ENGINEERING
Graphics and Design

PRADEEP JAIN
PhD (Pursuing), Mtech. (CAD/CAM), Btech. (M.E.)
Associate Professor
AKGEC, Ghazaiabd

A.P. GAUTAM
Mtech. (A & R), Btech. (M.E.)
Post Diploma (CAD/CAM/CAE & Mold Design)
Diploma (M.E. Production)
General Manager
Hitech Glazing, Ghaziabad
Formally, Assistant Professor, ABESIT, Ghaziabad

ANKITA MAHESHWARI
PhD (Pursuing), Mtech. (CAD/CAM), Btech. (M.E.)
Assistant Professor
AKGEC, Ghazaiabd

KHANNA BOOK PUBLISHING CO. (P) LTD.
Publisher, of Science, Technology and Engineering Books
4C/4344, Ansari Road, Darya Ganj, New Delhi-110002
Phone: 011-23244447-48 **Mobile:** +91-99109 09320
E-mail: contact@khannabooks.com
Website: www.khannabooks.com

Dear Readers,

To prevent the piracy, this book is secured with HIGH SECURITY HOLOGRAM on the front title cover. In case you don't find the hologram on the front cover title, please write us to at contact@khannabooks.com or whatsapp us at +91-99109 09320 and avail special gift voucher for yourself.

Specimen of Hologram on front Cover title:

Enlargement Of Overlay

Moreover, there is a SPECIAL DISCOUNT COUPON for you with EVERY HOLOGRAM.

How to avail this SPECIAL DISCOUNT:

Step 1: Scratch the hologram
Step 2: Under the scratch area, your "coupon code" is available
Step 3: Logon to www.khannabooks.com
Step 4: Use your "coupon code" in the shopping cart and get your copy at a special discount
Step 5: Enjoy your reading!

Price: ₹ 499.00

ISBN: 978-93-86173-47-8

Engineering Graphics and Design
by Pradeep Jain, Anand Gautam,
Ankita Maheshwari

First Edition: 2018
Reprint: 2021, **2024**

To view complete list of books,
Please scan the QR Code:

KPH

Published by:
Khanna Book Publishing Co. (P) Ltd.
CIN: U22110DL1998PTC095547

Visit us at: **www.khannabooks.com**
Write us at: **contact@khannabooks.com**

Printed in India by: Vaibhav Products, Delhi.

PREFACE

Designers use drawings when developing ideas. Well-prepared engineering drawings are used to communicate with manufacturers of components and devices. These drawings describe products and include instructions about manufacturing, assembly and operation. Clear, complete and accurate drawings help to prevent expensive and/or dangerous mistakes.

Engineering Graphics and Design is a graphic language incorporating the following:

- The ability to visualize or perceive a graphic;
- Knowledge and understanding of drawing principles and practices;
- Freehand, instrument and CAD drawings, and
- The Design Process.

The traditional first-year engineering graphics course has undergone significant change in the past quarter century. Although the emergence of computer-aided design (CAD) and the expansion of the graphics curriculum to include design are perhaps the most significant developments, more recent trends include a movement away from 2D CAD and toward 3D parametric solid modeling, an increased emphasis on freehand sketching at the expense of instrument drawing, a greater focus on the development of spatial visualization skills and an expansion of the curriculum to include the latest developments in design technology. All of this has occurred despite a strong countervailing trend to de-emphasize graphics in order to accommodate other material in the four-year undergraduate engineering curriculum.

The aim of this book, then, is to provide a clear, concise treatment of the essential topics included in a modern engineering graphics and design course.

Authors

CONTENTS

CHAPTER

1

INTRODUCTION TO ENGINEERING DRAWING

Learning Objectives

- Use of common drafting tools to construct engineering drawings
- Understand and apply proper dimensioning system in engineering drawings
- Construct, read, and understand the Title Block
- Utilize various line types and SI units to draw best descriptive drawings
- Use of various types of scales
- Understand fundamentals of geometrical constructions
- Learn to perform basic geometrical constructions

1.1 INTRODUCTION

"A picture is worth thousand words". For centuries drawings have been used as major mode of communication by marking message on sand or scratching drawings of animals, trees, and people on rock walls of caves. Artistic drawings reveal philosophical, aesthetic and abstract ideas which are totally different from technical/engineering drawings.

Technical communication is important while transferring ideas. Engineering drawing is a graphic language for expressing and documenting technical details. It is the universal language for and of engineers. Engineering drawing is a starting point for engineers of all the disciplines – Mechanical, Civil, Chemical, Computer Science, Electrical, Electronics and Communication, Mining, Agriculture etc. One technical drawing is equivalent to many sentences as the notations, diagrams, conventions etc. can be read and understood by a person having knowledge of these drawings. These drawings are used at various stages from conceptualization to designing, production, inspection, maintenance etc. while developing an engineering product. The engineering drawings depict lot of information such as the shape and size of an object, its dimensions, material details, date, and name of person who made the drawing, scale used etc. and provide road map to people involved in manufacturing. The drawings should be made properly because a small mistake made in the drawing can misguide the operator or workman and lead to failure of a system or a process.

The reason why manual drawings are introduced to students prior to computer aided drawings is to make students learn the standards, conventions, tools and help them visualize how drawings are made. This is similar to teaching calculations manually before introducing calculators or learning alphabets prior to reading books. Even though the use of computers makes the drawing process convenient and time saving still the drafting board and equipment cannot be replaced as a learning tool.

In every language one follows grammar. Similarly, engineering drawing must follow a grammar known as standards. These standards are followed internationally as they provide uniformity, easy collaborations and convenient exchange of technologies and technical data. This book provides the deep and detailed description of the grammar, symbols, conventions used in engineering drawing which would help students in making accurate, neat and legible drawings.

1.2 BUREAU OF INDIAN STANDARDS

International Standards Organization (ISO), Geneva has formulated and issued standards for engineering drawing internationally and modified standards to be followed in India are set up by Bureau of Indian Standards (BIS)." *Standardization is the process of formulating and applying rules for an orderly approach to a specific activity taking due account of functional conditions and safety requirements."*

Bureau of Indian Standards (under the Bureau of Indian Standards Act, 1986) establishes reviews or cancels Indian Standards in relation to any article or process.

Some of the standards followed in engineering drawing are:

Table 1.1: Standards followed in engineering drawing

Standard	Description
BIS-SP 46-1988	It is the standard for "Engineering Drawing Practice for schools and colleges". It was recently revised as SP – 46 – 2003
IS 10713:1983/ISO 5455:1979	Technical drawings-Scales
IS 8000(Part 3):1985/ISO 1660:1987	Technical drawings-dimensioning and tolerancing of profiles
IS 10711: 2014/ISO 5457:1999	Sizes and layout of drawing sheet
IS 11664:1986	Folding of drawing prints
IS 10209:1993	Terms related to projection method
IS 8976:1978	Guide for preparation and arrangement of sets of drawings and parts lists
IS 11663:1986	Conventional representation of common features and materials on technical drawings

IS 11655: 2014/ ISO 7200:1984	Technical product documentation – Data fields in title blocks and document headers
IS 11670:1993	Technical drawings – Abbreviations and Symbols for use in Technical drawings
IS 10719: 1983/ISO 1302:1978	Method of indicating surface texture on technical drawings
ISO 216:2007	Writing paper and certain classes of printed matter – Trimmed sizes – A and B series, and indication of machine direction
IS 8930(Part 1): 2014/ ISO 10209-1:2012	Technical product documentation – Vocabulary – Terms relating to technical drawings, product definition and related documentation
Is 15021(Part 2) : 2001/ ISO 5456-2:1996	Orthographic Projections
IS 15021 (Part 3) : 2001/ ISO 5456-3:1996	Technical Drawings – Projection Methods – Part 3: Axonometric Representations

1.3 DRAWING INSTRUMENTS AND HANDLING

The drawing instruments are used to record information on suitable paper or surface. Uses of reliable and accurate instruments ensure good results. The correct selection of instruments enhances the quality of drawing. Unsuitable equipment gives inaccurate results therefore while procuring special attention should be given to the accuracy and quality of instruments.

The list of instruments required for manual drawing are:

(*i*) Drawing Board
(*ii*) Drawing sheet
(*iii*) T-Square
(*iv*) Mini drafter
(*v*) Set squares
(*vi*) Instrument box
(*vii*) Paper fasteners
(*viii*) Scales
(*ix*) Protractor
(*x*) Pencils
(*xi*) Sharpeners and erasers

The detailed description of each component is given below:

(*i*) **Drawing Board:** It is a rectangular board made of best seasoned, soft wooden platens to allow easy insertion and removal of drawing pins. This material also helps in resisting daily wear and tear along with climatic changes. It is cleated at the back to prevent warping. One of the edge at which T-square is made to slide has a strip of hard ebony edge fitted up in a groove on shorter edge of the board.

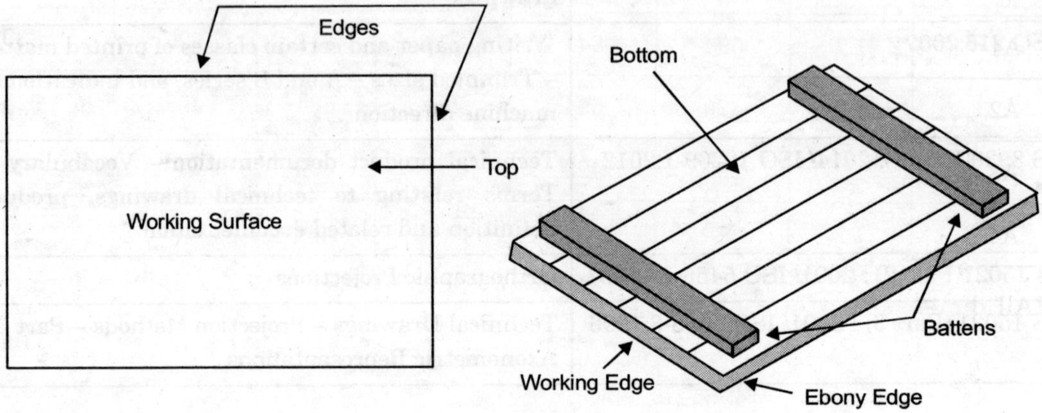

Fig. 1.1: Top and bottom of drawing board

*Handle the board properly and do not spoil the top surface.

These are the following standard sizes of the drawing board:

Table 1.2: Standard sizes of drawing board

Designation	Length * Width	Preferable sheet size to be used
D0	1500*1000	A0
D1	1000*700	A1
D2	700*500	A2
D3	500*500	A3

*D2 is for students. DO, D1 are for drawing offices

(*ii*) **Drawing sheet:** It is the medium on which drawings are made. Good quality papers with smooth surfaces should be used for drawings. The sheets are available in different sizes. A0 is the sheet with dimensions 1189*841. Each higher number sheet i.e., A0, A1, A2, A3 and so on is half in size of the next lower numbered sheet.

There are different sheets classified on the basis of their sizes (ISO-A):

Table 1.3: Standard sizes of drawing sheets

Designation	Figure	Trimmed Sheet (T)		Drawing Space		Untrimmed Sheet (U)	
		a1	b1	a2	b2	a3	b3
A0	1.2	841	1189	821	1159	880	1230
A1	1.2	594	841	574	811	625	880
A2	1.2	420	594	400	564	450	625
A3	1.2	297	420	277	390	330	450
A4	1.2	210	297	190	267	240	330

*All dimensions are in mm

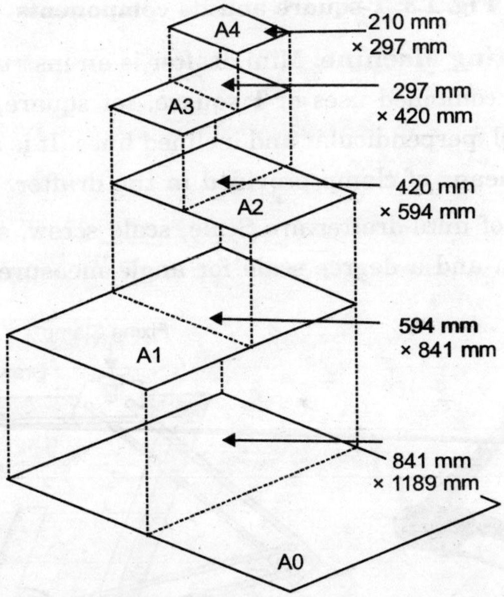

Fig. 1.2: Size of various drawing sheets

(iii) **T-square:** It should be made of seasoned hard quality wood. T-square is used by draftsmen as a guide for drawing horizontal lines on table. It is composed of a long strip called blade fastened rigidly to a shorter section called head and is positioned on drawing board in desired position by sliding working edge of its head along ebony edge of board. These are available in different sizes T0, T1, T2, T3 of lengths 1500, 1000, 700 and 500 respectively.

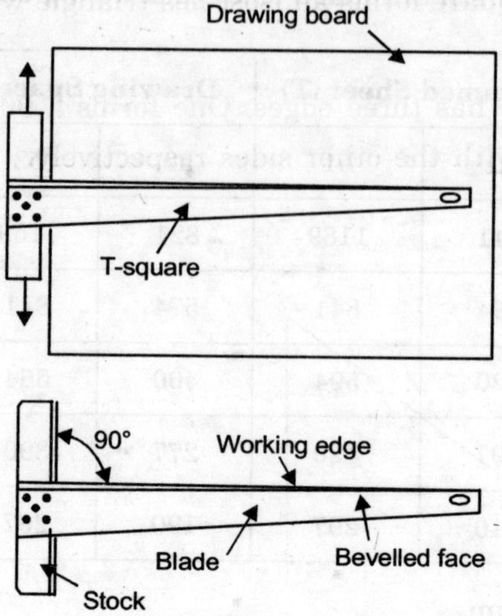

Fig. 1.3: T-square and its components

(*iv*) **Mini Drafer/Drafting Machine:** Mini drafter is an instrument to draw geometrical shapes and figures. It has combined uses of T square, set square, scale and protractor and can be used to draw parallel, perpendicular and inclined lines. It is mounted on top left corner of the drawing board by means of clamp provided in the drafter.

The main components of mini-drafter are Scale, scale screw, scale plate, steel bars, bar plate, clamping mechanism and a degree scale for angle measurement.

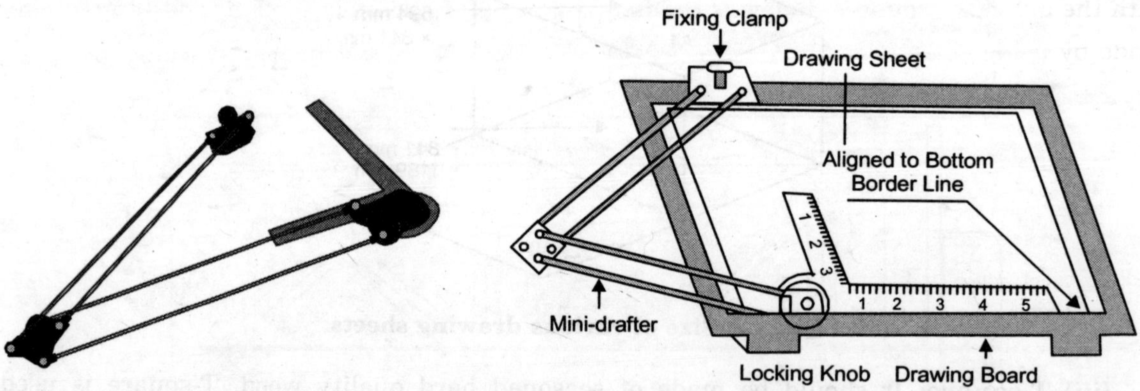

Fig. 1.4: Mini-drafter and positioning of mini-drafter on drawing board

(*v*) **Set Squares:** They are made of wood, tin or transparent acrylic and are available in 2 sets- 45° set square and 30°-60° set square. They are used in conjunction with T-square for drawing parallel straight lines, inclined lines, perpendicular lines and dividing a circle in 6, 8, 12 and 20 parts.

1. Forty-five degree set square forms an isosceles triangle where two angles are 45° each and the third angle is of 90°.

2. Thirty-sixty set square has three edges. One forms a 90° angle and other two edges form angles of 30° and 60° with the other sides respectively.

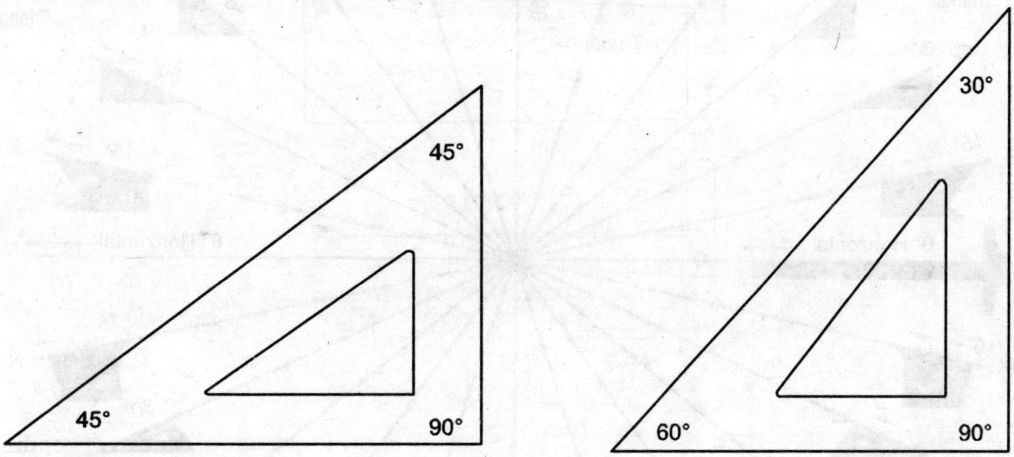

Fig. 1.5: Set squares

Set squares are available in various lengths and are used for drawing all straight lines (except horizontal lines). Perpendicular lines or lines inclined at 30°, 45°, 60°, 90° can be easily made. Various angles such as 15°, 75° and 105° and all the multiples of 15° can be made with the aid of set squares. Below is the figure showing all possible combinations and angles made by using set-squares.

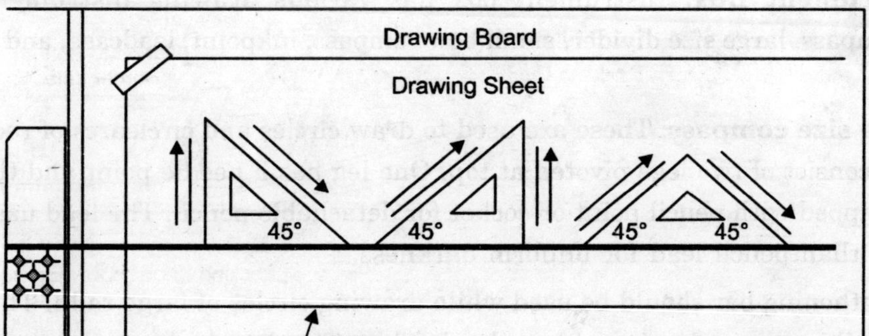

Fig. 1.6: Showing drawing lines at 45° and 90°

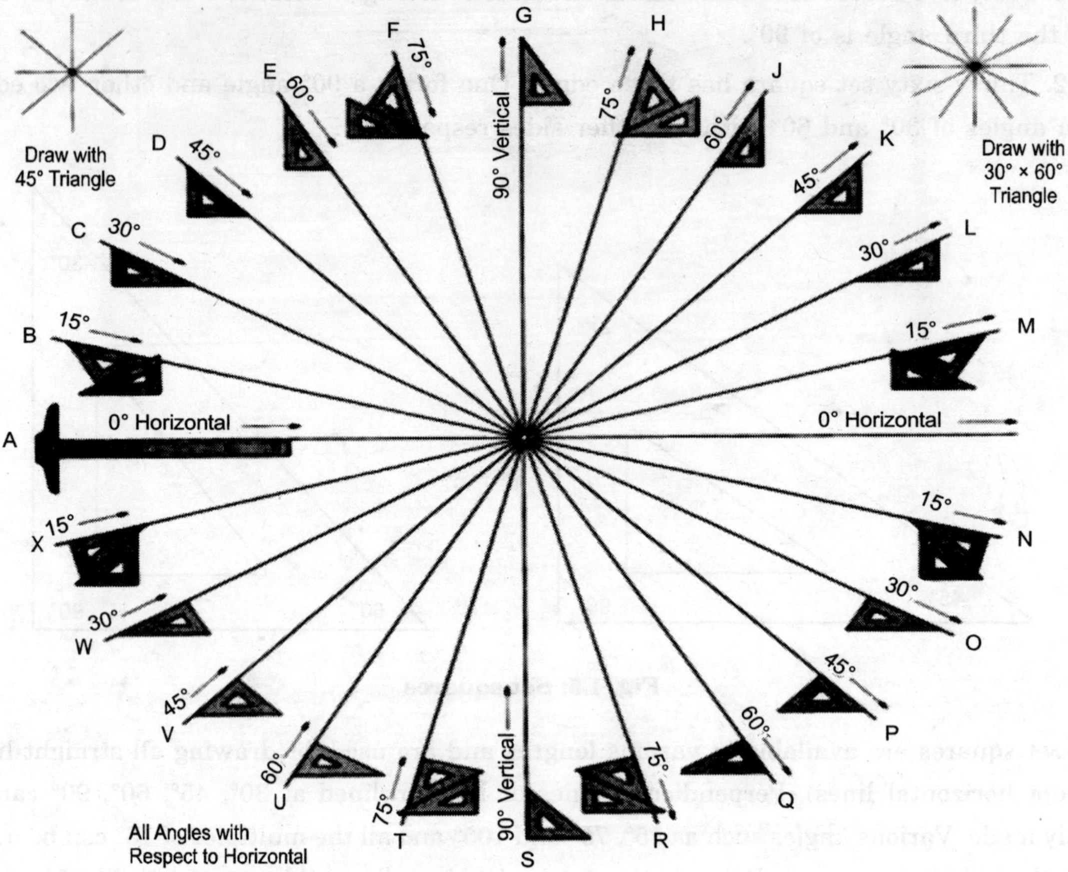

Fig. 1.7: All possible combinations of set-squares

(*vi*) **Instrument Box:** Instrument box has various drawing instruments namely: Large size compass, large size divider, small bow compass, inkpoint, leadcase, and lengthening bar.

(*a*) **Large size compass:** These are used to draw circles and circlearcs of required sizes. They consist of two legs pivoted at top. One leg has a needle point and the other one is equipped with pencil point or socket for detachable pencil. The lead used should be softer than pencil lead for uniform darkness.

(*b*) A lengthening bar should be used while drawing circles of large radii. The lower part of pencil is removed to insert lengthening bar. The detached pencil part is then fitted at the end of lengthening bar.

(*c*) **Divider:** Large size divider is used to transfer dimensions from one part to another, divide curved or straight lines into equal segments. Small spring bow divider is used to mark minute divisions and short equal distances such as off-set distances.

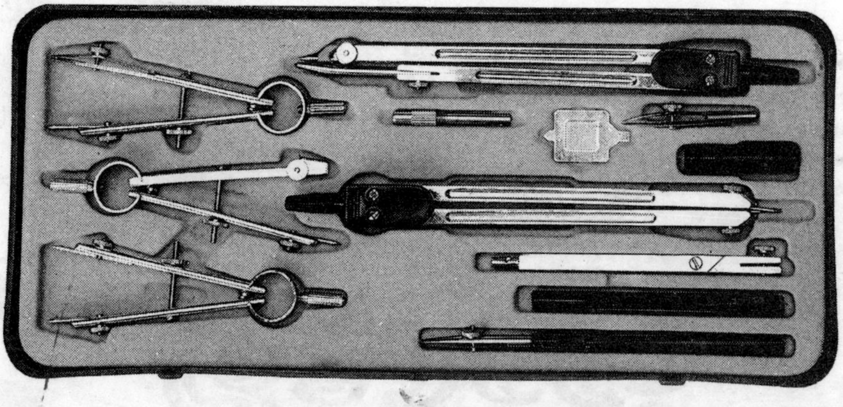

Fig. 1.8: Basic instrument box

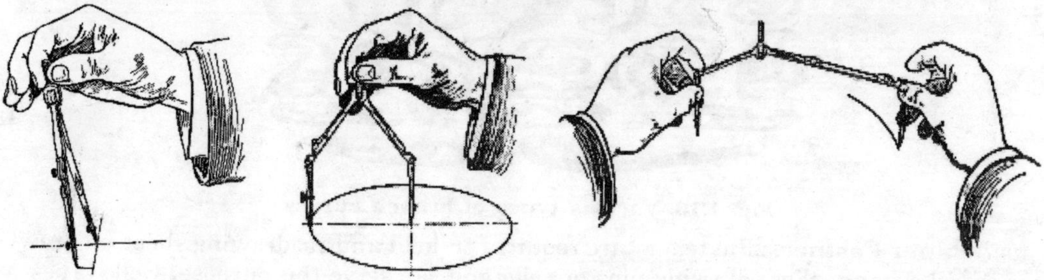

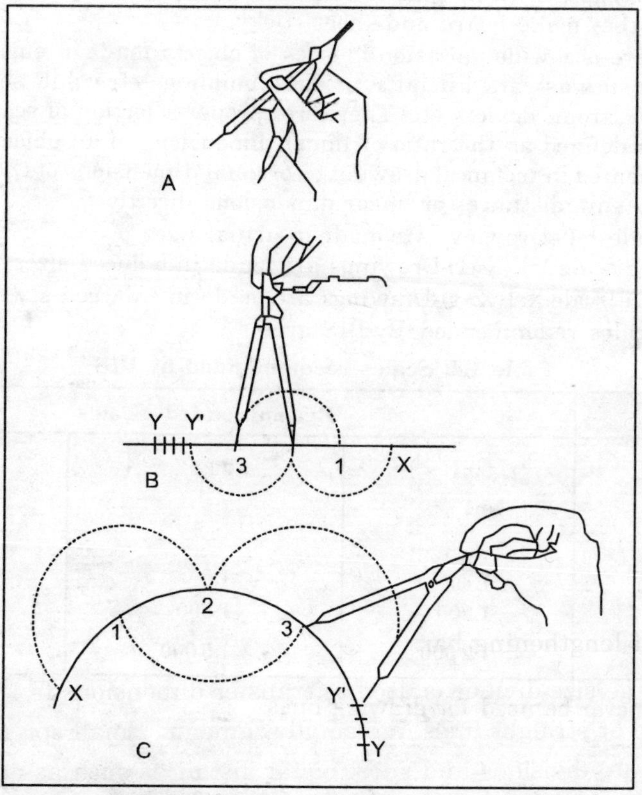

Fig. 1.9: Use of compass, divider and lengthening bar

Proper adjustment of needles helps in quick and easy adjustment with one hand. Second and third fingers are used to close the dividers.

*Test the tightness of joint so that it is firm while making the drawing. Keep both the needle points of divider at the same length. Adjust the legs of the divider properly so that holes are not made on the paper.

(*vii*) **French Curves/Flexible Curves/Irregular Curves:** French curves are used to draw (*i*) a smooth curve passing through a number of points (*ii*) curves where radius of curvature is not constant. These are made in various shapes and can be made out of wood, metal or highly transparent celluloid.

Fig. 1.10: Various types of French curves

(*viii*) **Paper Fastners:** Fasteners are required to hold and fix drawing sheet on drawing board. Steel clamps, clips, drawing pins or cellotape can serve the purpose. Cello tapes are usually avoided as they make board and sheet sticky.

(*ix*) **Scales:** There is a wide variation in sizes of objects made in engineering drawings. Some are very large such as cars, aircrafts, bridges, buildings etc. while some are very small such as watches, electronic devices etc. Therefore, proper selection of scales is required for accuracy. A scale is defined as the ratio of linear dimensions of an object or an element of the object as represented in technical drawing to original dimensions of the same. Scales can also be used to measure distances or linear dimensions directly.

1. Full size scale(1:1)-Drawings are made in actual sizes
2. Reduced size scale 1:X(X>1)-Drawings are made in reduced sizes
3. Enlarged size scale X:1(X>1)-Drawings are made in enlarged sizes

The common scales recommended By-BIS are:

Table 1.4: Scales recommended by BIS

Category	Recommended Scales		
Enlargement	50:1	20:1	10:1
Scales	5:1	2:1	
Full size			1:1
Reduction	1:20	1:5	1:10
Scales	1:200	1:500	1:100
	1:2 000	1:5 000	1:1 000

*Scale should never be used for drawing lines.

(*x*) **Protractor:** It is made of wood, tin or celluloid. The most common type is semi-circular. The circumference edge is divided in 180 divisions, 1° each and is readable from both the sides. They are used to draw angles or dividing circle into n number of equal parts.

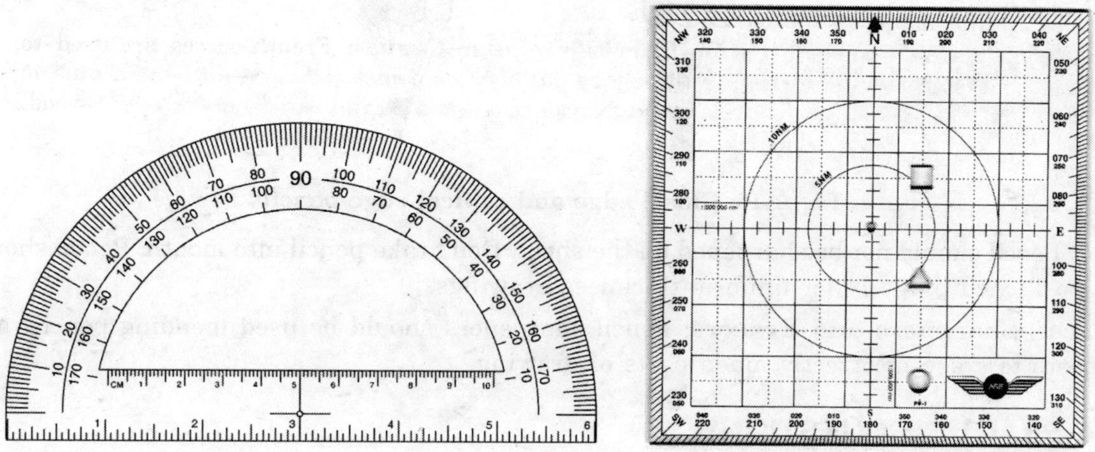

Fig. 1.11: Circular protractor and Square protractor

(*xi*) **Pencils:** Pencil or lead stick is a primary tool in engineering drawings. Different grades of pencils are used for different purposes. Pencil lead is made up of graphite mixed with varying quantities of clay to produce various levels of hardness.With low quality pencils proper thickness and uniformity of drawings cannot be achieved.

There are 18 grades of pencils in order of descending hardness: 9H, 8H, 7H, 6H, 5H, 4H, 3H, 2H, H, F, HB, B, 2B, 3B, 4B, 5B, 6B, 7B, 8B, 9B where 9H is the hardest and 9B is the softest. Pencils used for drawing purposes should be hard enough to be clearly visible but not very hard that is destroys the sheet by cutting grooves.

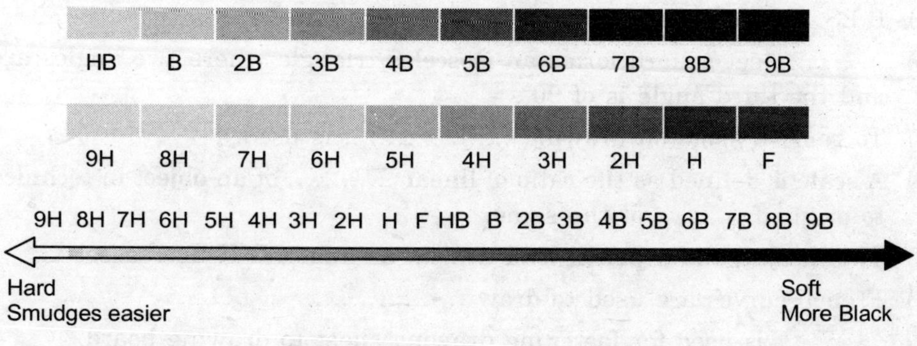

Fig. 1.12: Grades of pencils

There are various ways of mending the pencil out of which two are generally used:

(*i*) **Chisel edge pencil:** These are used for drawing straight lines. The edge is flat on each side just like the shape of a chisel.

(*ii*) **Conical pencil:** These are best for free hand work and used with medium grade pencils.

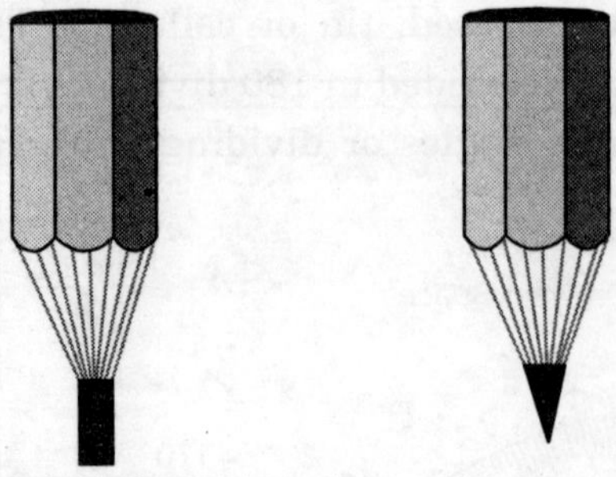

Fig. 1.13: Chisel edge and conical edge pencils

*Pencil should not be sharpened on the sheet. Don't take pencil into mouth. Pencil should be positioned properly for uniform thickness drawings.

(xii) **Sharpners and Erasers:** Pencil sharpeners should be used mending pencils and erasers to erase incorrectly made parts of drawing.

1.4 PRACTICE QUESTIONS

1. Define engineering drawing and its applications.
2. What are the standard sizes of drawing sheets according to BIS?
3. Make these figures on drawing sheet using mini-drafter.
4. Draw a circle of 40mm radius and by using set squares and T-square divide it into eight parts.
5. Draw a line of 200mm length and by using a divider, divide it into 12 parts of equal length.
6. Fill up the blanks:
   - *(i)* A lengthening bar should be used while drawing circles of ....
   - *(ii)* B.I.S. stands for ....
   - *(iii)* .... set square forms an isosceles triangle where two angles are 45° each and the third angle is of 90°.
   - *(iv)* To remove spots on drawing .... is used.
   - *(v)* A scale is defined as the ratio of linear .... of an object in technical drawing to original .... of the same.
   - *(vi)* The hardness of a pencil lead .... as numeral H decreases.
   - *(vii)* French curves are used to draw ....
   - *(viii)* ....is used for fastening drawing sheet to drawing board.

1.5 LAYOUT OF DRAWING SHEET

While preparing a drawing sheet a set of standards are to be followed. The drawing sheet consists of drawing space, title block and margins. The selection of proper scale and allotment of drawing sheet space for margins, title block, etc. is termed as layout of drawing sheet.

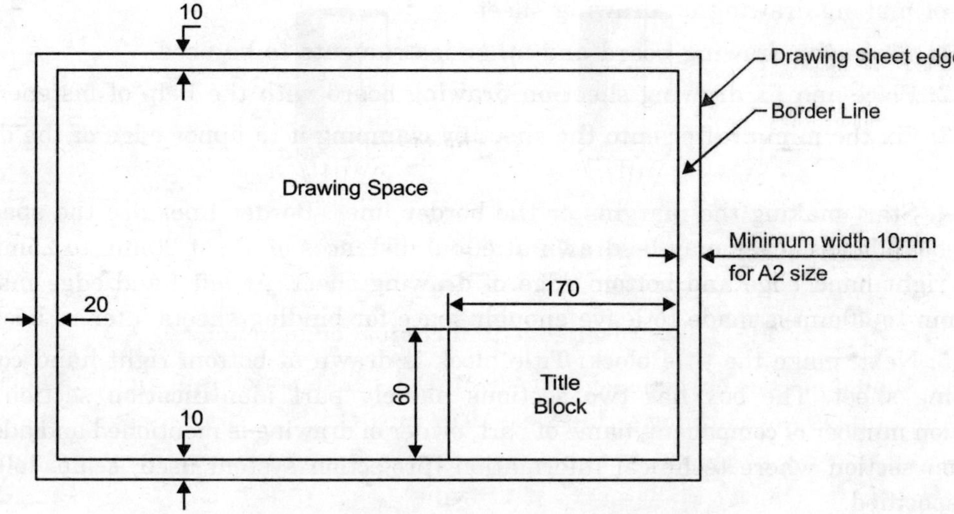

Fig. 1.14 (a): Layout of drawing sheet

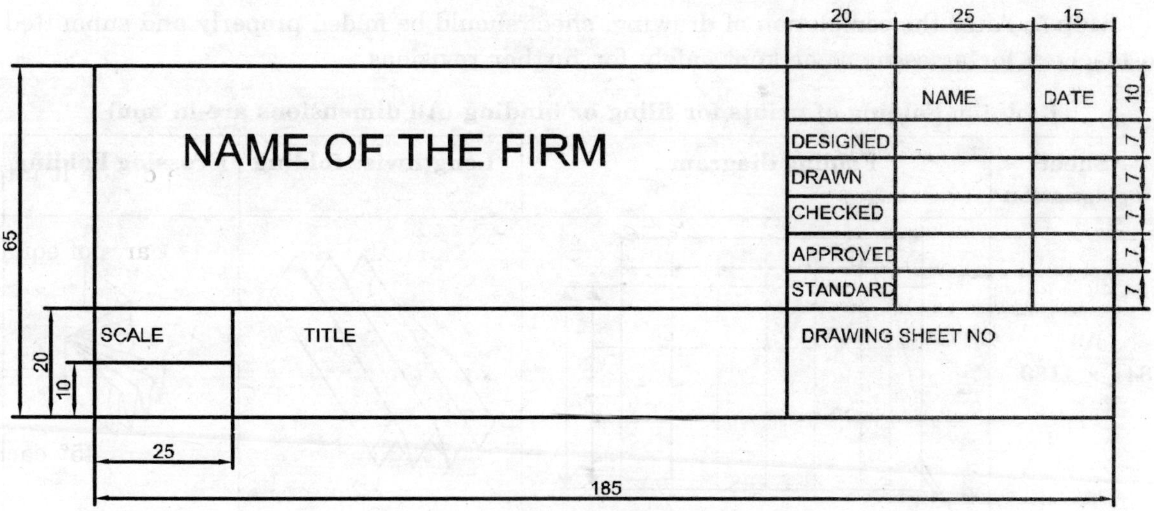

Fig. 1.14 (b)

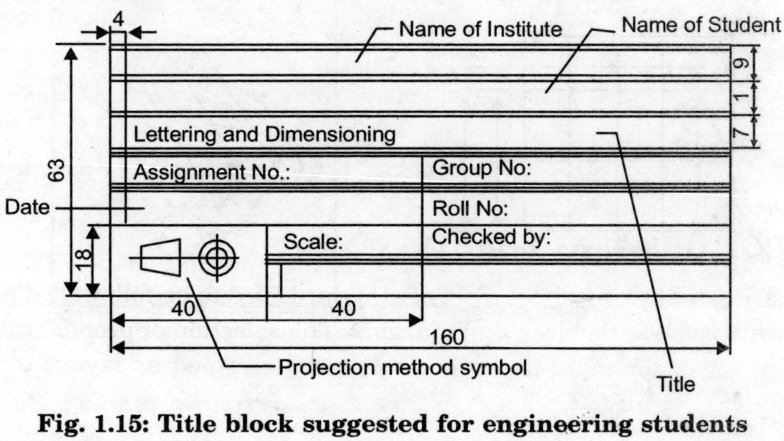

Fig. 1.15: Title block suggested for engineering students

Steps of making drawing on drawing sheet:

Step 1: Clean the drawing board and other instruments to be used.

Step 2: Place and fix drawing sheet on drawing board with the help of fasteners.

Step 3: Fix the mini drafter onto the sheet by clamping it to upper edge of the drawing board.

Step 4: Start making the margins or the border lines. Border lines are the spaces left at all the boundaries. They may be drawn at equal distances of about 20mm to 25mm from top edge, right hand edge and bottom edge of drawing sheet. At left hand edge margin of about 25mm to 40mm is made to leave enough space for binding sheets later in book form.

Step 5: Next, make the title block. Title block is drawn at bottom right hand corner of the drawing sheet. The box has two sections namely part identification section where identification number of component, name of part, owner of drawing is mentioned and additional information section where technical information (projection system used, scale, tolerances etc.) are specified.

Step 6: As per the questions given, make suitable drawings in drawing space.

Step 7: After the completion of drawing, sheet should be folded properly and submitted to teachers for assessment or kept safely for further revisions.

Table1.5: Folding of prints for filing or binding (All dimensions are in mm)

Sheet Designation	Folding diagram	Lengthwise folding	Crossing Folding
A0 **841 × 1189**	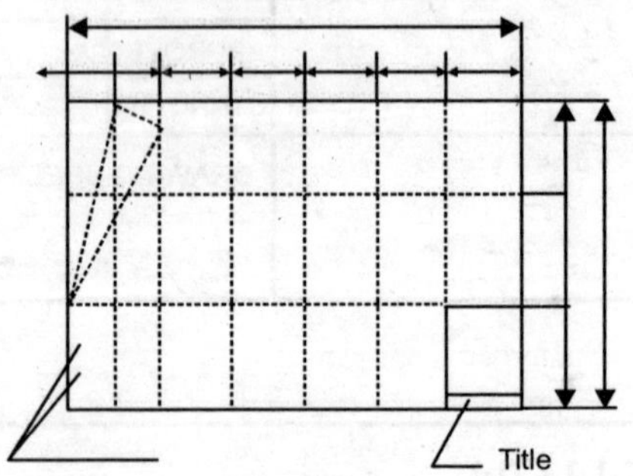	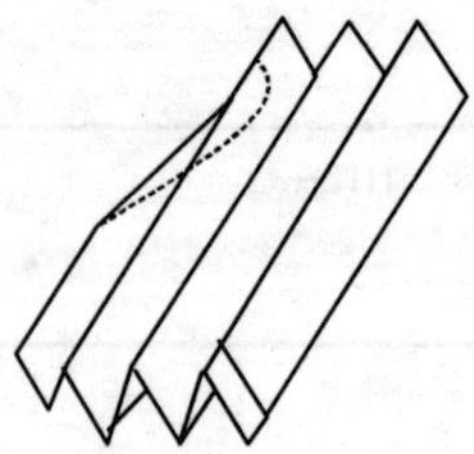	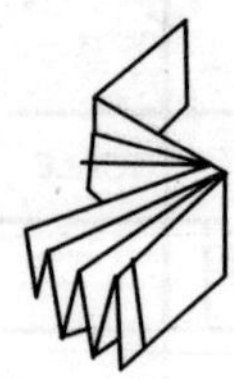
A1 **594 × 841**	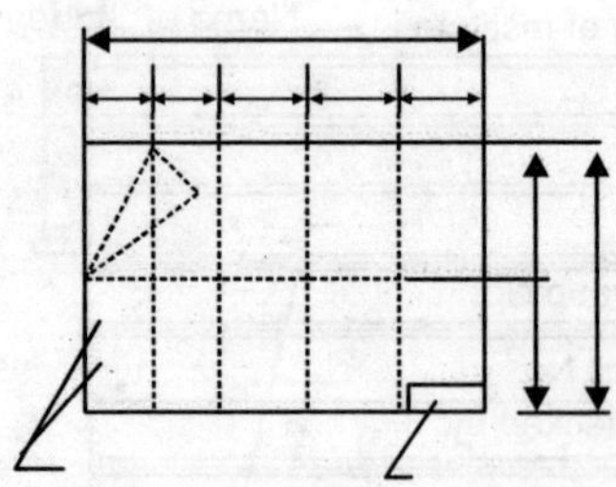	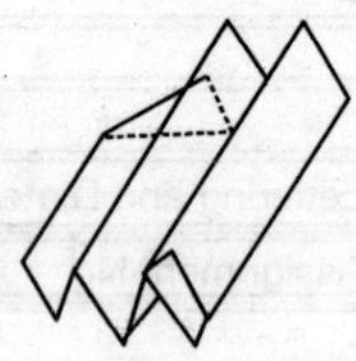	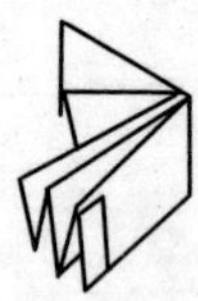

A2 **420 × 594**	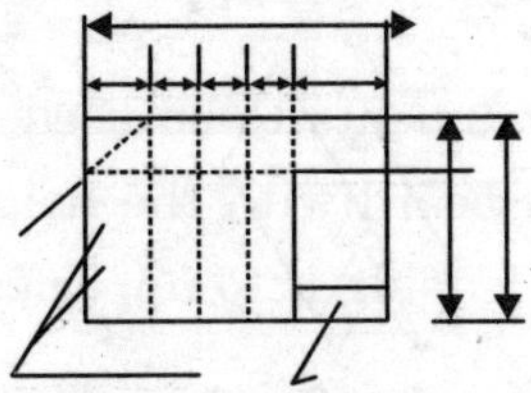	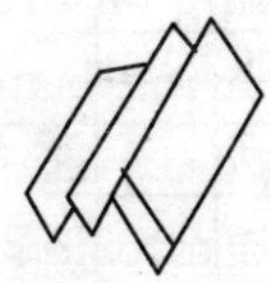	
A3 **297 × 430**	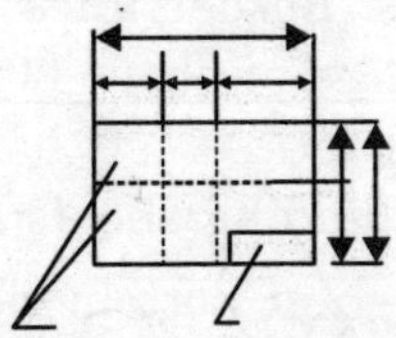		–

Table 1.6: Codes for folding drawing sheets in filing cabinet (IS:696-1972)(All dimensions are in mm)

Sheet Designation	Folding diagram	Lengthwise folding	Crossing Folding
A0 **841 × 1189**	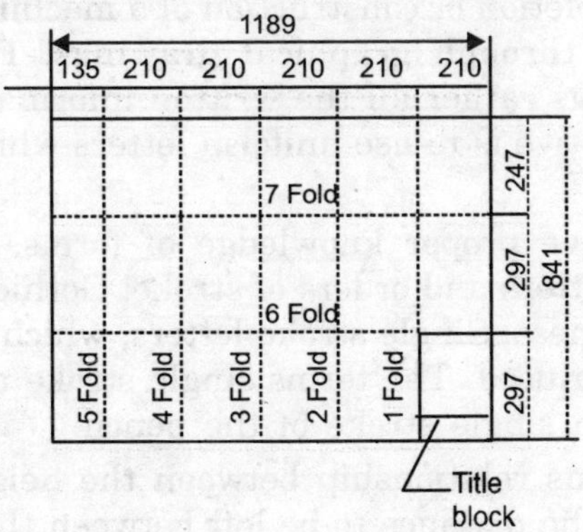	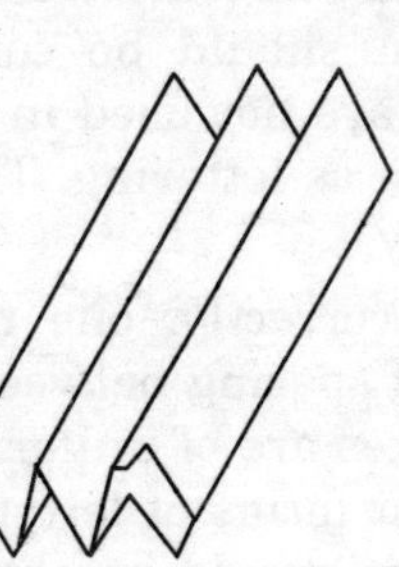	
A1 **594 × 841**	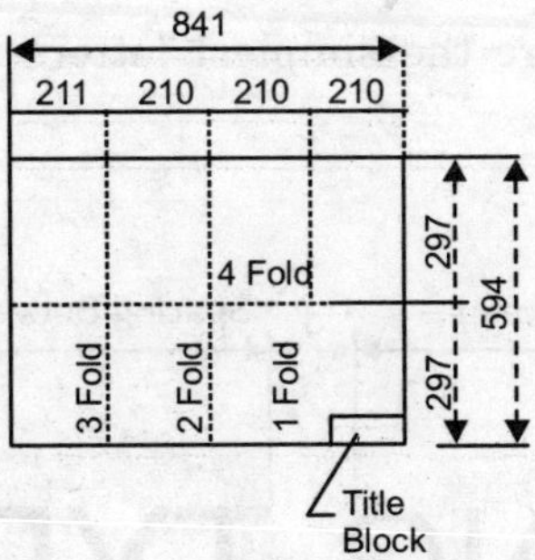	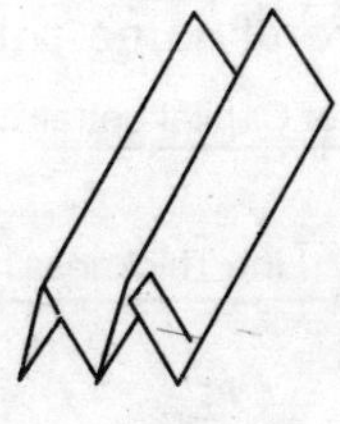	

A2 **420 × 594**	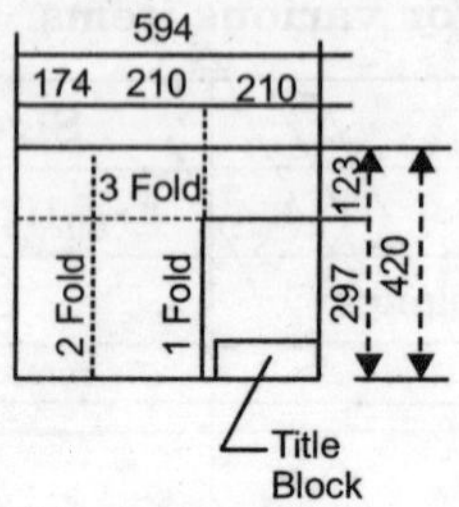		
A3 **297 × 430**	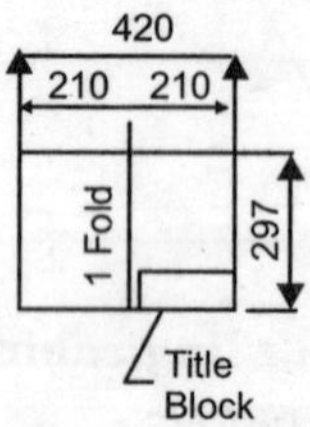		–

1.6 MANUAL TECHNICAL LETTERING

To impart necessary information for completion or construction of a machine or structure, complete information should be conveyed through graphical drawings. Free hand and ornamental writings are not used in drawings rather all the written information is in form of fixed format known as lettering. The objective is to use uniform letters which can be read and reproduced easily.

To write letters correctly, one must have proper knowledge of forms, proportions of letters, proportions of spacing between the letters and orders of stroke. Gothic writing is one in which all the strokes are of uniform thickness. Single stroke letters, which are preferably used, are the simplest forms of letter writing used. The terms single stroke means that the thickness of the letter should be obtained in single stroke of the pencil.

The general proportion of lettering means relationship between the height, width and spacing of each letter. By spacing we mean the distance to be left between the two adjacent letters. Good spacing is required for clarity of words.

1. Single-stroke vertical lettering: These are the simplest letters used in drawings. The alphabets made are of same thickness.

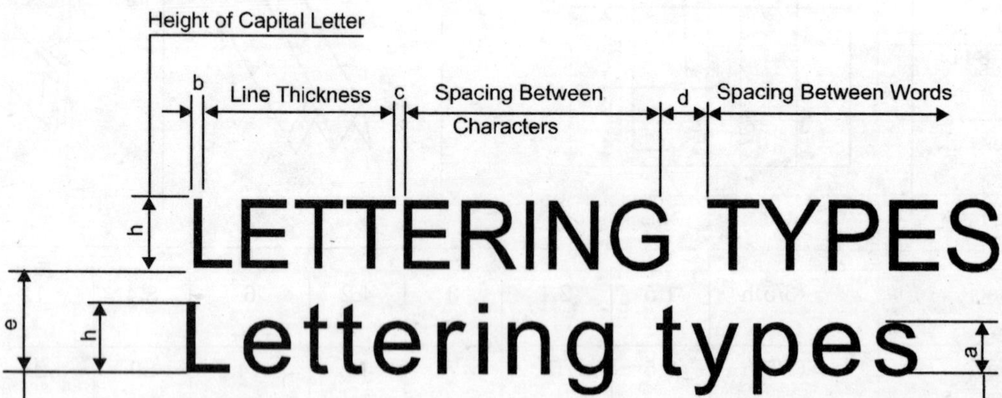

Table 1.7: Letter sizes recommended for various items

Sr. No.	Item	Size (mm)
1.	Name of the company	10, 14, 20
2.	Drawing numbers, letters denoting section planes	10, 14
3.	Title of the Drawing	7, 10
4.	Sub-titles and heading	5, 7
5.	Dimensioning, Notes, Schedules,	3.5, 7
6.	Alteration entries and tolerances	3.5

Table 1.8: Specifications for type A lettering

Specifications	Value	Size (mm)						
Capital letter height	h	2.5	3.5	5	7	10	14	20
Lowercase letter height	a = (5/7)h	-	2.5	3.5	5	7	10	14
Thickness of lines	b = (1/14)h	0.18	0.25	0.35	0.5	0.7	1	1.4
Spacing between characters	c = (1/7)h	0.35	0.5	0.7	1	1.4	2	2.8
Min. spacing b/n words	d = (3/7)h	1.05	1.5	2.1	3	4.2	6	8.4
Min. spacing b/n baselines	e = (10/7)h	3.5	5	7	10	14	20	28

Table 1.9: Specifications for Type B lettering

Specifications	Value	Size (mm)						
Capital letter height	h	2.5	3.5	5	7	10	14	20
Lowercase letter height	a = (7/10)h	-	2.5	3.5	5	7	10	14
Thickness of lines	b = (1/10)h	0.25	0.35	0.5	0.7	1	14	2
Spacing between characters	c = (1/5)h	0.5	0.7	1	1.4	2	2.8	4
Min. spacing b/n words	d = (3/5)h	1.5	2.1	3	4.2	6	8.4	12
Min. spacing b/n baselines	e = (7/5)h	3.5	5	7	10	14	20	28

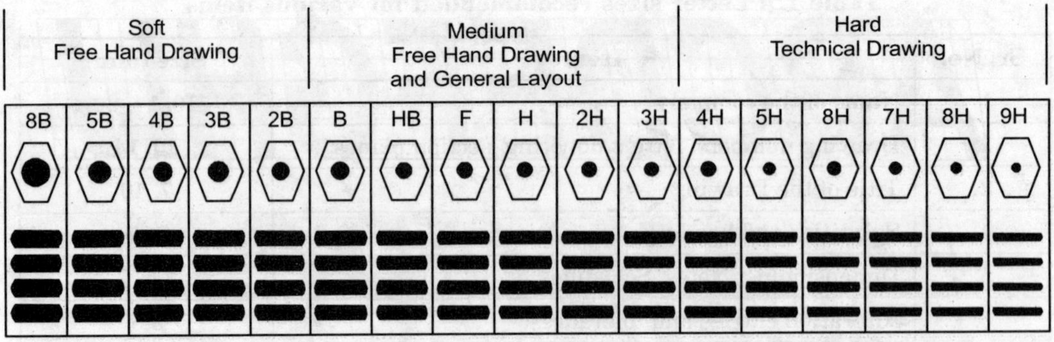

Fig. 1.16: Grades of pencils

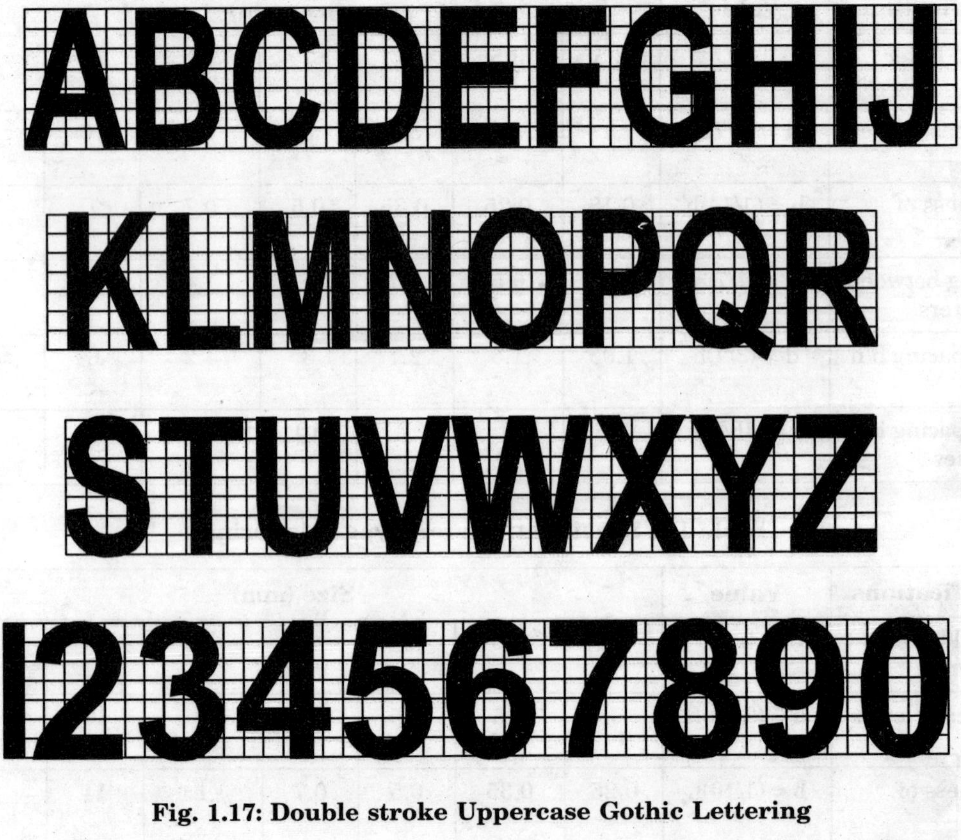

Fig. 1.17: Double stroke Uppercase Gothic Lettering

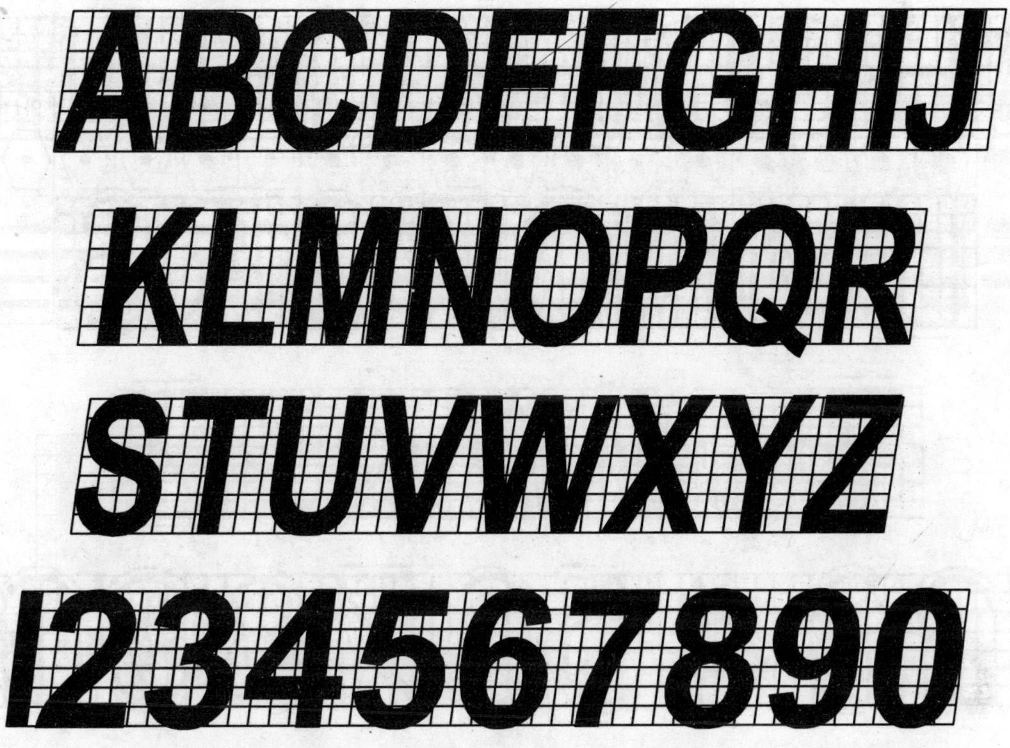

Fig. 1.18: Double stroke inclined (75°)

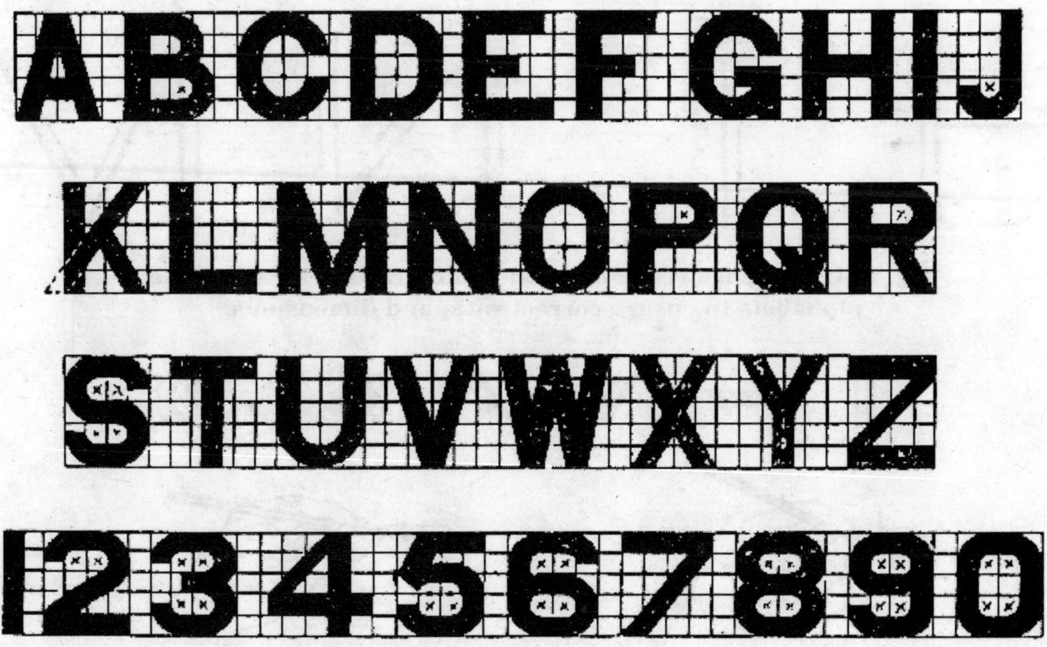

Fig. 1.19: Uppercase lettering Type 2

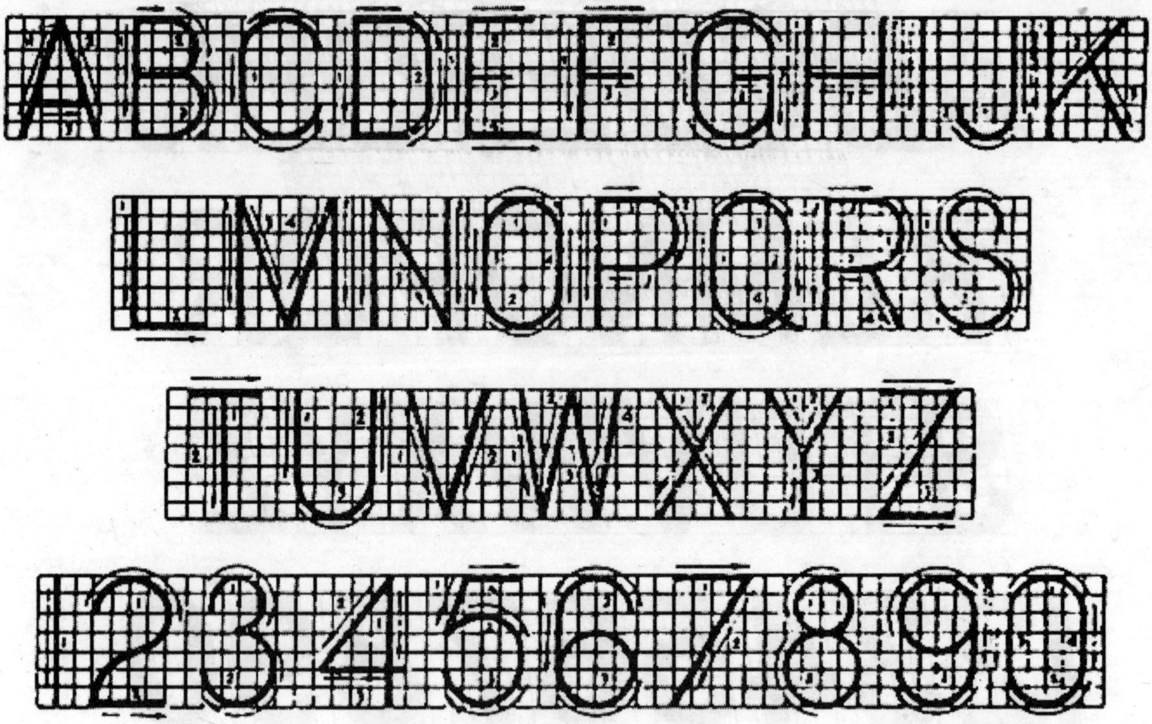

Fig. 1.20: Single stroke vertical lettering

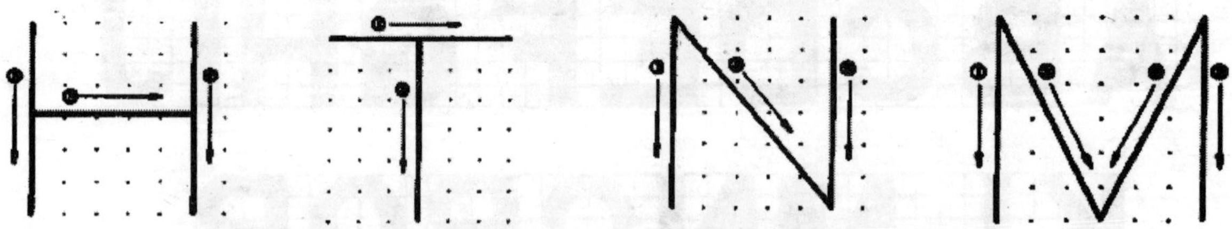

Fig. 1.21: Sequence recommended while drawing straight line alphabets to ensure correct ratio and dimensions.

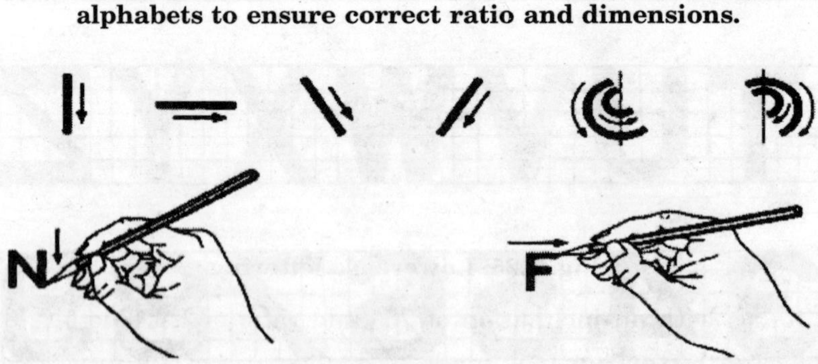

Fig. 1.22: Order and direction of stroke

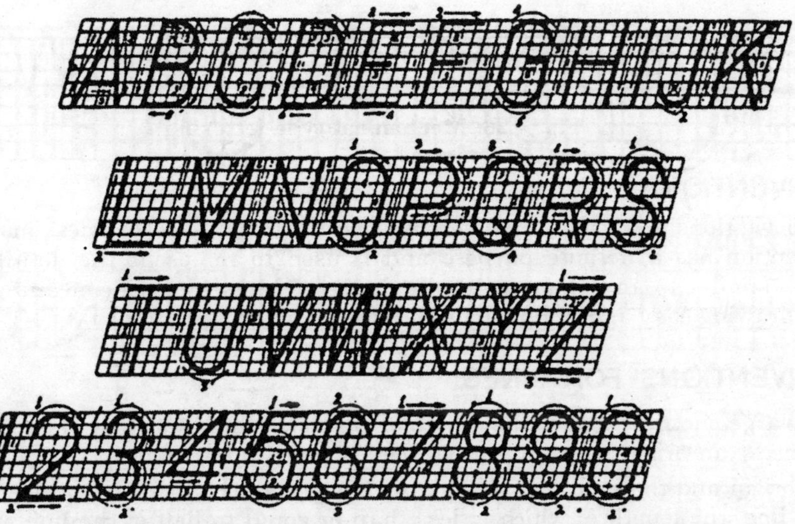

Fig. 1.23: Single stroke inclined Gothic lettering

Fig. 1.24: Lowercase lettering

Lower case is used in some drawings and extensively in map drawing.

Freehand lettering: The small size alphabets in italics written freehand are called lower case italic.

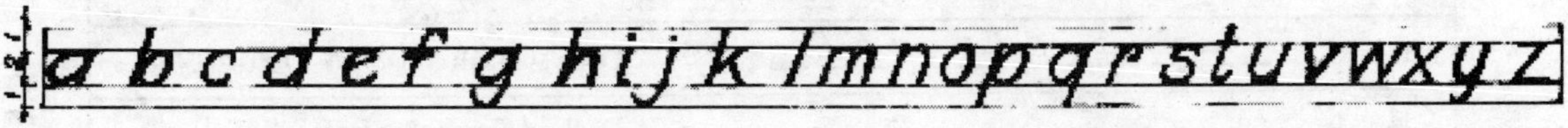

Fig. 1.25: Lowercase lettering

6 mm high italics with an inclination of 75° and ratio of 1:2:1 (letters-6 mm)

Mechanical type lettering: It is made with the ratio of 8:9 in vertical single stroke. (h = 6 mm).

Fig. 1.26: Mechanical type lettering

1.7 CONVENTIONS

The conventions are useful in differentiating different types of lines, materials etc. as each convention has a definite purpose and is used to represent the details of an object correctly.Its similar to any language where each word has a specific meaning. Conventions make the drawing easy and simple to understand and draw.

1.8 CONVENTIONS FOR LINES

Line is a geometrical object, the length of which is more than half of the line width and which connects an origin with an end without or with interruptions.

* The origin and the end may coincide with one another as in the case of a line forming a circle. A line, the length of which is less than or equal to half of the line width, is called a dot.

Table 1.10: Basic types of lines

No.	Representation	Description
1.	________________________	Continuous Line
2.	------------------------------	Dash ed Line
3.	— — — — — — — —	Dash es Space Line
4.	--.--.--.--.--.--.--.--.--.--	Long Dashed Dotted Line
5.	--..--..--..--..--..--..--	Long Dashed Doubledott ed Line
6.	—...—...—...—...—...—	Long Dashed Triple-dotted Line
7.	....	Dotted Line
8.	— — — — — — — — — —	Long Dashed Short Dashed Lin e
9.	— - - — - - — - - — - -	Dash ed Dotted Line
10.	- - . - - . - - . - - . - - .	Double Dash ed Dotted Line
11.	——.——.——.——.——	Double Dash edDotted line
12.	—..—..—..—..—..—..	Dash ed Double Dott ed Line
13.	——..——..——..——..	Double Dash ed Double Dotted Line
14.	—...—...—...—...—...	Dash ed Triplicate Dotted Line
15.	——...——...——...——...	Doub le Dash ed Triplicate Dotted Line

Table 1.11: Types of lines and their applications

S.No.	Type of Line	Illustration	Application
A	Continuous thick		Visible outlines
B	Continuous thin		Dimension lines, leader lines, extension lines, construction lines, outlines of adjacent parts, hatching and revolved section
C	Continuous thick wavy		Irregular boundary lines, short break lines
D	Short dashes medium		Hidden outlines and edges
E	Long chain thin	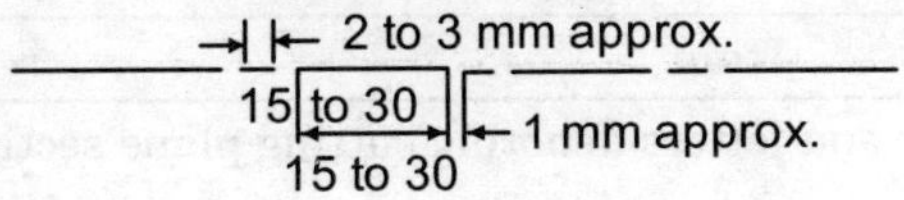	Centre lines, locus lines, extreme positions of the movable parts, pitch circles and parts situated in front of the cutting planes
F	Long chain thick at ends & thin elsewhere	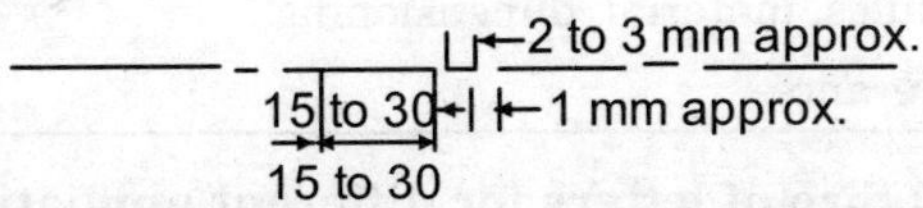	Cutting plane lines
G	Long chain thick	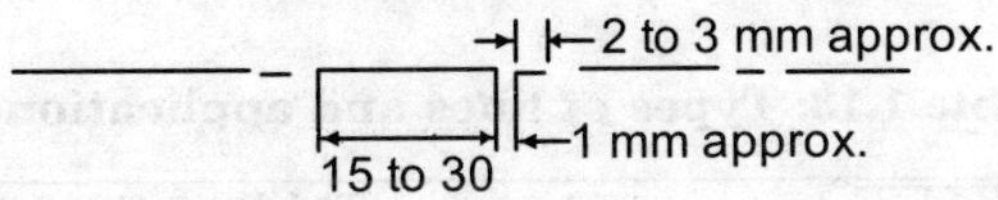	To indicate surfaces which are to receive additional treatment
H	Ruled line and short zigzag thick	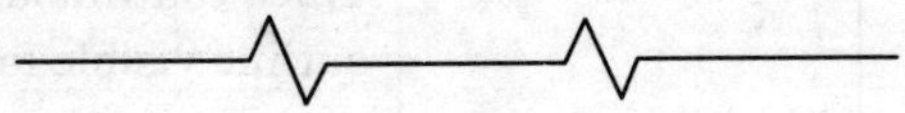	Long break lines

Table 1.12: Conventions used for different types of material

S.No.	Materials	Convention
1..	Steel, cast iron, copper aluminium, and its alloys, etc.	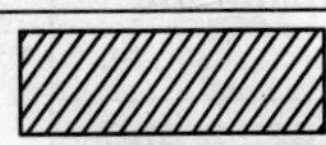
2.	Lead, zinc, tin, white metal, etc.	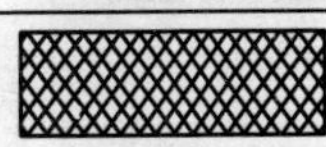
3.	Brass, bronze, gun metal, etc.	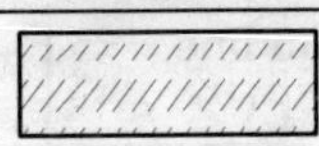
4.	Glass	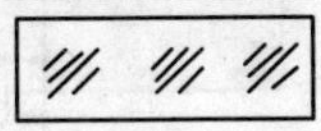
5.	Porcelain, stone, ware, marble, slate, etc.	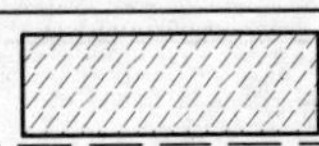

6.	Asbestos, fell, paper, mica, cork, rubber, leather, wax, insulating materials	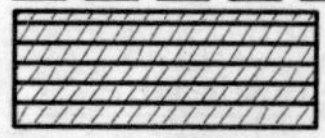
7.	Wood, plywood, etc.	
8.	Earth	
9.	Brick work, masonry, fire bricks, etc.	
10.	Concrete	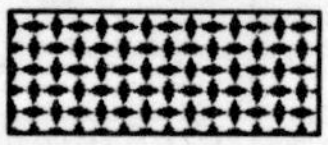
11.	Water, oil, petrol, kerosene, etc.	

Item	Size b (mm)
Drawing number in title block and letters denoting cutting plane section.	10, 12.
Title of drawing.	6, a.
Subtitle and headings.	3, 4, 5, 6.
Notes, such as legends, schedules, material, dimensioning.	3, 4, 5.
Alteration entries and the tolerance.	2, 3.

Fig. 1.27: Size of letters for different appliations

Table 1.13: Types of lines and applications

Visible line	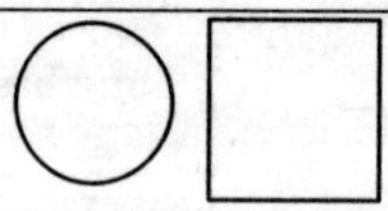	Thick continuous line used to show outline visible features on drawing
Center line	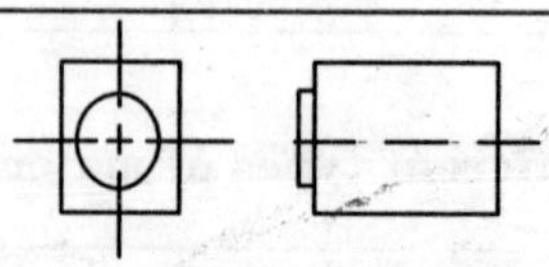	Long and short dashes line used to show center line or locus line of objects
Extension line	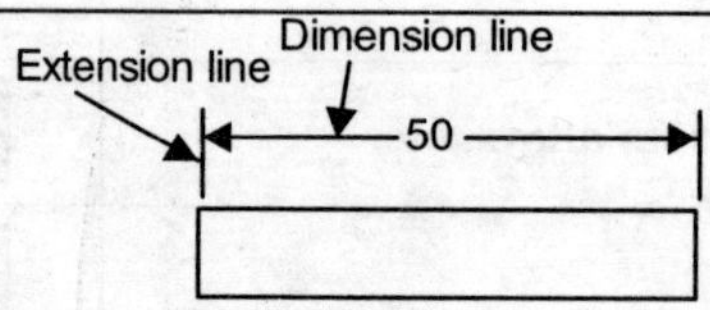	Thin continuous line for dimensioning objects
Dimension line		Thin continuous line for giving dimensions. Terminates with arrow heads
Section line	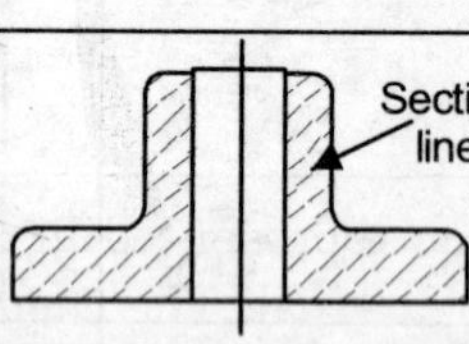	Thin continuous line for sectioning an object

Projector line	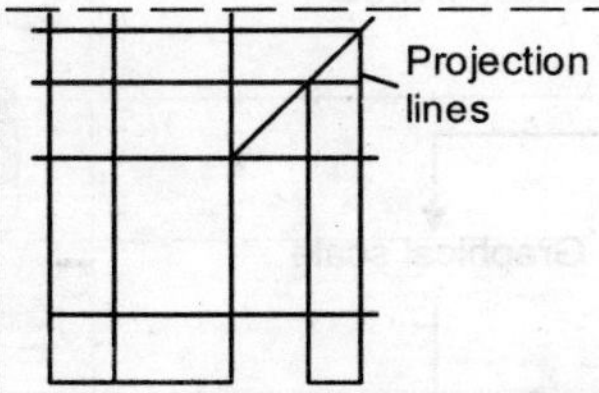	Thin continuous line for drawing projectors
Construction line	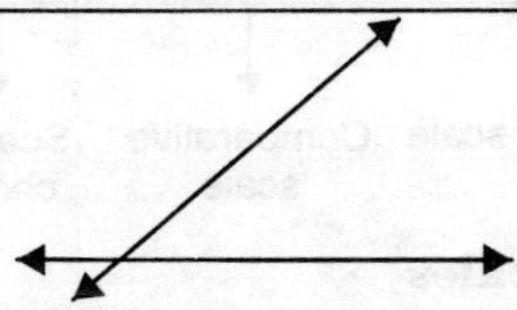	Thin continuous line for constructing an object
Hidden line	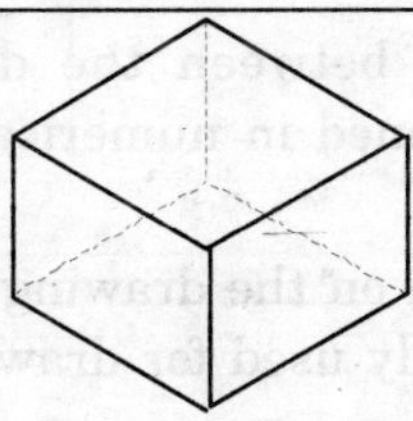	Short dashes line to show invisible or hidden parts on the drawing
Cutting plane line	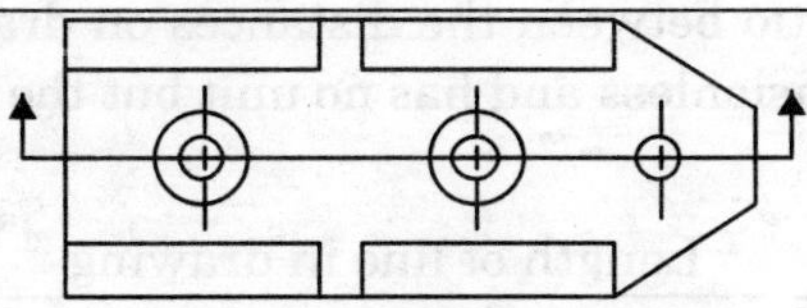	Thick long lines at ends with thin long and short lines at center to show cutting plane.
Long break lines	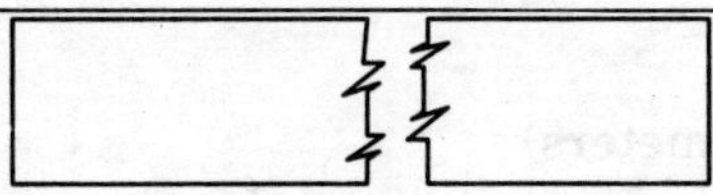	Freehand zig-zag lines at suitable intervals to show long breaks
Leader line	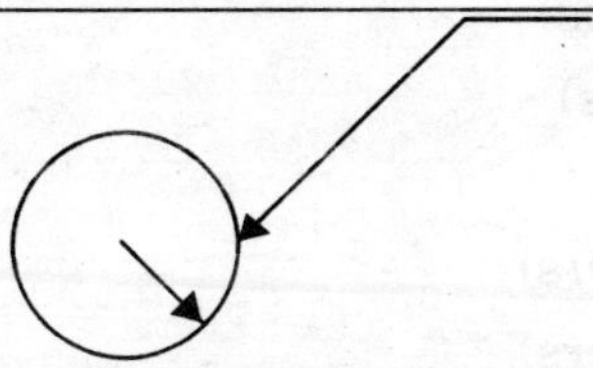	Continuous narrow line , establishes the connection between the features of a graphical representation and additional alphanumeric and/or written instructions
Reference line	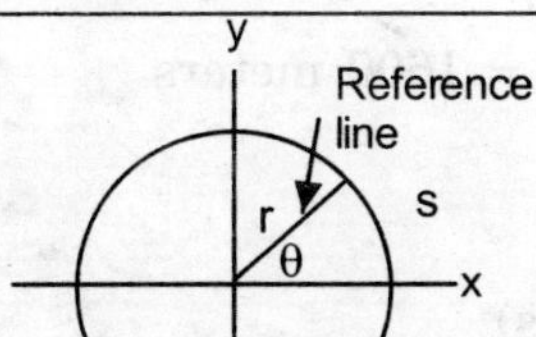	Continuous narrow line connecting with the leader line horizontally or vertically and on or at which the additional instructions are indicated

1.9 TYPES OF SCALES

Scale is the ratio of linear dimension of an object as represented in original drawing to the real linear dimension of the same element of object in real. The choice of scale depends upon the complexity of object, size of object depicted and the purpose of representation. The choice should always allow clear and easy depiction of object for correct interpretation. Various types of scales are:

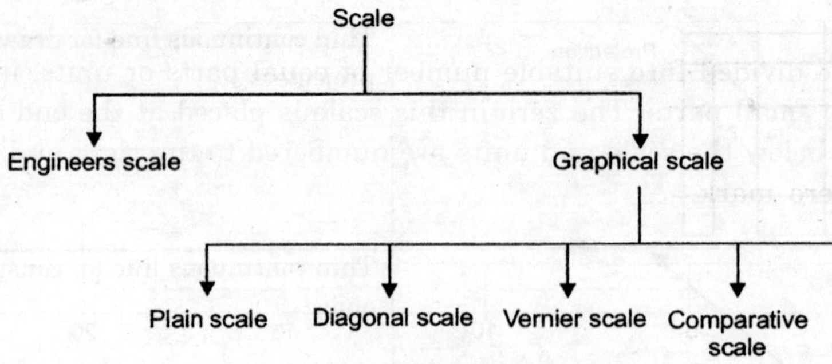

Fig. 1.28: Types of scales

Engineer's scale: In engineer's scale, the relation between the dimension on the drawing and the actual dimension of the object is mentioned in numerical values. Such as 5 mm = 1 m.

Graphical scale: In graphical scale, the scale is drawn on the drawing itself. Out of five types of graphical scales plain and diagonal scales are only used for drawing purposes.

Representative fraction (R.F.) is the ratio between the distances on drawing sheet to the corresponding actual distance. R.F. is dimensionless and has no unit but the unit of numerator and denominator should be same.

$$\text{Representative fraction} = \frac{\text{Length of line in drawing}}{\text{Actual length of line on object}}$$

Certain points to remember:

1 cm (centimeter) = 10 mm (millimeters)
1 dm (decimeter) = 10 cm (centimeters)
1 m (meter) = 10 dm (decimeters)
1 dam (decameter) = 10 m (meters)
1 hm (hectometer) = 10 dam (decameters)
1 km (kilometer) = 10 hm (hectometers)
1 mile = 1.6 km (kilometers) = 1600 meters
1 yard = 3 feet
1 feet = 12 inches
1 inch = 2.54 cm (centimeters)

1.10 TYPES OF GRAPHICAL SCALES:

The types of graphical scales are:

- Plain scale
- Diagonal scale
- Vernier scale
- Comparative scale
- Scale of chords

1.10.1 Plain Scale

It consists of a line divided into suitable number of equal parts or units, in which first part is subdivided into small parts. The zero in this scale is placed at the end of first main unit. RF is mentioned below the scale and units are numbered to the right and subdivisions to the left from the zero mark.

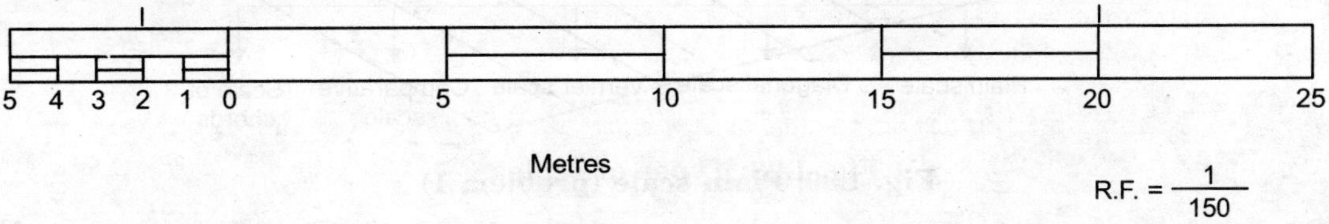

Fig. 1.29: Plain scale

How to construct Plain scale:

- Find the Representative fraction.
- Find the length of scale(L.S.) using the formula:

  Length of scale = R.F. X maximum distance to be measured
- Draw a straight line equal to length of scale and complete a rectangle by drawing height of suitable length.
- Divide it into equal number of parts and mark zero at the end of first main division.
- Make sub-divisions on first main division as per the requirement.
- Number the units to the right and sub-divisions to the left from the zero mark.
- Write the names of units and sub-divisions at respective ends.

Problem 1: Construct a plain scale of R.F. 1/5 to show centimeters and decimeters. The scale should be readable to 1 meter. Also, show length of 7.6 decimeters on the scale.

Solution. Given: R.F. = 1/5

Length of scale = RF × maximum length of scale

= 1/5 × 10 × 10 (1 m = 10 dm, 1 dm = 10 cm)

= 20 cm (Length of line to be drawn on sheet)

- Draw a line of 20cm and divide it into 10 equal parts each representing 1 decameter.
- Put zero at the end of first main division and mark 1, 2, 3, 4, 5, 6, 7, 8 and 9 at each subsequent division to the right.
- Divide first division into 10 equal parts each representing 1 centimeter.
- Write centimeters on left and decimeters on right.
- To mark distance of 7.6 decimeters take 7 decimeters from main scale and 6 subdivisions equal to .6 decimeters or 6 centimeters from the scale made on left side of zero mark.

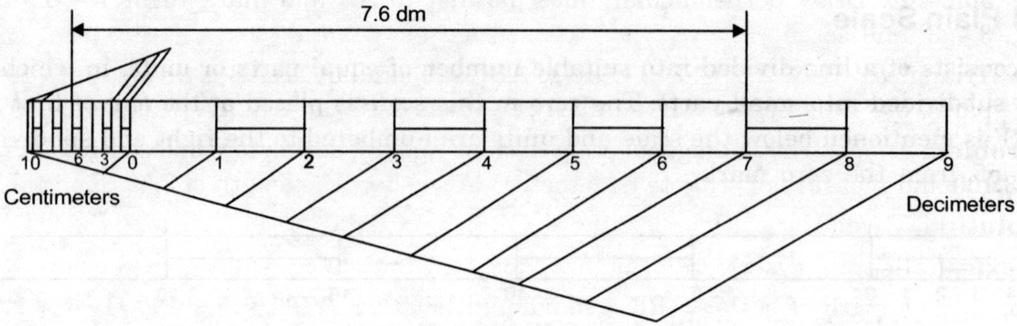

Fig. 1.30: Plain scale (problem 1)

1.10.2 Diagonal Scale

A diagonal scale consists of a set of parallel lines crossing obliquely with other set of lines such that their intersections create subdivisions. It is mainly used when measurements are required in three units such as cm, m and km. The units of measurement created could not be read on plain scale. The scale has two parts: the lower part which is a plane scale and the upper part which has oblique lines as diagonals. On this scale it is possible to read up to 1/100th of a meter which is not possible on plane scale.

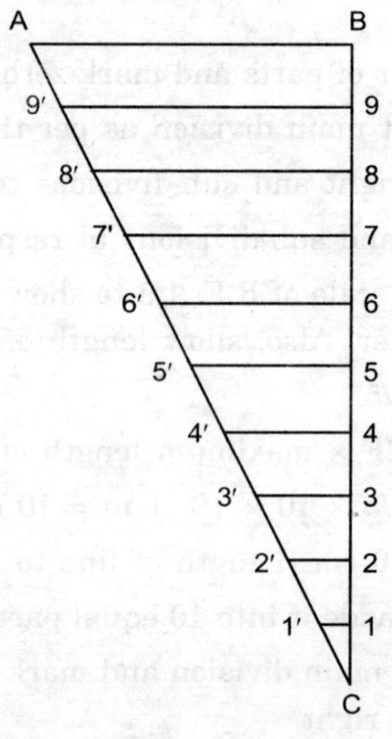

Fig. 1.31: Diagonal scale

How to construct a diagonal scale to obtain small divisions of short lines:

- Let a line AB be divided into 10 equal parts.
- At one end B, draw a line perpendicular to AB and mark off ten divisions (equal distances of any length) along it starting from B to C.

- Join AC. Draw perpendicular lines parallel to AB and mark them 1 – 1′, 2 – 2′, 3 – 3′ and so on.
- Triangles 1′ 1 C, 2′ 2 C, 3′ 3 C up to ABC are similar. As AB is divided into 10 equal parts, 5 – 5′ is half of AB, similarly 2 – 2′ is 2/10th of AB, 7 – 7′ is 7/10th of AB.

Problem 2: Construct a diagonal scale to show meters, decimeters and centimeters. Take RF = 3:100 and measure of scale to be 5 meters. Also show the length of 3.69 m on the scale.

Solution. Given data: RF = 3/100

Maximum length of scale = 5 meters

So, Length of scale = RF × maximum length = 3/100 × 5 × 100 (1 m = 100 cm)

= 15 cm

- Draw a line of length 15cm and divide it into 5 equal parts each representing 1 meter.
- Divide the first division into 10 equal subdivisions. Each subdivisionn showing 1 decimeter.
- At the starting point, erect a perpendicular of suitable length and complete the rectangle. Divide the line drawn into 10 equal parts and draw horizontal lines along each subdivision throughout the rectangle.
- Draw perpendiculars at meter divisions 0, 1, 2, 3 and 4.
- Join the top most left point with point 9 (first sub division on horizontal scale on the left side of zero mark.
- Draw lines parallel to line drawn in earlier step through the remaining points 8, 7, 6, 5, 4, 3, 2, 1 and 0.
- Now, mark a distance of 3 meters 6 decimeters and 9 centimeters on the scale.

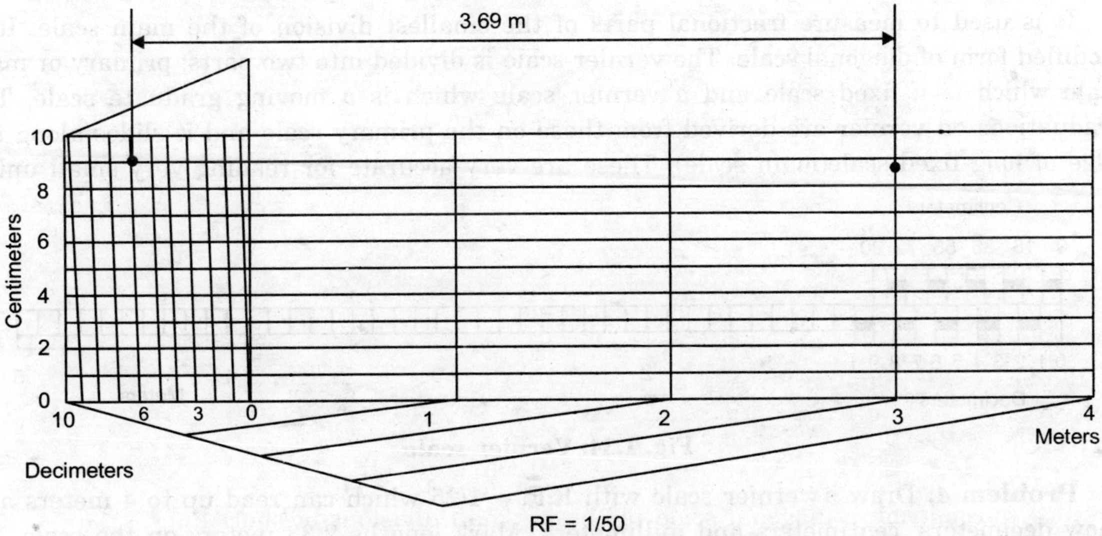

Fig. 1.32: Scale drawn for problem 2

Problem 3: Two points are at a distance of 14 cm on a drawing. But the real distance between these points is 20 km. Draw a diagonal scale that can read up to 25 km and mark a distance of 17 kilometers and 6 hectometers on the scale.

Solution. Representative fraction = 14 cm/ 20 × 1000 × 100 cm

= 7/1000000 (1 km = 1000 m, 1 m = 100 cm)

Maximum length of scale = 25 km

Length of scale = 7/1000000 × 25 × 1000 × 100 = 17.5 cm

Let there be 5 divisions on horizontal scale, each representing 1 km.

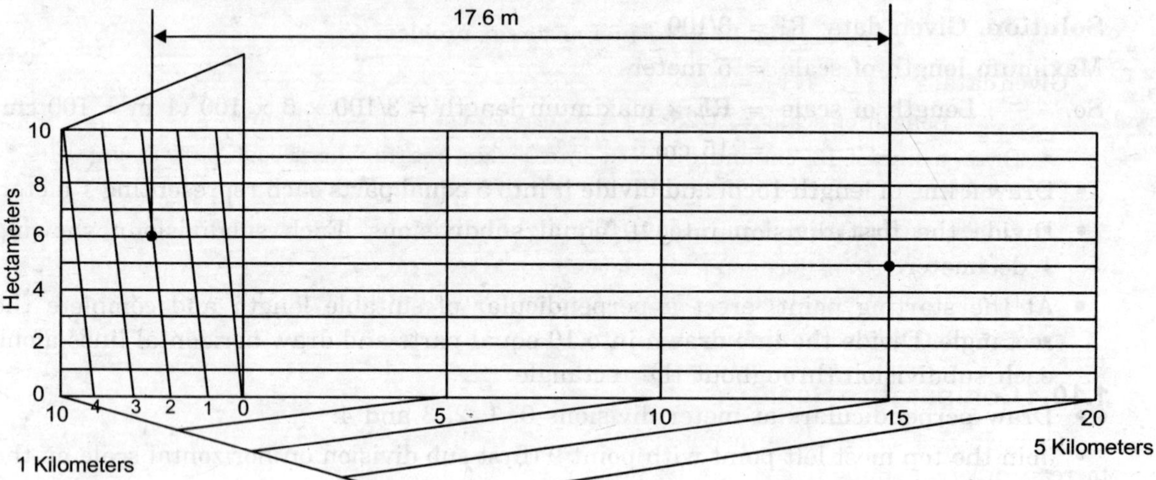

Fig. 1.33: Scale drawn for problem 3

1.10.3 Vernier Scale

It is used to measure fractional parts of the smallest division of the main scale. It is modified form of diagonal scale. The vernier scale is divided into two parts: primary or main scale which is a fixed scale and a vernier scale which is a moving graduate scale. The graduations on vernier are derived from those on the primary scale and it slides along the edge of long fixed scale(main scale). These are very accurate for reading very small units

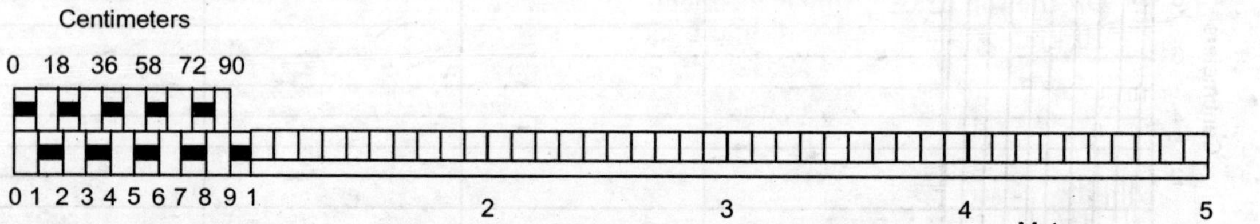

Fig. 1.34: Vernier scale

Problem 4: Draw a vernier scale with R.F. = 1/25 which can read up to 4 meters and show decimeters, centimeters and millimeters. Mark lengths 2.33 meters on the scale.

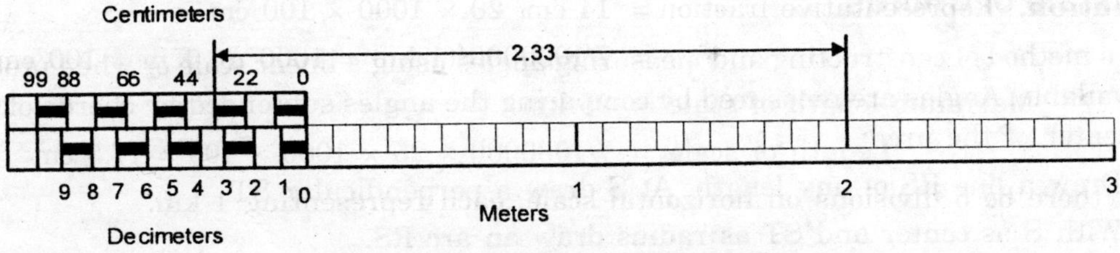

Fig. 1.35: Vernier scale for problem 4

Given data: R.F. = 1/25

Length of scale = 1/25 × 4 × 100 = 16 cm (1 m = 100 cm)

- Draw a line of 16 cm and divide it into 4 equal parts where each part is equal to 1 meter.
- Divide each of these 4 parts into 10 equal subparts to show decimeter.
- Take 11 parts of decimeter length each equal to 1 cm on left side of zero and divide them into 10 equal parts. Each part shows length of 11 cm or 1.1 dm.
- Complete the scale and mark 2.33 dm.

1.10.4 Comparative Scale

Comparative scale or corresponding scales consists of two scale of same RF but graduated to read different units. They may be plain scales or diagonal scales or vernier scales or separately constructed scales one above the other. They are used to compare distances expressed in different systems such as kilometers and miles.

Problem 5: A distance of 40 miles is represented by a line 8 cm long. Construct a plain scale that reads 80 miles. Construct a comparative scale to read upto 120 km (1 m = 1.609 km)

(i) Miles scale - Given that 40 miles = 8 cm so 80 miles will read 16 cm.

RF = 8/40 × 1609 × 1000 × 100 = 1/804500

Divide a line of 18 cm into 8 equal parts. Sub divide first part to measure miles.

(ii) Kilometers scale - Length of scale = 1/804500 × 120 × 1000 × 100 = 14.90 cm

On the top draw a line of 14.9 cm and divide it into 12 equal parts each representing 10 km.

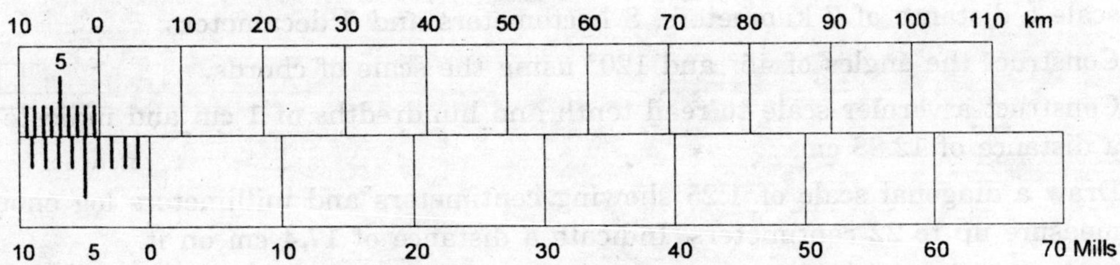

Fig. 1.36: Comparative scale

1.10.5 Scale of Chords

It is a method of constructing and measuring angles using a linear scale or when protractor is not available. Angles are measured by comparing the angles subtended by chords of an arc at the center of the arc.

- Draw a line RS of any length. At S draw a perpendicular ST.
- With S as center and ST as radius draw an arc RS.
- Divide the arc in 9 equal parts each division of 10° and mark them as 10, 20, 30, 40 and so on.
- With R as center and R-10, R-20, R-30 ... as radius turn down the divisions to line through R and complete the scale.
- Construct linear scale by drawing rectangle and mark the divisions with zero below R and number the divisions subsequently as 10°, 20°, 30° and so on.

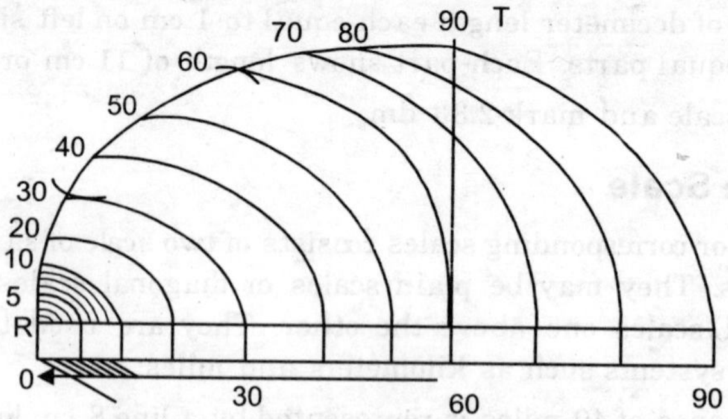

Fig. 1.37: Scale of chords

1.11 PRACTICE PROBLEMS

1. What is a scale? Why are they used in engineering drawing?
2. Differentiate between plain scale and diagonal scale?
3. An area of 144 sq. cm on a map represents an area of 36 square km on the field. Find the RF of the scale for this map and draw a diagonal scale to show kilometers, hectometers and decameters and to measure up to 10 kilometers. Indicate on the scale a distance of 7 kilometers, 8 hectometers and 5 decameters.
4. Construct the angles of 45° and 120° using the scale of chords.
5. Construct a vernier scale to read tenth and hundredths of 1 cm and indicate on it, a distance of 12.88 cm.
6. Draw a diagonal scale of 1:25 showing centimeters and millimeters log enough to measure up to 22 centimeters. Indicate a distance of 17.4 cm on it.
7. Draw a plain scale to show decimeters & centimeters and to measure upto 1 m.

1.12 FUNDAMENTALS OF PLANE GEOMETRIC CONSTRUCTIONS

Engineering drawing is made by constructing combinations of primitive geometric forms such as points, lines and planes. These building blocks define the shape and position of object in space. The engineers must be familiar with geometries and geometrical constructions as they are used frequently in engineering drawings.

Primitives:

(*i*) **Point:** A point represents a location in space or on a drawing and has no depth, width or height. It can be represented by intersection of two lines, a crossbar on line or simply a cross.

Table 1.14: Methods of drawing points

A point denoted by cross	
Point at the center of a circle	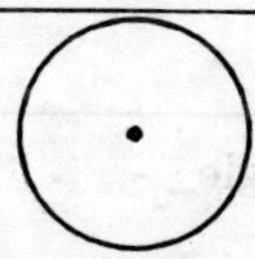
Point at intersection of two lines	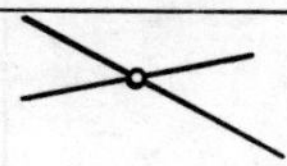
Point at end or midpoint of a line	

(*ii*) **Line:** It is a geometric primitive of zero thickness which has length and direction.

Table 1.15: Methods of drawing lines

Parallel lines	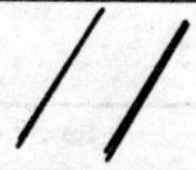
Intersecting lines	
Tangent lines	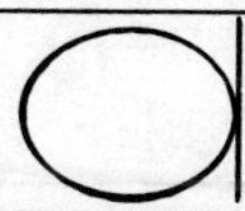
Non parallel lines	
Perpendicular lines	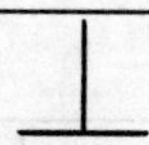

(*iii*) **Angle:** An angle is the inclination between two intersecting lines.

Table 1.16: Types of angles

Acute angle		Less than 90°

Right angle	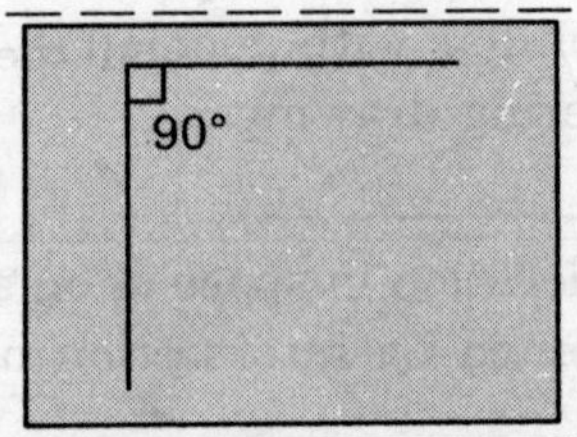	Exactly 90°
Obtuse angle	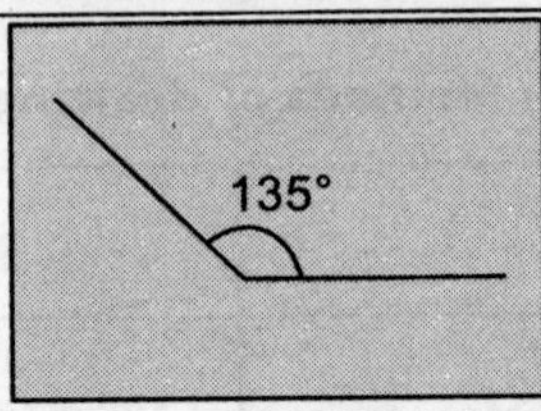	Greater than 90° but less than 180°
Straight angle	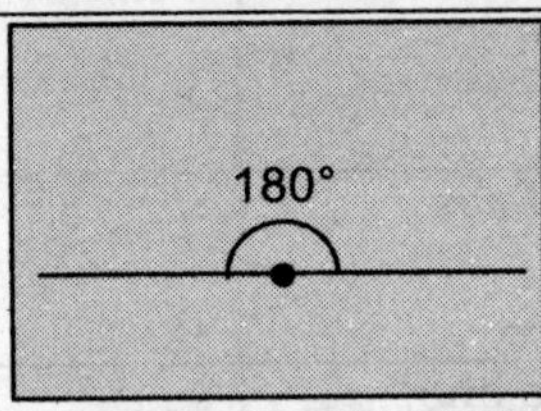	Exactly 180°
Reflex angle	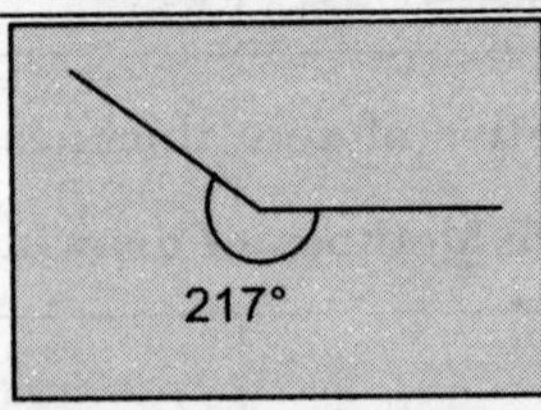	Greater than 180°
Full angle	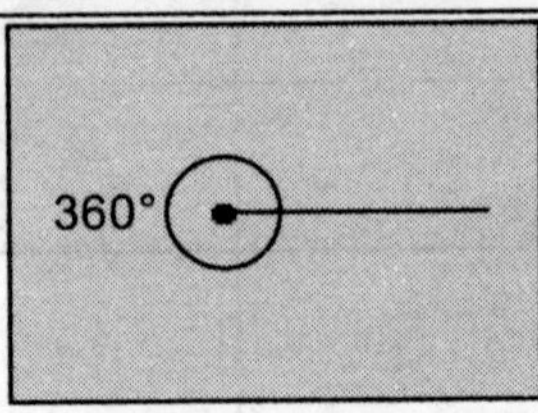	Exactly 360°

(*iv*) **Triangle:** A triangle is a closed plane figure with three straight sides and their interior angles sum is equal to 180°.

Table 1.17: Types of triangles

Name	Example	Definition
Equilateral triangle	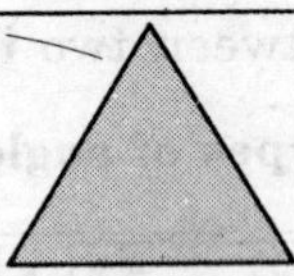	All three sides are of equal length
Isosceles triangle	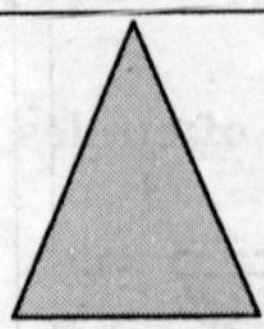	Two sides and the angles opposite to them are same.

Scalene triangle	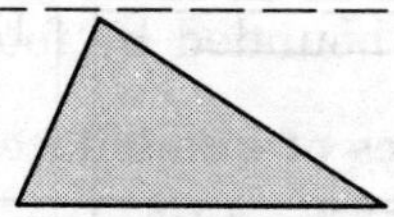	All the sides and angles are different.
Acute angled triangle	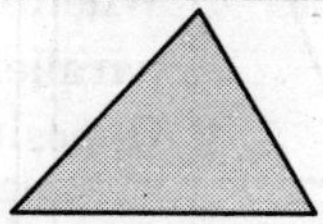	All the angles are less than 90°(acute)
Right angled triangle	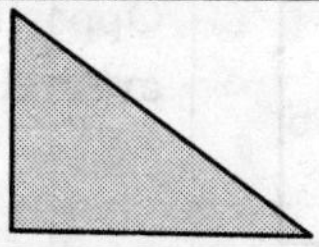	One of the angle is 90°
Obtuse angled triangle	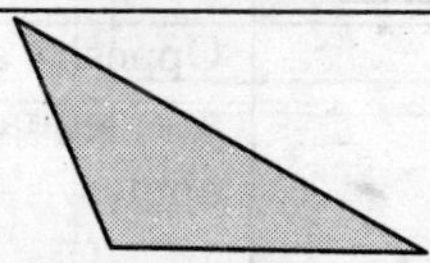	One angle is greater than 90°(obtuse)

(*v*) **Circle:** A circle is a closed curve with all points on the circle at same distance from the center point. The fixed distance from the center to circumference is called radius. The half of circle is termed as semi-circle.

The *diameter* is twice of the radius and is a straight line drawn through the center and meeting at both ends on the circumference.

A *sector* is the part of circle bounded by an arc and two radii.

A *chord* is a straight line whose ends terminate on the circumference of the circle.

A *quadrant* is a sector with central angle of 90°

A *segment* is the smaller portion of circle separated by the chord.

Concentric circles are two or more circles with the same center point.

Eccentric circles are two or more circles without a common point.

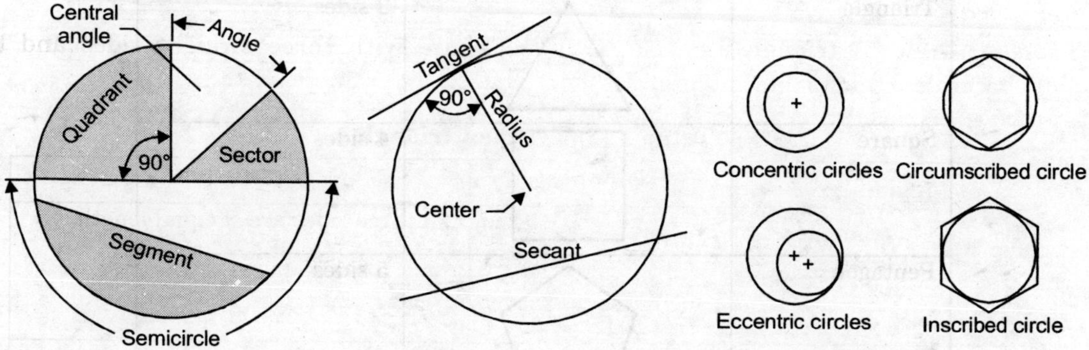

Fig. 1.38: Components of circles and its types

(*vi*) **Quadrilateral:** It is a plane figure bounded by four sides.

Table 1.18: Types of quadrilaterals

Parallelogram	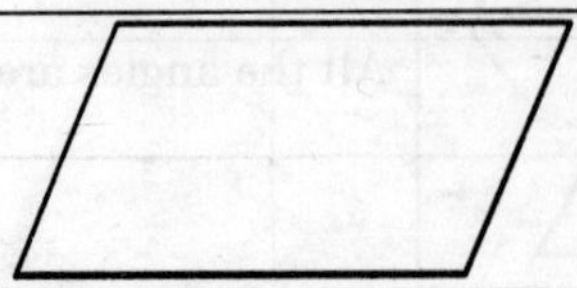	When opposite sides are equal and parallel. Opposite angles are equal.
Rectangle	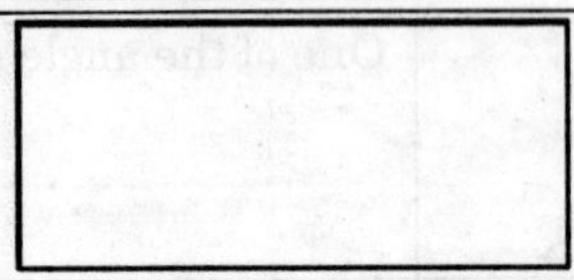	Opposite sides are equal. All angles are right angles.
Square	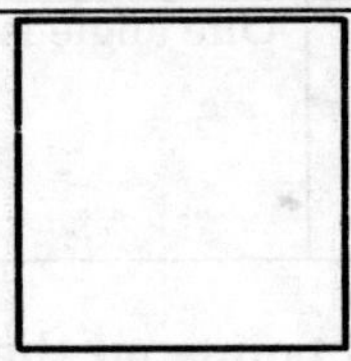	Opposite sides are parallel. All sides are equal. All angles are right angles.
Rhombus	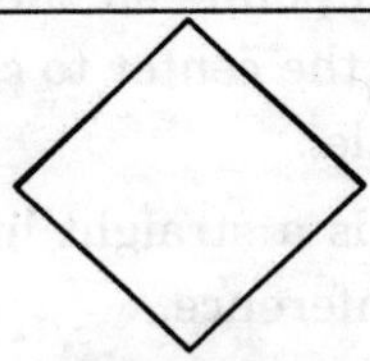	All sides are equal. Opposite sides are parallel. Opposite angles are equal. Diagonals bisect each other but angles are not right angles.
Trapezoid	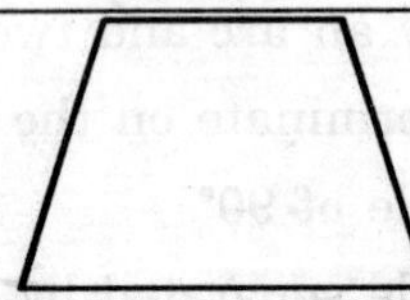	Two sides are parallel

(*vii*) **Polygon:** A polygon is a plane bounded by three or more straight lines.

Table 1.19: Types of polygon

Triangle	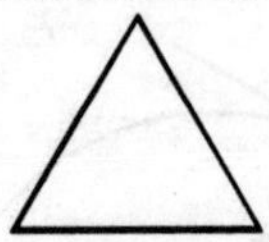	3 sides
Square	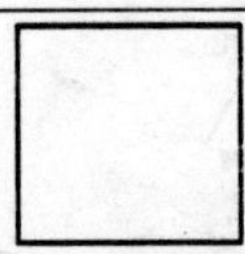	4 sides
Pentagon	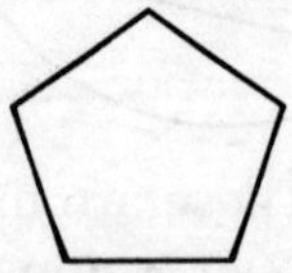	5 sides
Hexagon	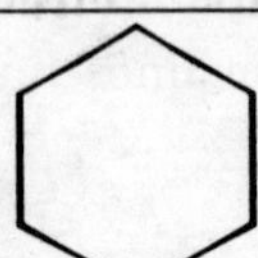	6 sides

Heptagon	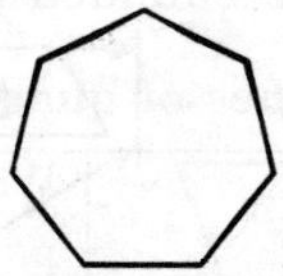	7 sides
Octagon	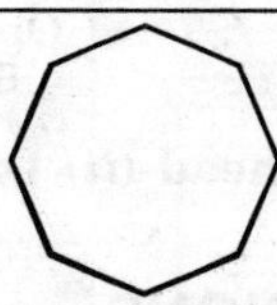	8 sides
Nonagon	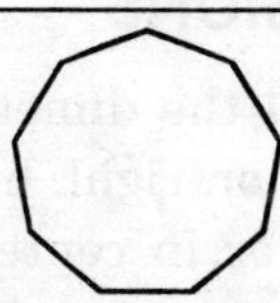	9 sides
Decagon	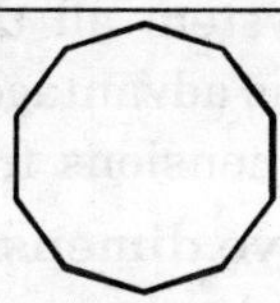	10 sides
Undecagon	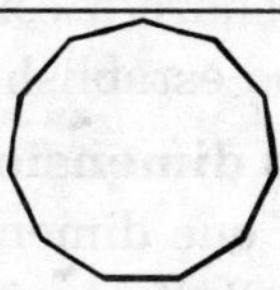	11 sides
Dodecagon	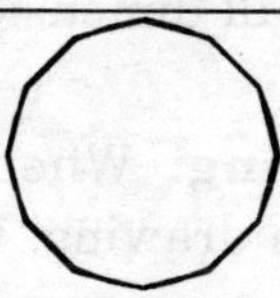	12 sides

1.13 DIMENSIONING

A complete drawing must convey the designer's intent clearly to the end user and must be properly dimensioned to show shape and size of an object and also give information about length, breadth, distances, material finish, material requirements etc.

1.13.1 Features of Dimension Lines

(*i*) **Dimension Line:** It is the thin continuous lines that show the extent and direction of the dimension. Arrow heads are placed at the ends of dimension lines and dimension is mentioned either at the top or in the middle of dimension line.

(*ii*) **Extension Line:** These extended lines are drawn perpendicular to the dimension without touching the object lines. These are drawn after giving a gap of 1 or 2 mm from the point and should extend 2 or 3 mm beyond the dimension line.

(*iii*) **Arrow head:** Each end of dimension lines has an arrow head. The length of arrow head should be 3 times the spread.

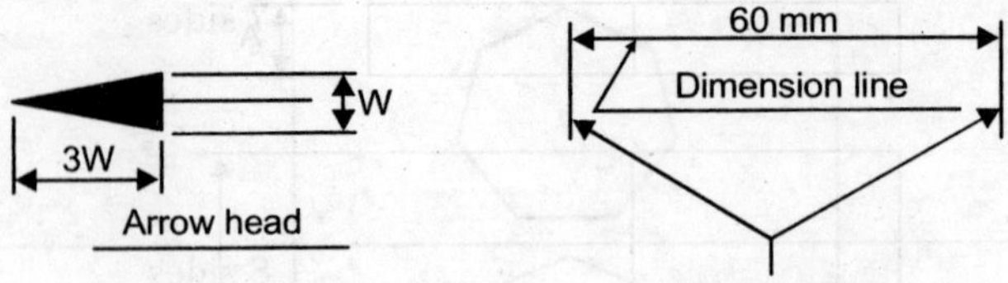

Fig. 1.39: (*i*) Arrow head (*ii*) Dimension line

1.14 SYSTEM OF PLACING DIMENSIONS

(*i*) **Aligned system:** In this system all the dimensions are placed in such a way that they can be read from the bottom or right hand edge of the drawing sheet. The dimension values are placed above or in center of dimension line.

(*ii*) **Unidirectional system:** In this system all the dimensions can be read from the bottom of drawing sheet. The system is advantageous in dimensioning of large drawings where it is inconvenient to read dimensions from right hand side.

(*iii*) **Parallel dimensioning (Progressive dimensioning):** All the dimensions are shown from a common reference line and share a common extension line. This method is adopted when dimensions have to be established from a particular datum surface

(*iv*) **Chain dimensioning (Continuous dimensioning):** All the dimensions are aligned in such a way that an arrowhead of one dimension touches tip-to-tip the arrowhead of the adjacent dimension. The overall dimension is placed outside the other smaller dimensions.

(*v*) **Parallel Combined dimensioning:** When chain dimensioning and parallel dimensioning are used on the same drawing, the method of dimensioning is called combined dimensioning.

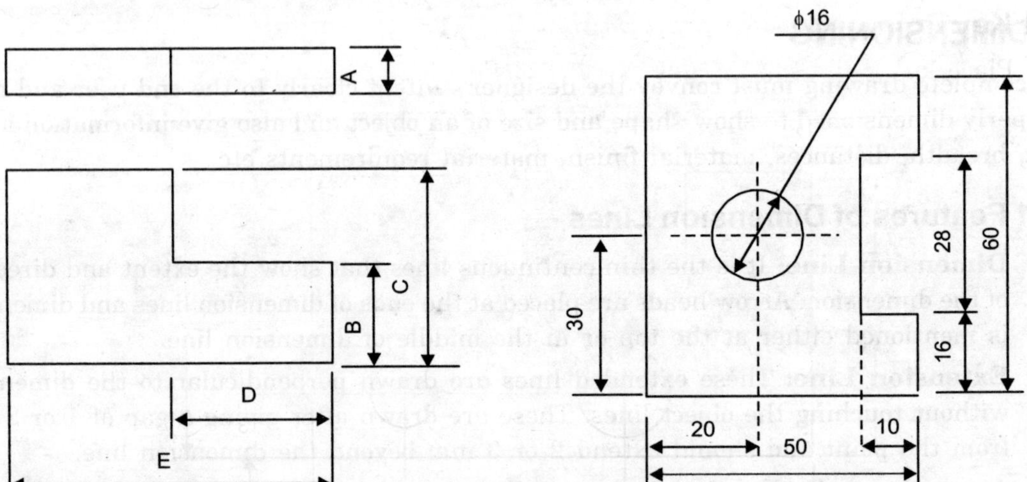

Fig. 1.40: Aligned Dimensioning

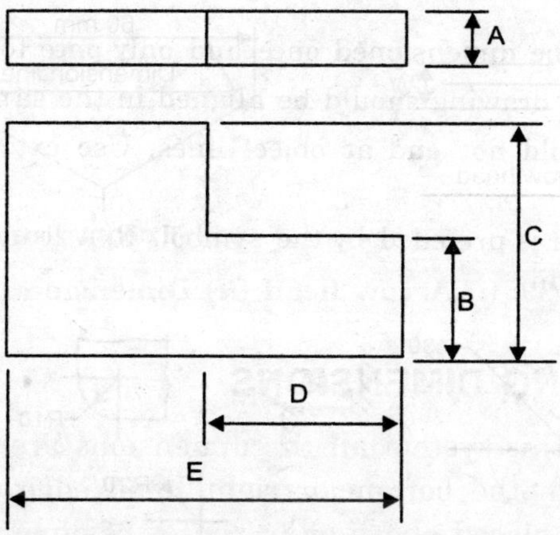

Fig. 1.41: Unidirectional dimensioning

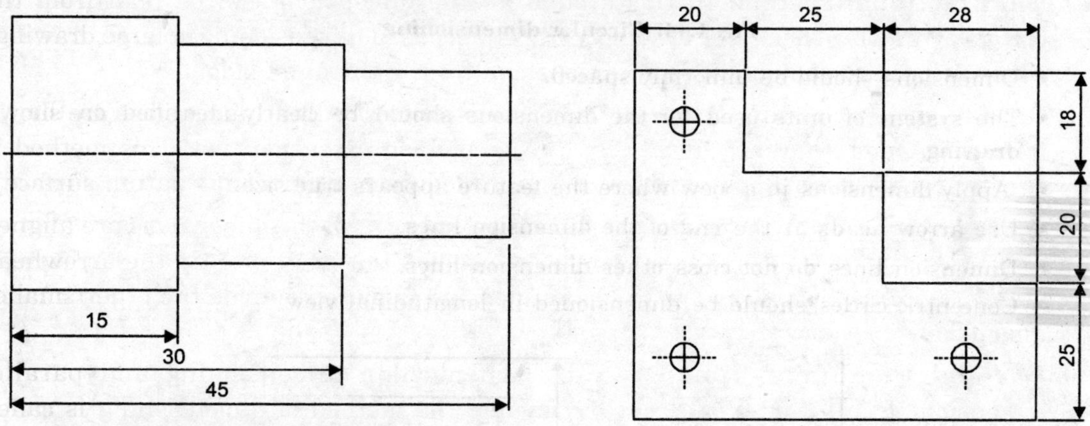

Fig. 1.42: (*i*) Parallel dimensioning (*ii*) **Chain dimensioning**

1.14.1 Key Points to be Kept in Mind while Dimensioning

- Place larger dimensions towards the outside.

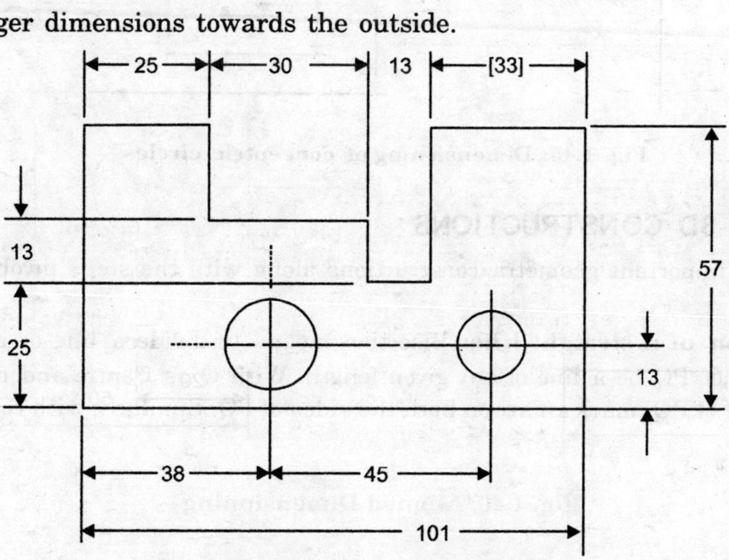

Fig. 1.43: Example of dimensioning

- Each feature should be dimensioned once and only once to avoid confusion.
- The dimensions on a drawing should be aligned in the same direction.
- Dimension lines should not end at object lines. Use extension lines to show the dimension instead.
- The radius dimension is preceded by the symbol, R. A leader line is commonly used for diameters and radii.

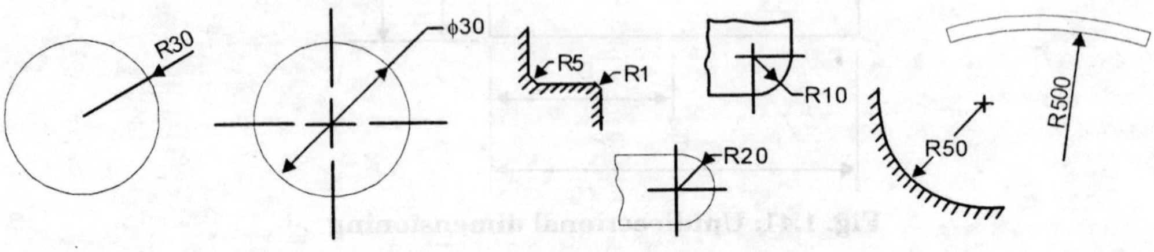

Fig. 1.44: Circular dimensioning

- Dimensions should be uniformly spaced.
- The system of units used for the dimensions should be clearly identified on the drawing,
- Apply dimensions in a view where the feature appears true size.
- Use arrow heads at the end of the dimension lines.
- Dimension lines do not cross other dimension lines.
- Concentric circles should be dimensioned in longitudinal view.

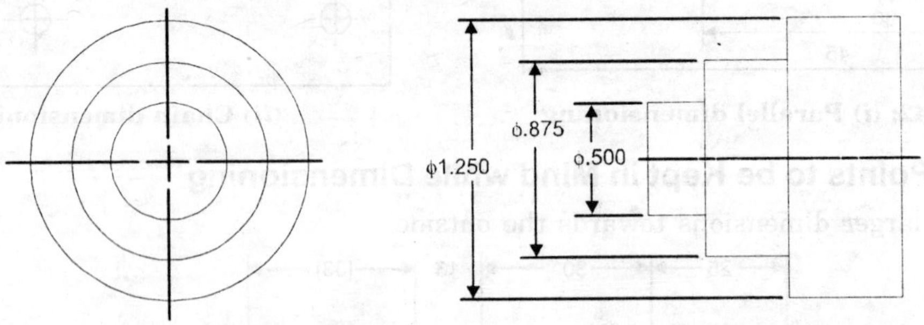

Fig. 1.45: Dimensioning of concentric circles

1.15 2D AND 3D CONSTRUCTIONS

Some of the important geometric constructions along with the steps involved are given below:

(*i*) **Bisection of a straight line:** Bisection means to divide a line or an arc in two halves. Let PQ be a line of any given length. With Q as Centre and radius greater than half of PQ, mark an arc on both the sides of PQ. Similarly, with center as P and same radius, draw the arcs on both sides of PQ intersecting the arc made earlier with Q as center. Join the points where intersection of arcs meets.

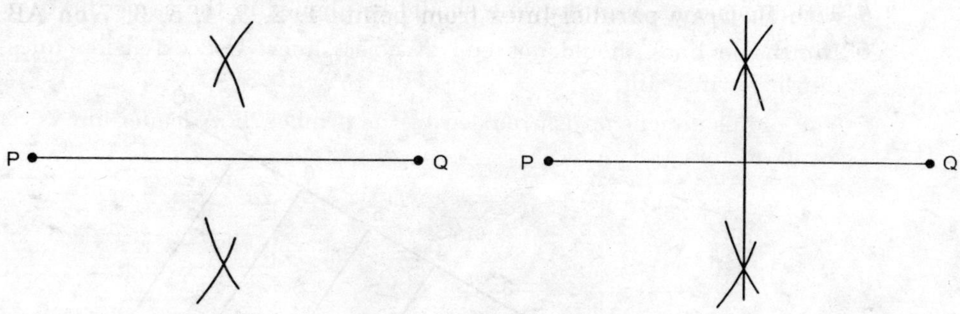

Fig. 1.46: Bisection of a straight line

(*ii*) **Drawing of perpendiculars to a given point outside the line:** Draw a line and a point P which is not on the line. Put the compass on point P and stretch it unless it pass the line. Draw an arc on left and right such that they intersect the line. Keeping the point of intersection on left stretch the compass untill it is more than half the distance between intersections. Draw an arc on the side opposite of point P. Keeping the compass length same put the point on intersection point at right and draw an arc on opposite side of P. Connect P with the intersection of arcs.

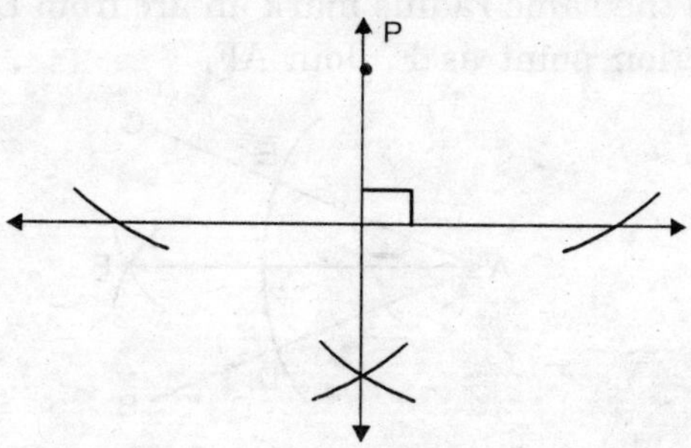

Fig. 1.47 Perpendiculars to a given point outside the line

(*iii*) **Drawing of parallel lines to a given line:** Let AB the given line and F given point. With F as center and any radius draw an arc CE cutting AB at E. Taking E as center and same radius, draw an arc FD at D. With E as center and same radius, draw an arc intersecting CE at G. Draw a line through F and C.

Fig. 1.48: Parallel lines to a given line

(iv) **To divide straight line into a number of parts:** Take a straight line AB and draw a line AC at an acute angle. From A cut 7 equal divisions on AC using divider. Join 7 with B. Draw parallel lines from points 1, 2, 3, 4, 5, 6, 7 on AB at 1′, 2′, 3′, 4′, 5′, 6′, B.

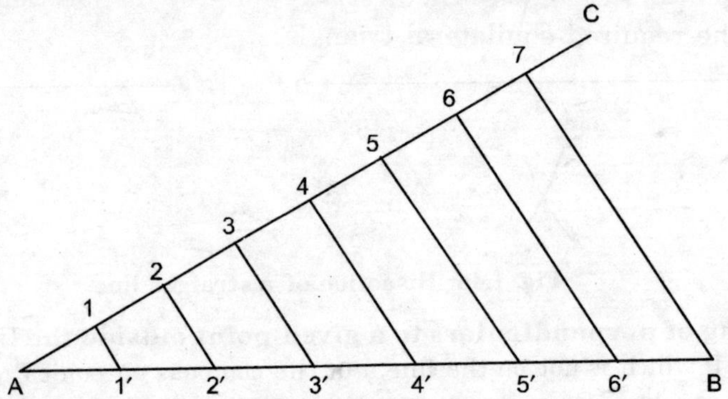

Fig. 1.49: To divide straight line into a number of parts

(v) **To bisect an angle:** Let BCA be the angle between AB and AC. With A as center cut an arc ED intersecting AC at E and AB at D. With E as center and some radius mark an arc. With the same radius mark an arc from D intersecting the previous arc. Mark the intersection point as F. Join AF.

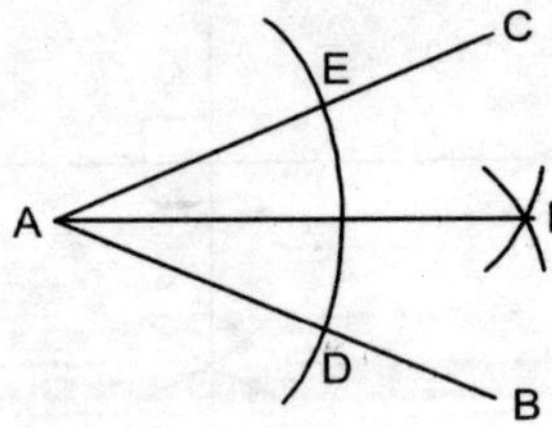

Fig. 1.50: To bisect an angle

(vi) **To find the centre point of a given arc:** Let AB be the arc. Select two chords CD and EF on AB. Draw the perpendicular bisectors of CD and EF. Join them to obtain point O which is the center of arc.

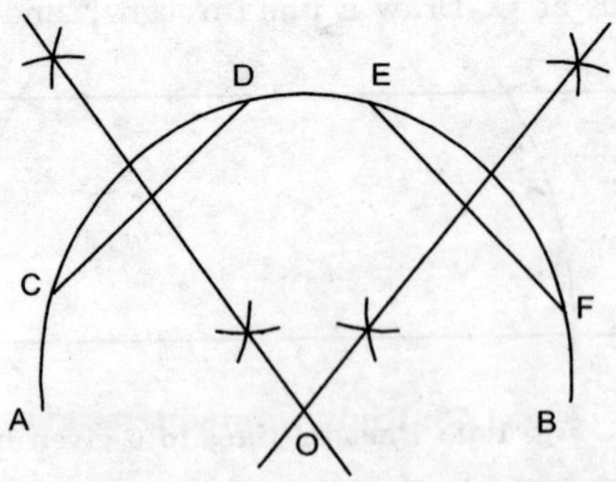

Fig. 1.51: To find the center point of a given arc

(*vii*) **To construct an equilateral triangle when one of the side is given:** Draw a horizontal straight line PQ of given distance. Place your compass on point P and measure the distance to point Q. Swing an arc of measured size above or below the line segment. Keeping the span of compass same, place the compass on point Q and make an arc, intersecting with the first arc. Join the intersection point R with P and Q. PQR is the required equilateral triangle.

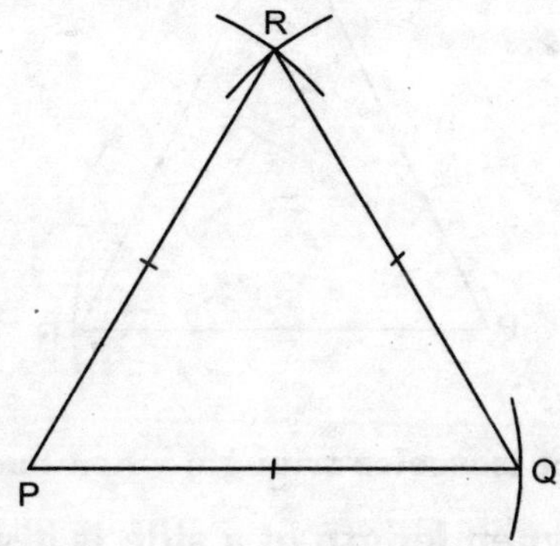

Fig. 1.52: To construct an equilateral triangle when one of the side is given

(*viii*) **To construct a right angled triangle:** Let AB the hypotenuse and CD be perpendicular distance from a right angle to hypotenuse be given. Draw AB as hypotenuse and at perpendicular distance draw a line GH parallel to AB. Bisect AB and describe a semi-circle on it passing through GH. Let the point of intersection be C. Join AC and AB.

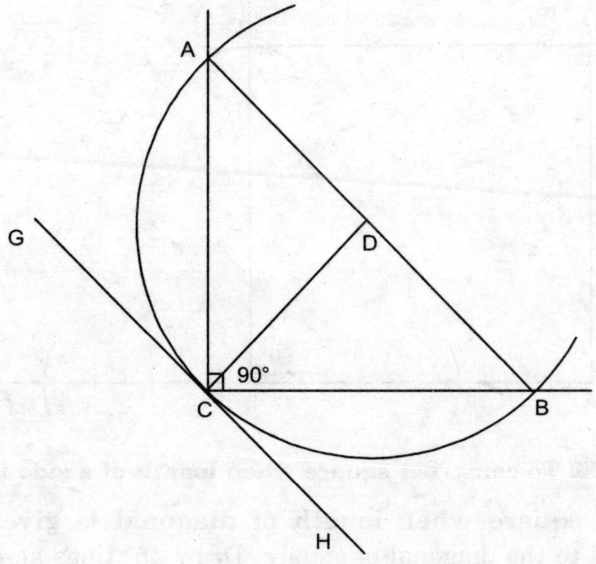

Fig. 1.53: Right angled triangle

(*ix*) **To construct isosceles triangle when base and a side is given:** Draw the base PR of given length. As two sides of isosceles triangle are same, take distance equal to the given side in compass and with P as centre draw an arc on either side of the line segment. With the same dimension, draw an arc intersecting the previously made arc with Q as center. Mark the intersection point as Q. Join PQ and RQ.

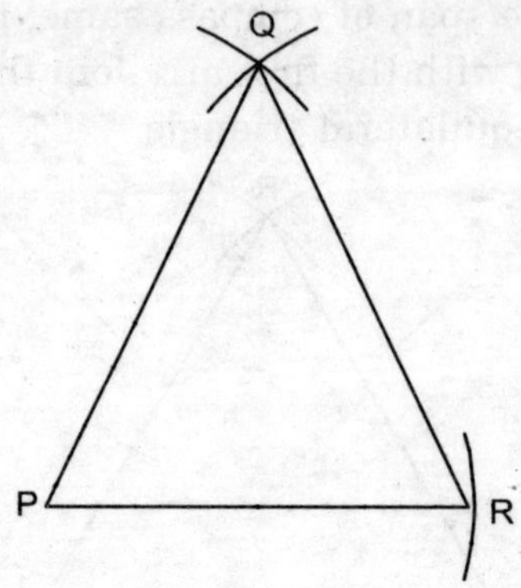

Fig. 1.54: To construct isosceles triangle when base and a side is given

(*x*) **To construct square when length of a side is given:** All the sides of a square are equal and all the angles are of 90°. Make angle ABC of 90°. Taking the compass chord distance equal to length of square mark arcs on line segments AB and BC with B as center. From point C mark an arc. Again mark an arc with A as center intersecting previously made arc. Mark he intersection of both the arcs as D. Join AD and CD.

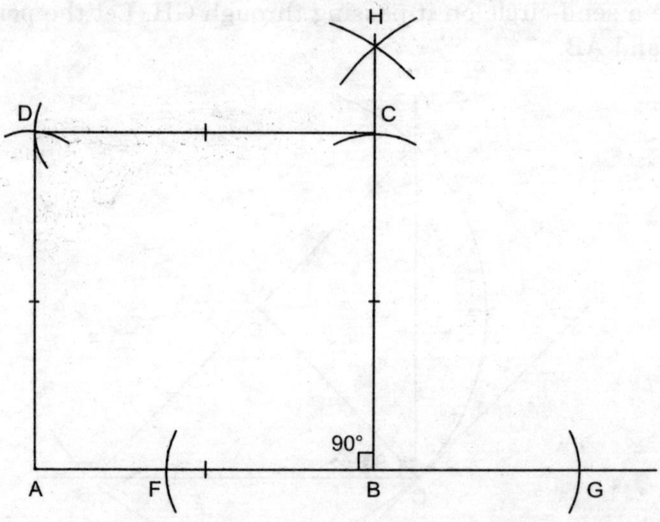

Fig. 1.55: To construct square when length of a side is given

(*xi*) **To construct square when length of diagonal is given:** Draw a circle with diameter equal to the diagonal of square. Draw 45° lines across the center and join the points where lines inclined at 45° intersect of circle.

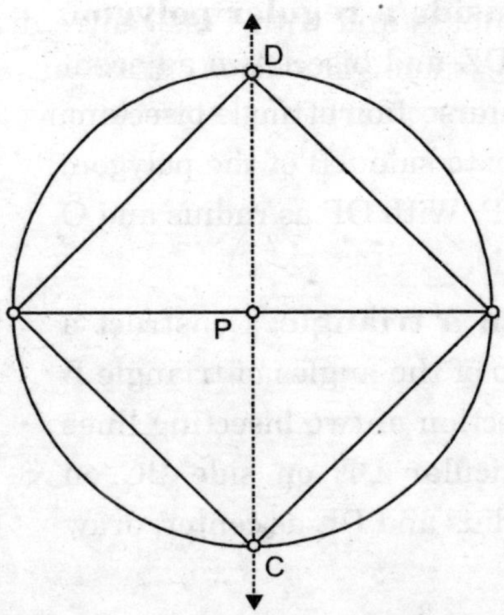

Fig. 1.56: To construct square when length of diagonal is given

(*xii*) **To construct hexagon, when one side is given:** Draw a circle with radius equal to side of hexagon using a compass. Move the compass at the top of circle and without changing angle of compass mark an arc on the edge of circle. Move the compass to the arc made and again make an arc on the edge of circle. Continue making the marks unless you reach the first mark. Join the intersections of arcs to form a hexagon.

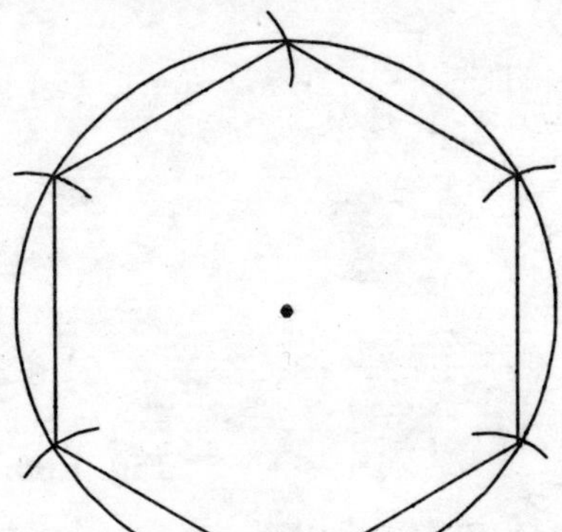

Fig. 1.57: To construct hexagon, when one side is given

(*xiii*) **To construct regular pentagon when one side is given:** Draw vertical and horizontal lines. With the intersection of lines as center draw a circle. Without adjusting the compass, place the point of the compass on the circle where it crosses the horizontal line. Now draw arcs on the previous circle above and below and connect those points. Now center the compass on the crosshair made from the bisector and draw an arc from the top of the circle down to the horizontal line. Do not close the compass. Putting the point of the compass at the top of the circle draw an arc from where the last arc intersected the horizontal line out to the circle. Now move around the circle using each arc as the center of the next arc. Draw lines from each intersection to form a pentagon.

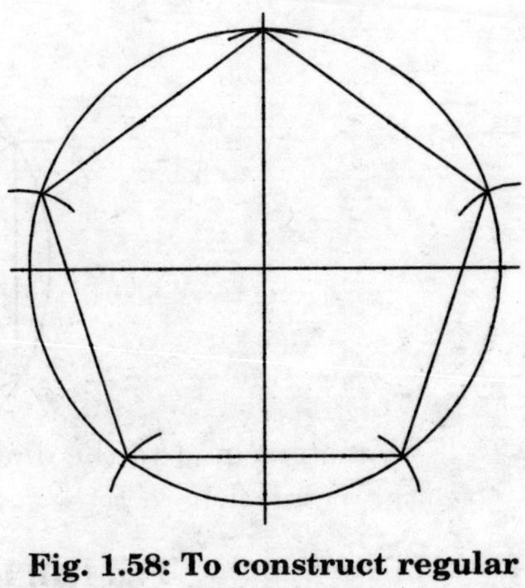

Fig. 1.58: To construct regular pentagon when one side is given

(*xiv*) **To inscribe a circle inside a regular polygon:** Draw the polygon ABCDE and bisect two adjacent angles A and B. At the intersection of angle bisectors, O draws a perpendicular to side AB of the polygon. Draw a perpendicular OP. With OP as radius and O as center draw the circle.

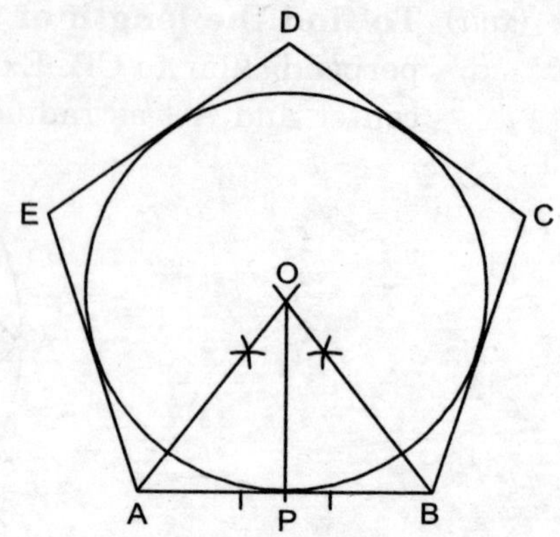

Fig. 1.59: To inscribe a circle inside a regular polygon

(*xv*) **To inscribe a circle in a triangle:** Construct a triangle ABC. Bisect two of the angles of triangle B and C. Mark the intersection of two bisecting lines as D. Draw a perpendicular DE on side BC on triangle. Taking D as radius and DE as center, draw a circle.

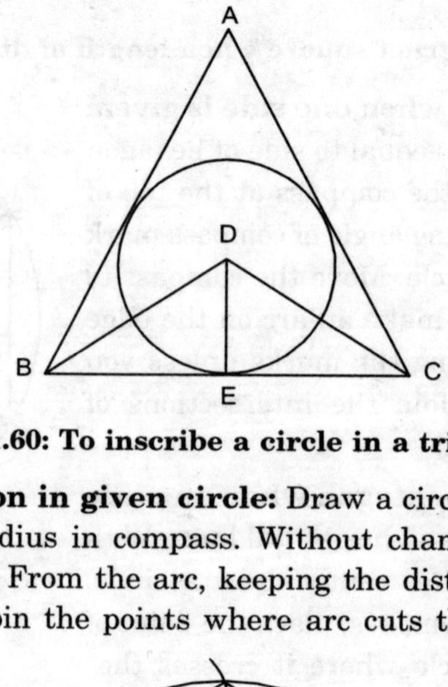

Fig. 1.60: To inscribe a circle in a triangle

(*xvi*) **To inscribe a hexagon in given circle:** Draw a circle of given radius and take the dimension equal to radius in compass. Without changing the distance of compass, mark an arc on circle. From the arc, keeping the distance same, mark another arc. Repeat the process. Join the points where arc cuts the circle.

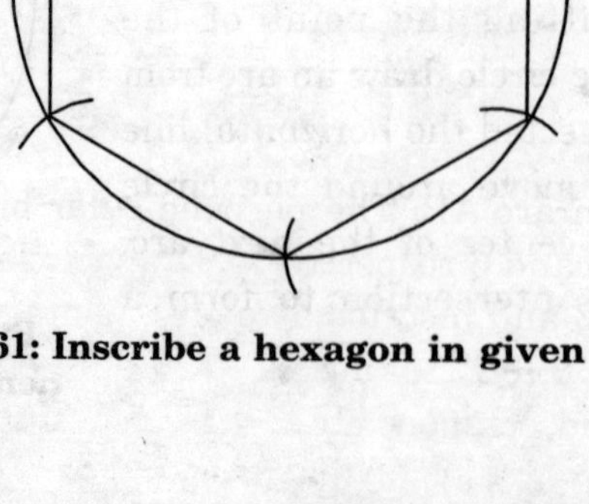

Fig. 1.61: Inscribe a hexagon in given circle

(*xvii*) **To find the length of an arc:** Let AB be the arc with center C. Draw AB and BE perpendicular to CB. Extend AB to a point G such that BG is half of AB. With G as center and AG as radius draw an arc AF. Take line BF is the length of arc AB.

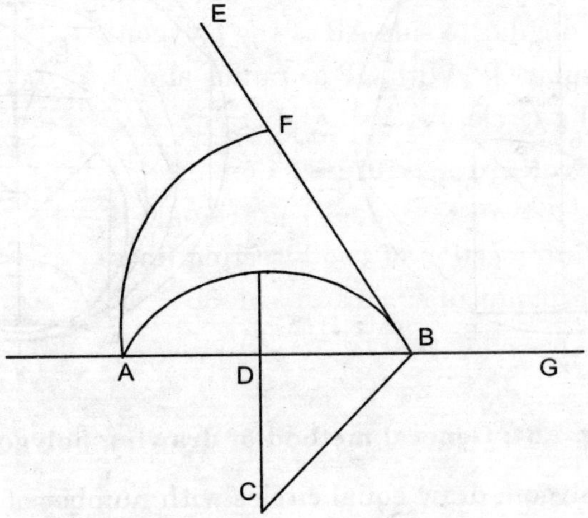

Fig. 1.62: Finding the length of an arc

(*xviii*) **To draw common tangents to two circles of different radii:** A common tangent can be tangent either internally or externally. A common internal tangent is a common tangent that intersects the segment that joins the center of the two circles. A common external tangent is a common tangent that does not intersect the segment that joins the centers of the two circles.

Draw two circles, C1 and C2, each outside the other. Let the centers be O and Q. Join OQ. Draw perpendiculars OB and CQ intersecting circles at B and C. Join CB. It is the required tangent.

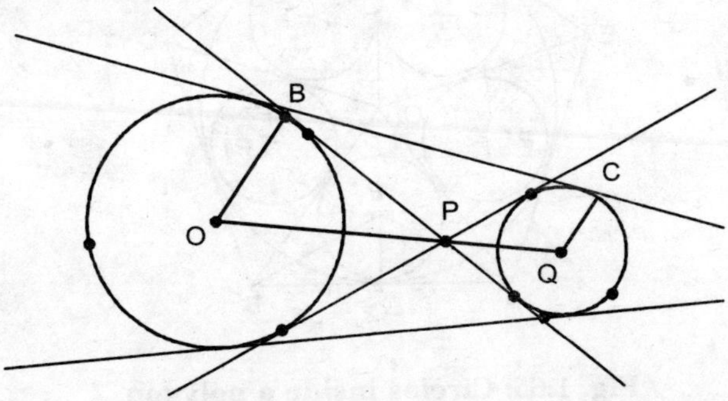

Fig. 1.63: Common tangents to two circles of different radii

(*xix*) **General method of drawing polygon:** Draw AB which is the given length of polygon. At point B, draw a perpendicular BP = AB. Join AP. Taking B as center and AB as radius, draw an arc AP. The perpendicular bisector of AB meets the line AP and arc AP at points 4 and 6 respectively. Draw circles with centers as 4, 5 and 6 and radii as 4B, 5B and 6B and inscribe a square, pentagon, & hexagon in the respective circles. Mark point 7, 8, etc. taking 6 – 7, 7 – 8, etc. = 4 – 5 to get the centers of circles of heptagon and octagon, etc.

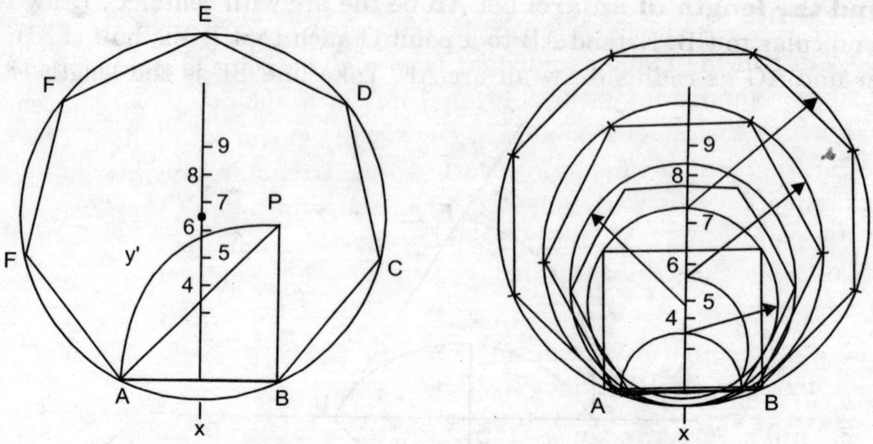

Fig. 1.64: General method of drawing polygon

(*xx*) Inside a regular polygon, draw equal circles with number of circles equal to the sides of the polygon such that each circle touches one side of the polygon and two of the other circles

Draw a regular pentagon ABCDE and bisect all the angles of the polygon, meeting at a point O. This divides the polygon into five triangles. In each triangle inscribe a circle touching the side of triangle.

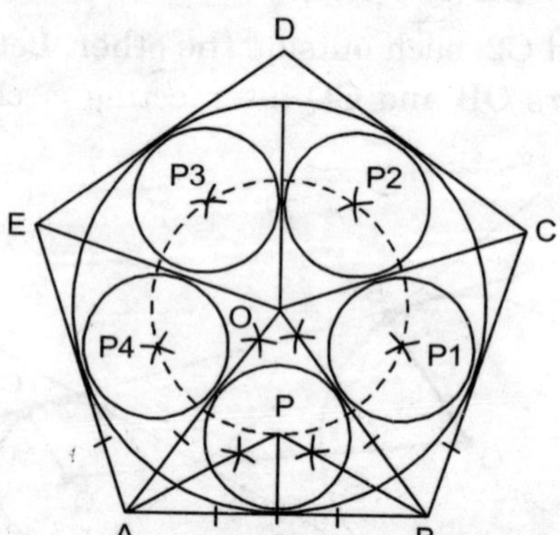

Fig. 1.65: Circles inside a polygon

1.16 INTRODUCTION TO CONIC SECTIONS, CURVES AND LOCI OF POINTS

A locus is the set of all points satisfying some condition or a set of rules. Locus in Latin means "position" or "location".

1.16.1 Loci of Points

The locus of points in the plane equidistant from a given point is a circle, and the set of points in three-space equidistant from a given point is a sphere. If we join each of the positions through which the moving point passes, we will get the path followed by the point. If the movement of the point is random, its path will be irregular, and we shall not be able to predict, for example, how its path will look when the point stops. Therefore, the path of a randomly moving point cannot be determined. On the other hand, if the point's movement obeys a given set of rules, the resulting path will be predictable and can be a circle, a straight line, an ellipse, and so on.

(*i*) **Locus as a straight line:** For example a point moves such that its distance *d* from a fixed straight line AB is always fixed.

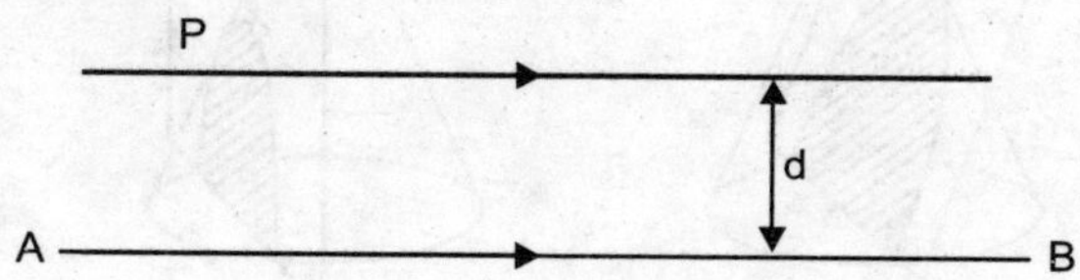

Fig. 1.66: Locus as a straight line

(*ii*) **Locus as a circle:** For example let a point P move in such a way that it is always at a constant distance *r* from a point O which is fixed.

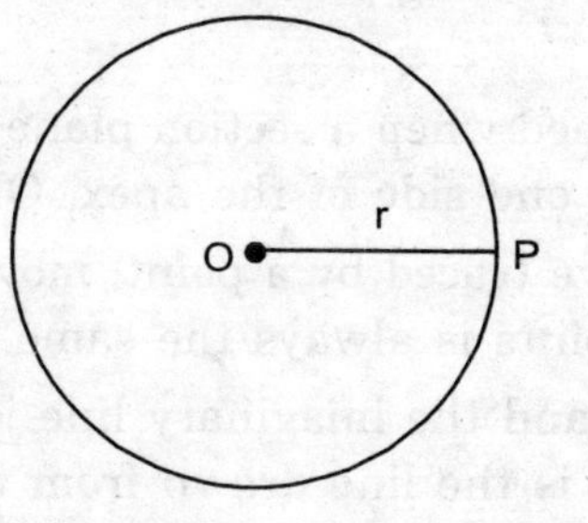

Fig. 1.67: Locus as a circle

1.17 CONIC SECTIONS

The conic sections (or conics) - the ellipse, the parabola and the hyperbola play a very important role in the application of mathematics to engineering. The conics were first studied by the Greek mathematician Apollonius more than 200 years BC. The conics are the curves which are obtained when a double cone is intersected by a plane. This section covers the nature of curves, their geometrical properties and construction methods.

Conic sections are always "smooth". Conic is defined as the locus of a point moving in a plane such that the ratio of its distance from a fixed point and a fixed straight line is always constant.

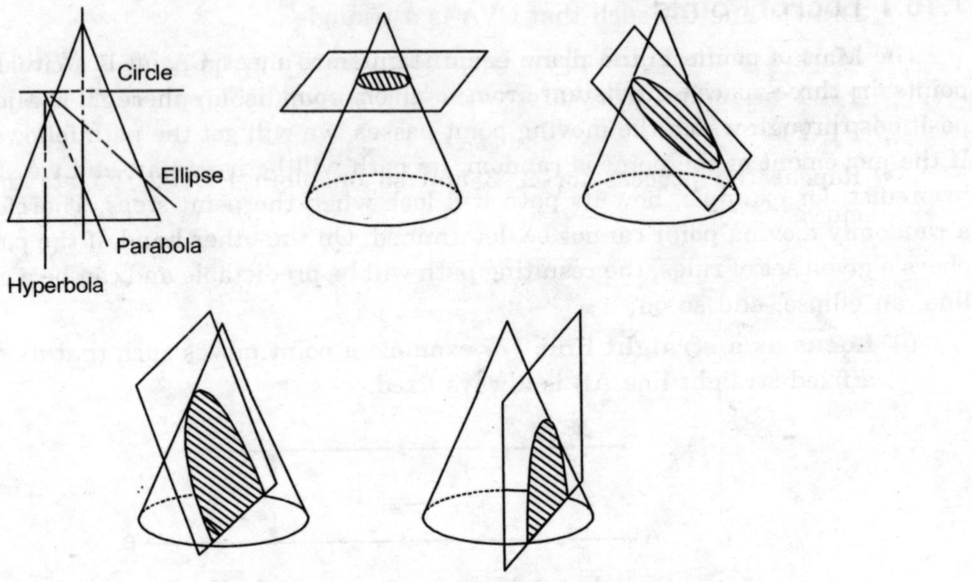

Fig. 1.68: Showing a triangle and cutting planes used to form circle, ellipse, parabola and hyperbola

On the basis of position of the cutting plane relative to the axis of cone, there are three types of conic sections:

(*i*) **Ellipse:** An ellipse is obtained when a section plane which is inclined to the axis, cuts all the generators of the cone on one side of the apex. OR

An ellipse is defined as a curve traced by a point, moving in a plane such that the sum of its distances from two fixed points is always the same.

Apex is the top point of cone and the imaginary line joining apex and center of base of cone is called the *axis*. *Generator* is the line drawn from the apex to circumference of base circle.

Methods for Generating Ellipse

1. **Focus-Directrix or Eccentricity Method:** Method of constructing conics when the distance of the focus from the directrix and its eccentricity are given.

The fixed point is called *focus* F and fixed straight line is called *directrix*. *Eccentricity*(e) is the ratio of shortest distance from focus to the shortest distance from directrix. For ellipse $e < 1$.

Problem 1: Construct an ellipse when distance of focus from directrix is 70 mm and eccentricity is 3/4.

Solution: Draw directix DD′ and draw a perpendicular at any point A.

- Mark focus F at distance of CF = 70 mm.
- Divide CF in 7 equal parts and mark vertex V at 4th part from C.
- From V draw a perpendicular, VF = VA.

- Draw a line CA such that CVA is a triangle.
- Mark a point 1 and draw perpendicular to meet CA at 1′. With F as center and 1 – 1′ as radius, draw arcs to cut the perpendicular through 1. Mark the points as P1, P1′.
- Repeat the process for 2, 3 and so on. Join P1, P2, P3 etc to obtain smooth curve.

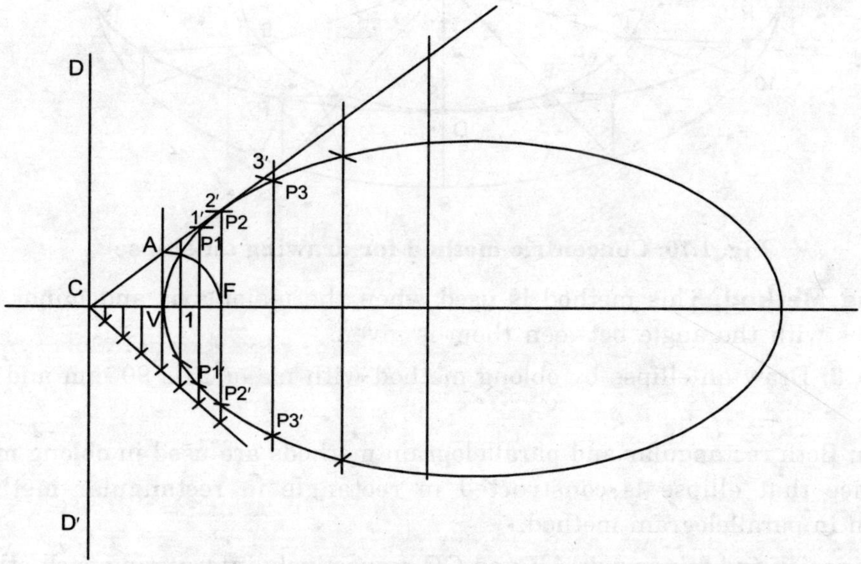

Fig. 1.69: Focus-Directrix Method for drawing an ellipse

2. **Concentric Method:** This method is used when the major axis and minor axis of an ellipse are given. Major axis is the longest diameter of an ellipse.

Minor axis is the shortest diameter of an ellipse.

Problem 2: Draw ellipse by concentric circle method. Take major axis 70 mm and minor axis 40 mm long.

Solution: AB is major axis and CD is minor axis.

- Draw AB and CD intersecting each other at right angles at point Q.
- With Q as center and AB and CD as diameters draw two circles.
- Divide the major circle in 12 equal parts and mark as A, 1, 2, 3, 4 …. 10, B.
- Draw radial lines joining 12 points with center Q and cutting the minor circle and dividing it in 12 parts. Name the points as 1, 2 and so on.
- Through 1 of major circle draw a line parallel to minor axis and through 1 of minor circle draw a line parallel to major axis. Mark the intersecting point as P1 which is the required point on the locus of ellipse.
- Repeat the process and obtain 12 points and make an ellipse by joining A, P1, P2, C, P4, P5, B, P6 …. A.

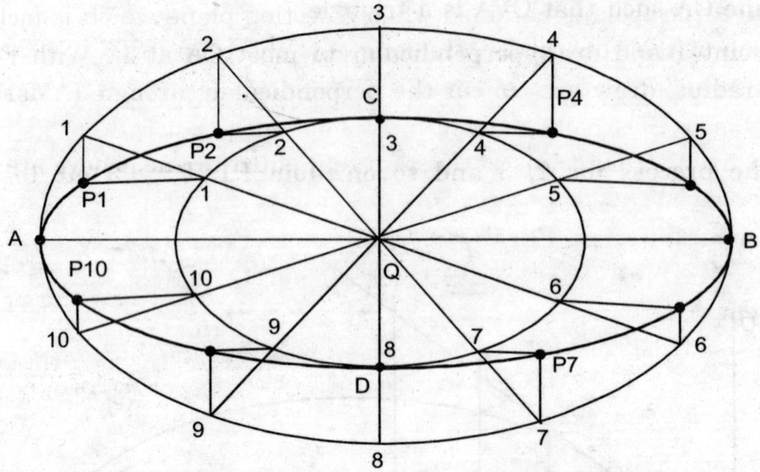

Fig. 1.70: Concentric method for drawing an ellipse

3. **Oblong Method:** This method is used when the major axis and minor axis or the conjugate axes with the angle between them is given.

Problem 3: Draw an ellipse by oblong method with major axis 90 mm and minor axis 50 mm.

Solution: Both rectangular and parallelogram methods are used in oblong method with only difference that ellipse is constructed in rectangle in rectangular method and in parallelogram in parallelogram method.

- Draw major and minor axis AB and CD respectively intersecting each other at right angles at O
- Construct an oblong (rectangle or parallelogram) PQRS.
- Divide AO and AP in same number of equal parts. 4 parts in this case.
- Join all vertical points to upper end of minor axis and all horizontal points to lower end of minor axis. Mark the intersections as P1, P2, P3, P4.
- Similarly mark other points and draw the curve to form an ellipse.

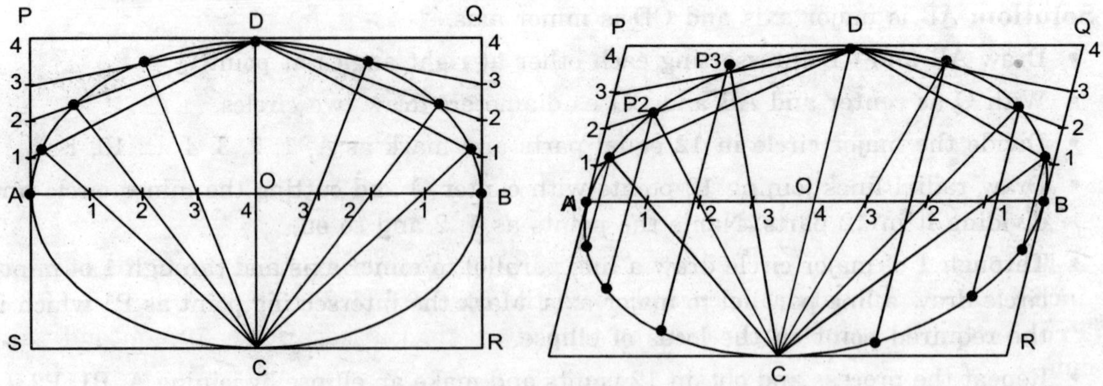

Fig. 1.71: Oblong method for drawing ellipse

(*ii*) **Parabola:** A parabola is obtained when a section plane which is inclined to the axis but parallel to one of the generators of a cone cuts the cone.

OR

A parabola is locus of a point moving in a plane such that ratio of its distances from a fixed point and a fixed straight line is constant and equal to one. There are a large number of applications for parabolic shapes such as searchlight mirrors, telescopic mirrors etc.

Methods of constructing Parabola

1. **Rectangle Method:** This method is applicable when the axis and the base of a parabola are given.

Problem 4: Construct parabola by rectangle method if the dimensions of rectangle are 100 mm and 150 mm.

Solution: Draw a rectangle of given size and divide it into two equal vertical parts.

- Divide AC and CB in 6 equal parts.
- Join point D with divisions on vertical divisions.
- Mark points of intersections of D with 1, 2, 3, 4, 5, 6 and divisions on AC and CB as P1, P2 up to P5. Similarly, locate P1′, P2′, P3′, P4′, P5′.
- Join A, P1, P2, P3, P4, P5, D, P5′, P4′, P3′, P2′, P1′, B. It is the requited parabola.

*If rectangle is used instead of parallelogram, the method of construction will be known as parallelogram method.

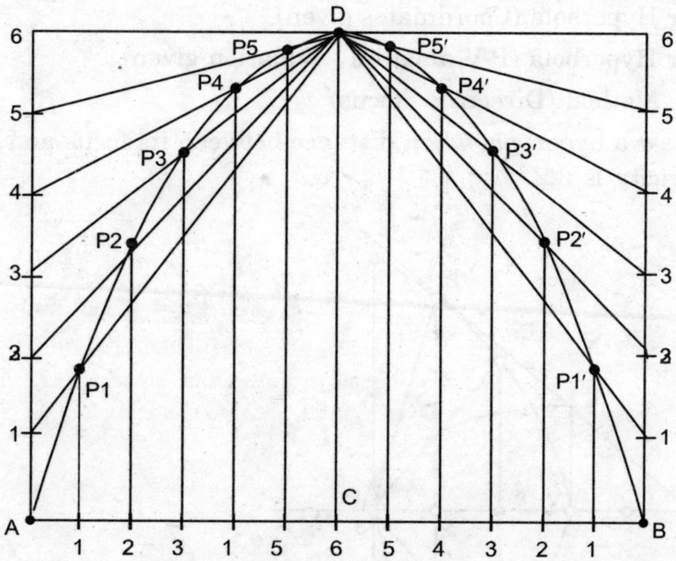

Fig. 1.71: Rectangle method for drawing parabola

2. **Method of Tangents (Triangle Method):** This method is used when the base and the inclination of tangents at open ends with the base are given.

Problem 5: Construct a parabola by tangent method, if base RS is 70 mm and CD is 35 mm.

Solution: Draw the base 70 mm and CD as 35 mm

- Extend CD such that CD = CO. Join O with R and S.

- Divide RO and OS in equal parts, say 3 and mark them as shown.
- Join the corresponding points 1 – 1′, 2 – 2′, 3 – 3′.
- Draw a smooth curve from R and tangential to lines 1 – 1′, 2 – 2′, 3 – 3′.

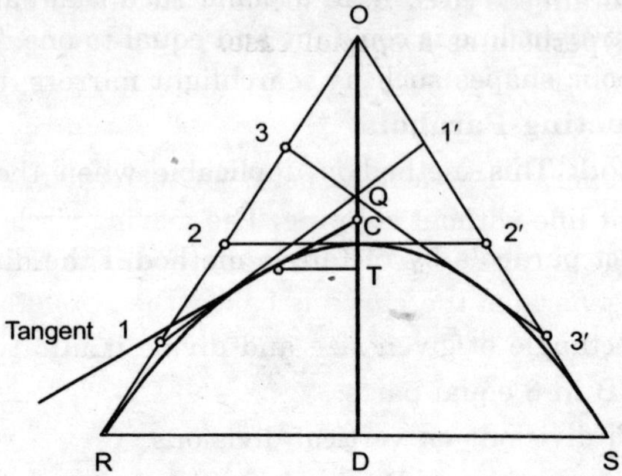

Fig. 1.72: Triangle method for drawing parabola

(*iii*) **Hyperbola:** A hyperbola is obtained when a section plane which makes a smaller angle with the axisthan the one made by generator of a cone. It can be constructed by following methods:

1. Rectangular Hyperbola (Coordinates given)
2. Rectangular Hyperbola (P-V diagram - Equation given)
3. Basic Locus Method (Directrix - focus)

Problem 6: Draw a hyperbola when distance between its focus and directrix is equal to 70 mm and eccentricity is 3/2.

Solution:

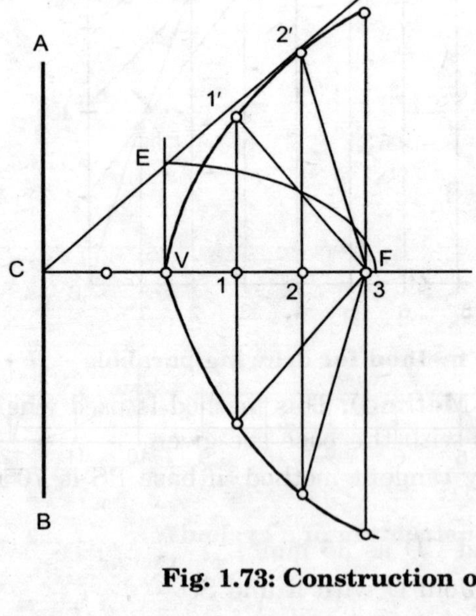

Fig. 1.73: Construction of a Hyperbola

- If the section plane passes parallel to base of cone, *circle* is obtained and if section plane passes through the apex of cone such that it is perpendicular to base a *triangle* is obtained.
- From ellipse we obtain the circle as a special case, and from the hyperbola we obtain the rectangular hyperbola as a special case.

1.18 CYCLOID

A cycloid is a curve generated by a fixed point on the circumference of the circle as the circle rolls along a straight line without slipping. The moving circle is called the "Generating circle" and the straight line is called the "Directing line" or the "Base line". The point on the Generating circle which generates the curve is called the "Generating point".

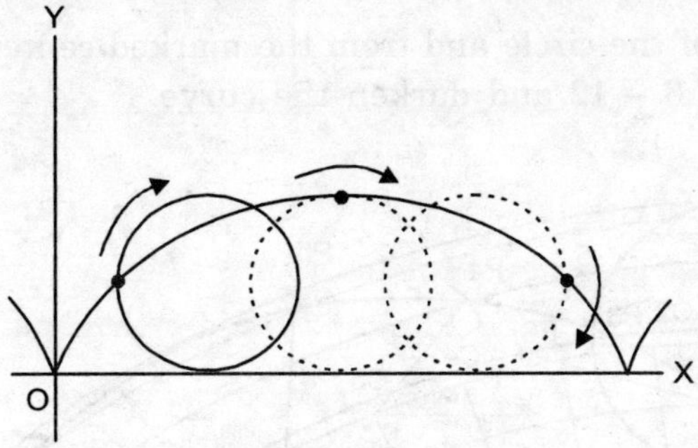

Fig. 1.74: Cycloid

Problem 7: Draw a cycloid if the diameter of rolling circle is 50 mm.

Solution: Draw a circle of radius 25 mm and base line OC equal to circumference of the circle which is equal to $\pi d = \pi \times 50$.

- Divide the circumference of circle and OC in 12 equal parts.
- Through divisions on circle draw perpendiculars to intersect the line OC 12 at C1, C2 and so on.
- Set the compass the radius of the circle and centers as C1, C2, C3 – C12 to cut the arcs on the lines from circle through 1, 2, 3, ... etc.
- Join the points to form the cycloid.

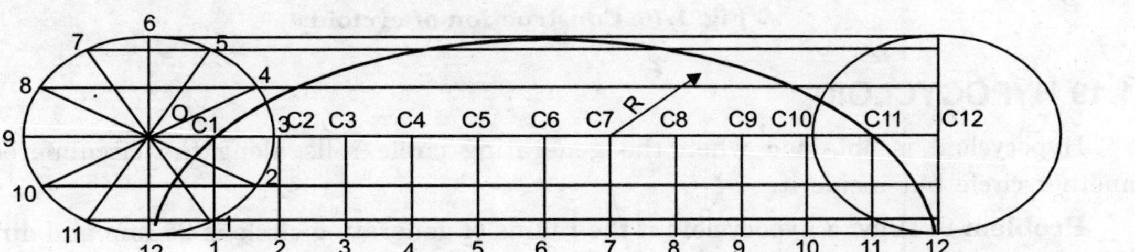

Fig. 1.75: Construction of a cycloid

Epicycloid: The cycloid is called Epicycloid when the generating circle rolls along the circumference of another circle outside it.

Problem 8: Draw an epicycloid if the radius of generating circle is 25 mm and directing circle is 75 mm.

Solution: Draw a circle of radius 25 mm and divide the circle in 12 equal divisions.

- From point P on circle draw a line equal to 75 mm.
- With O as center and radius OP draw a part of directing circle.
- With O as center and radius OC draw an arc. Divide angle POA in 12 equal parts and mark the positions of the circle- centers (C1, C2 – C12) as the circle rolls on the base surface.
- Project the positions of the point from the circle.
- Using the radius of the circle and from the marked centers C1, C2 – C12 cut off the arcs through 1, 2, 3 – 12 and darken the curve.

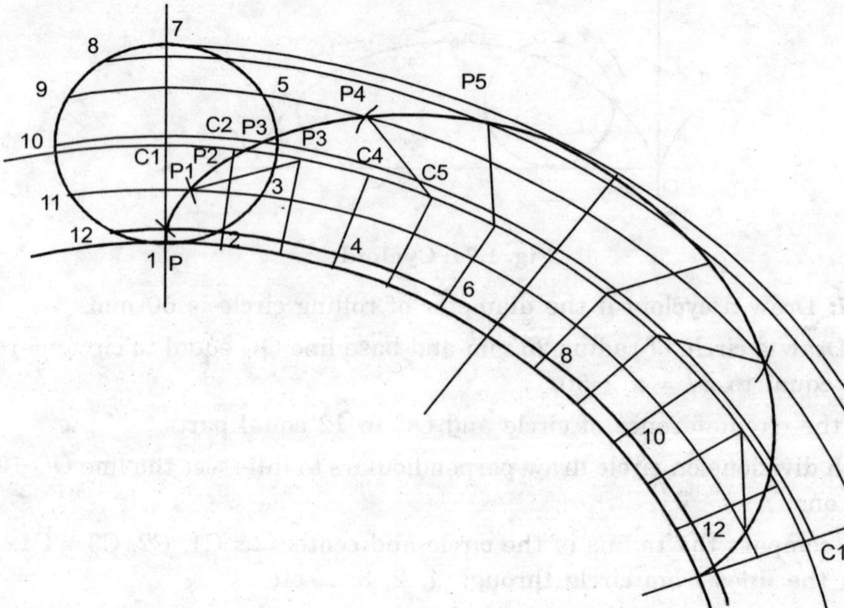

Fig. 1.76: Construction of cycloid

1.19 HYPOCYCLOID

Hypocycloid is obtained when the generating circle rolls along the circumference of another circle but inside it.

Problem 9: Draw a hypocycloid if the radius of generating circle is 25 mm and directing circle is 75 mm.

Solution: Draw a circle of radius 25 mm and divide the circle in 12 equal divisions.

- $\theta = r/R * 360°$

- Take a point O′ as center and radius 75 mm, draw an arc PQ subtending angle θ O′.
- On O'P mark PC1 radius of rolling circle. With O′ as center and radius *r* draw the rolling circle.
- Divide the rolling and generating circles into 12 equal parts.
- With C1, C2 etc. as centers and radius equal to *r*, draw arcs to intersect the arcs made with O′ as center through 7 – 12.
- Draw a smooth curve.

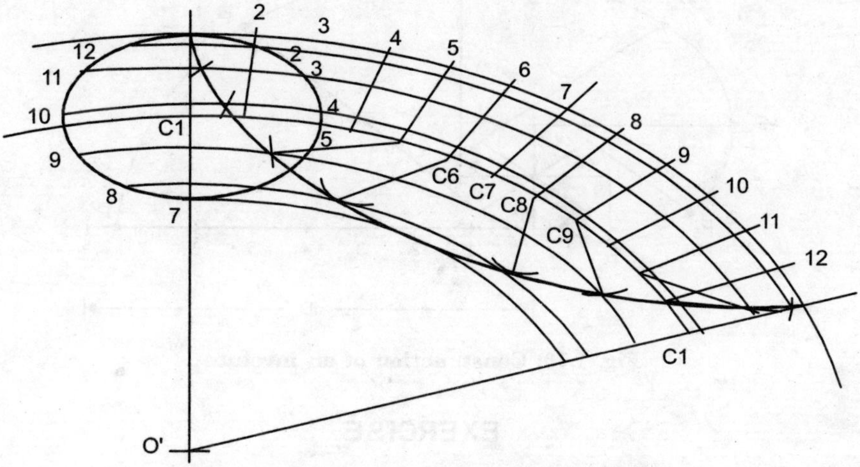

Fig. 1.77: Construction of hypocycloid

1.20 INVOLUTE

An involute is defined as the path of a point on a straight line which rolls without slipping along the circumference of a cylinder. It can also be defined as a curve traced by the free end of a thread unwound from a circle or a polygon in such a way that the thread is always tight and tangential to the circle or side of the polygon.

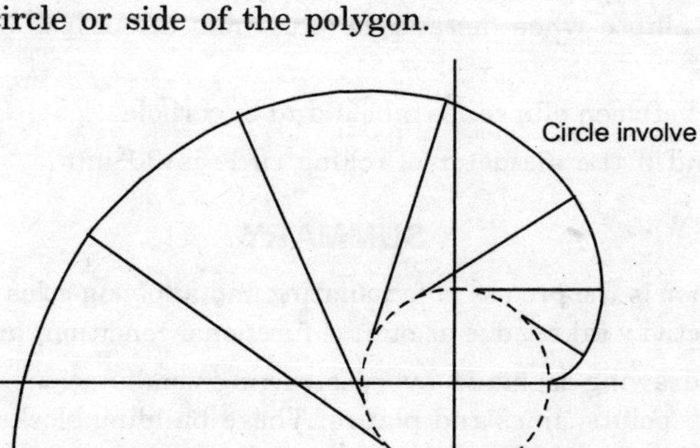

Fig. 1.78: Involute

Problem 10: Draw an involute of a circle of 40 mm diameter.

Solution: Draw the base circle of diameter 40 mm and divide it into 8 equal parts.

- Draw a tangent from point P equal to circumference of circle and divide in 8 equal parts.
- Draw the tangents at 7, 6, 5, 4, 3, 2, 1 of circle.
- Mark points P_1, P_2 such that P_7O = 01, $6P_6$ = 02 and so on

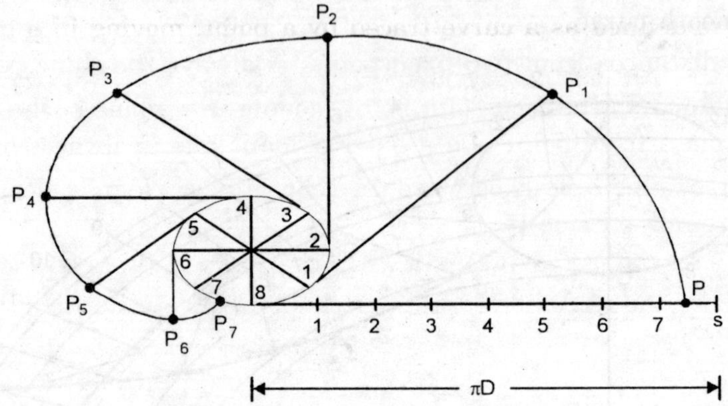

Fig. 1.79: Construction of an involute

EXERCISE

1. Define involute, cycloid and epicycloid.
2. Draw ellipse by concentric circle method. Take major axis 75 mm and minor axis 35 mm long.
3. What are the systems of placing dimensions?
4. Draw a hypocycloid if the radius of generating circle is 20 mm and directing circle is 70 mm.
5. Inscribe a hexagon in given circle of radius 40 mm.
6. Construct an ellipse when distance of focus from directrix is 60 mm and eccentricity is 3/4.
7. Differentiate between ellipse, parabola and hyperbola.
8. Draw a cycloid if the diameter of rolling circle is 30 mm.

SUMMARY

- Standardization is the process of formulating and applying rules for an orderly approach to a specific activity taking due account of functional conditions and safety requirements.
- Engineering drawing is made by constructing combinations of primitive geometric forms such as points, lines and planes. These building blocks define the shape and position of object in space.

- In Single-stroke vertical lettering the alphabets made are of same thickness.
- Each feature in an object should be dimensioned once and only once to avoid confusion.
- The dimensions on a drawing should be aligned in the same direction.
- Vernier scale is used to measure fractional parts of the smallest division of the main scale.
- A locus is the set of all points satisfying some condition or a set of rules.
- An ellipse is defined as a curve traced by a point, moving in a plane such that the sum of its distances from two fixed points is always the same.
- Conic is defined as the locus of a point moving in a plane such that the ratio of its distance from a fixed point and a fixed straight line is always constant.
- The cycloid is called Epicycloid when the generating circle rolls along the circumference of another circle outside it.
- An involute is defined as the path of a point on a straight line which rolls without slipping along the circumference of a cylinder.

CHAPTER

2

PROJECTIONS

Learning Objectives

- Explain the concept and types of projections
- Understand the difference between first angle and third angle projection
- Create multi-view drawings using first and third angle projections
- Visualize and draw projections of points
- Visualize and draw projections of lines
- Visualize and draw projections of planes

2.1 INTRODUCTION

In the universal graphic language shape and size of an object is described by technical drawing. To represent true shape and size of an object, straight lines from various points on the contour of object are drawn on the paper. The image formed by joining the points on paper in the right order is called projection of the object.

2.2 TYPES OF PROJECTION

Depending upon orientation of the object, location of the sight, and direction of lines of view relative to picture plane projections can be classified as:

Axonometric Projection: It is the projection obtained when projectors are parallel to each other but perpendicular to the plane of paper.

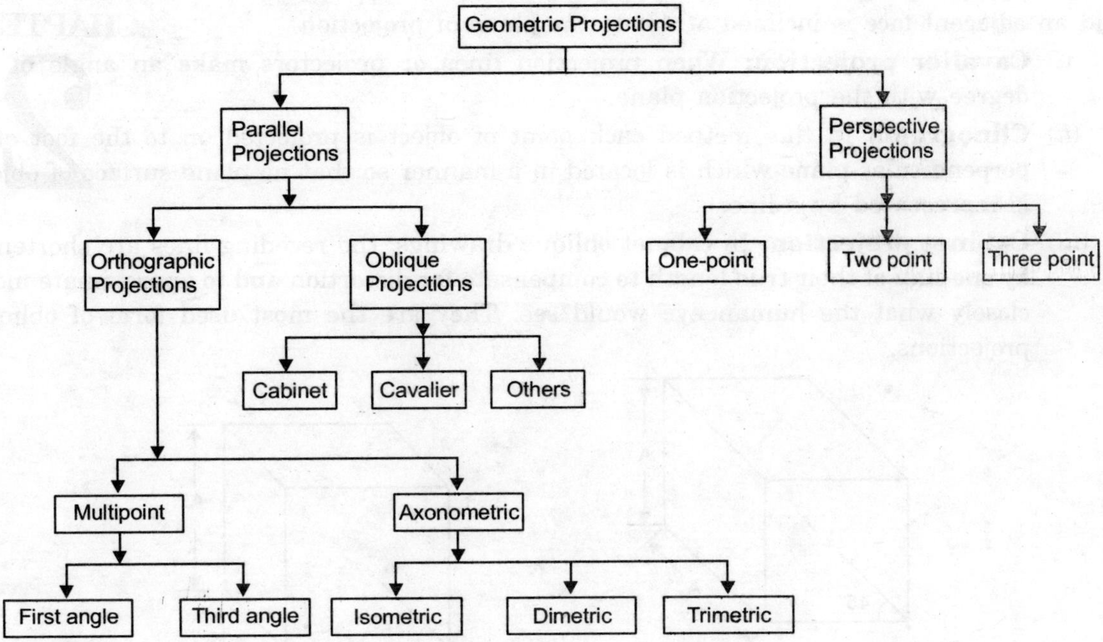

Fig. 2.1: Showing types of projections

(*i*) **Isometric projection:** It is the projection obtained when the angles between projections of axes are equal. It is most commonly used and easiest to draw and dimension.

(*ii*) **Dimetric projection:** In dimetric projection, view direction is such that two of the three axes appear equally shortened and angles of presentation are determined according to the angle of viewing. When two of the three angles are equal, the projection is termed as a dimetric projection and is easier to produce than trimetric drawings.

(*iii*) **Trimetric projection:** In this, all the three axes of space appear unequally foreshortened from the direction of viewing. The scale along each of the three axes and the angles among them are different from each other.

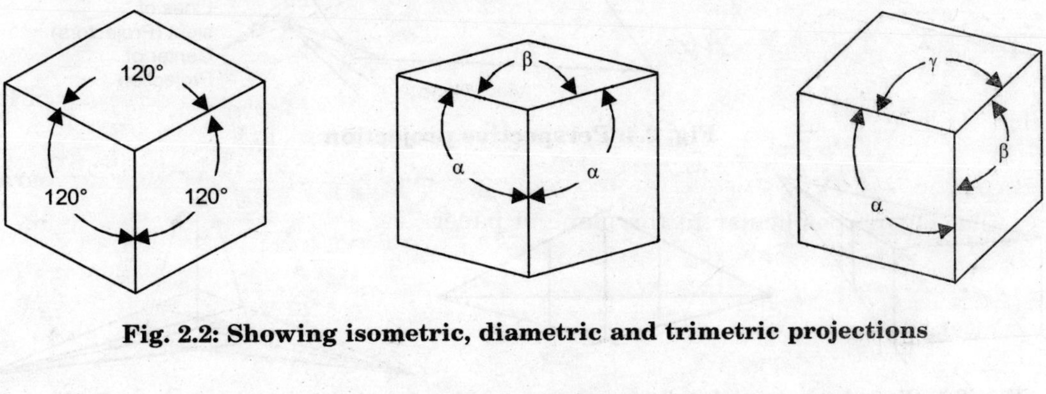

Fig. 2.2: Showing isometric, diametric and trimetric projections

(*b*) **Oblique projection:** It is the projection obtained when one face of object is parallel and an adjacent face is inclined at 45° to the plane of projection.

(*i*) **Cavalier projection:** When projection lines or projectors make an angle of 45 degree with the projection plane.

(*ii*) **Clinograph:** In this method each point of object is projected on to the foot of a perpendicular plane which is located in a manner so that no plane surface of object is represented by a line.

(*iii*) **Cabinet projection:** In cabinet oblique drawings, the receding lines are shortened by one-half of their true length to compensate for distortion and to approximate more closely what the human eye would see. They are the most used form of oblique projections.

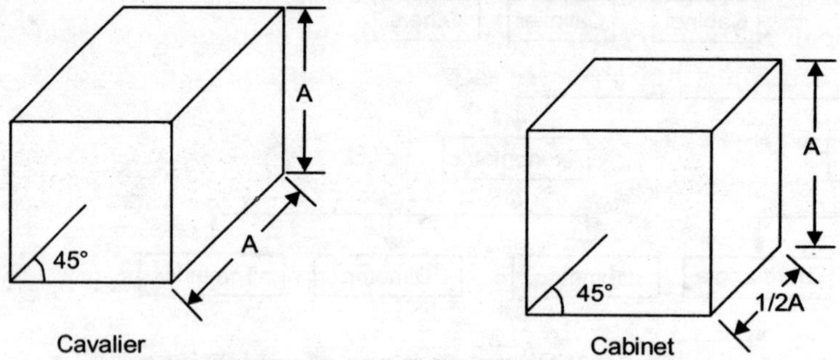

Fig. 2.3: Cavelier and cabinet projections

(*c*) **Perspective projection:** This type of projection is obtained when projectors converge to a point on the plane. It does not show true size of the object.

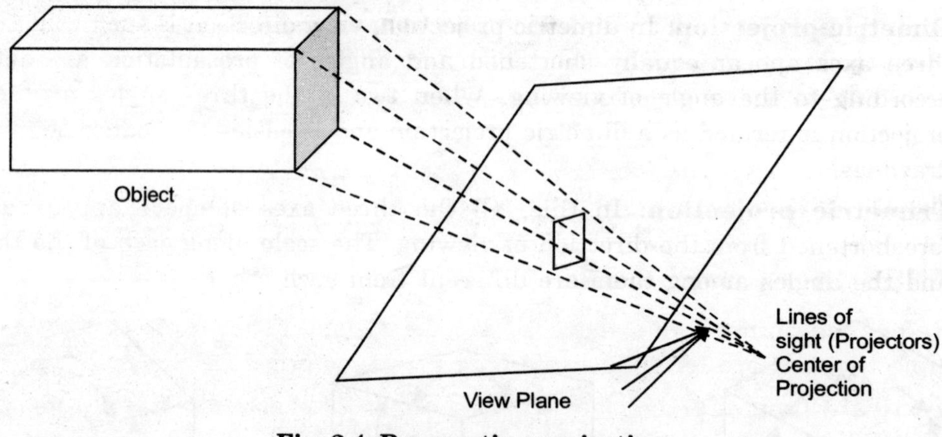

Fig. 2.4: Perspective projection

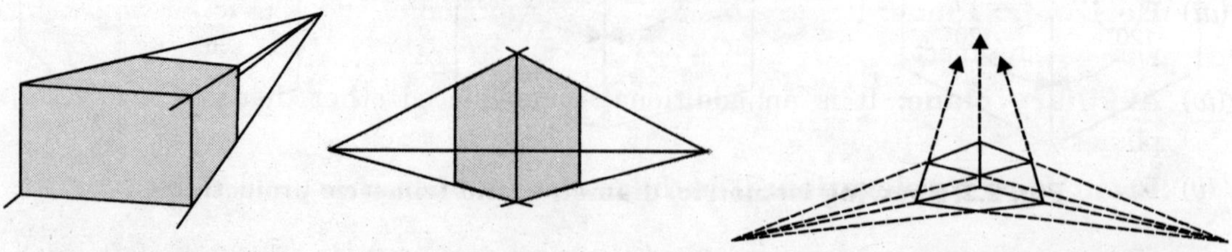

Fig. 2.5: Showing one point, two points and three points perspective projections

(*i*) **One point projection:** In this case one face of the object is kept parallel to Perspective Picture plane which is a vertical projection plane used to obtain the object's Perspective. The edges perpendicular to this face appears to be converging to a Vertical plane.

(*ii*) **Angular Perspective (2-point):** In this case one edge of the object is kept parallel to the PPP. Each edges perpendicular to them converges to one of the two Vertical planes.

(*iii*) **Oblique Perspective (3-point Perspective):** Neither an edge nor a face of the object is kept parallel to PPP. Edges converge to one of the three Vertical planes.

2.3 ORTHOGRAPHIC PROJECTION

Orthographic projection system is a method of representing exact or true shape of a 3D object on 2D sheet in two or more views. To draw orthographic projections object to be projected, plane of projection, observer's eye and projectors are required.

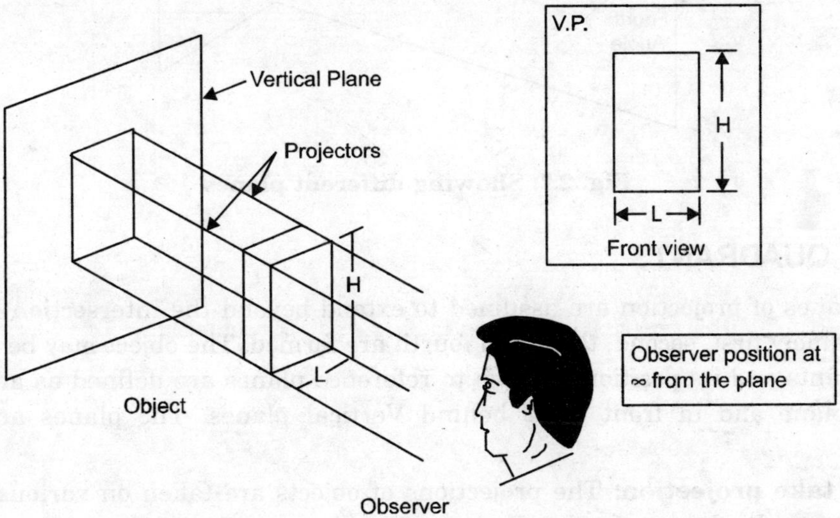

Fig. 2.6: Showing orthographic projection of a rectangular slab in front view

2.4 PLANES OF PROJECTION

(*i*) **Reference plane:** Two planes intersecting at right angles to each other are used for projection and are termed as reference planes or principal planes of projection.

(*ii*) **Vertical Plane:** It is the plane which is vertical and is used for projecting front view of an object.

(*iii*) **Horizontal Plane:** It is the plane which is horizontal and is used for projecting top view of an object.

(*iv*) **Auxiliary plane:** It is an additional plane placed other than the two reference planes.

(*v*) **Front view or elevation:** It is the front view of an object and is projected on VP.

(vi) **Top view or plan:** The projection of an object on horizontal plane is termed as plan or top view.

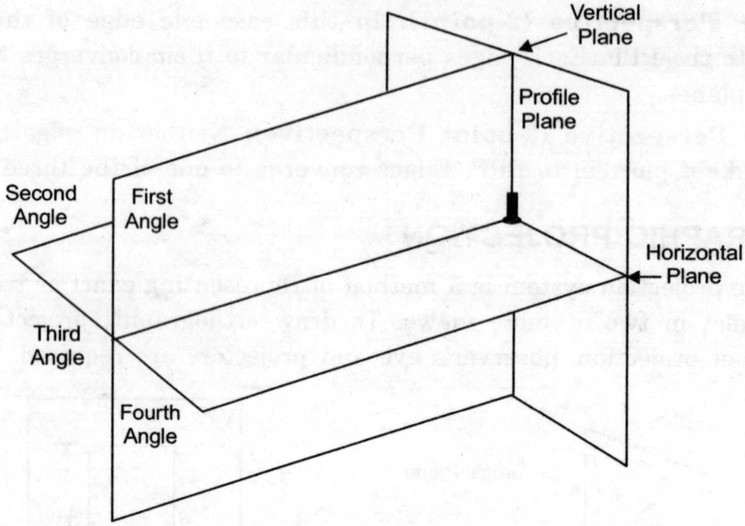

Fig. 2.7: Showing different planes

2.5 FOUR QUADRANTS

When planes of projection are assumed to extend beyond the intersection line, the four quadrants namely first, second, third and fourth are formed. The object may be placed in one of the quadrants and its position relative to reference planes are defined as above or below Horizontal plane and in front of or behind Vertical planes. The planes are considered transparent.

How to take projection: The projections of objects are taken on various planes from looking from front and from the top. To bring the views on flat surface, planes are rotated. In 1st and 3rd quadrants vertical plane is rotated outwards and in 2nd and 4th quadrants vertical plane is rotated inwards. Vertical plane is kept fixed in all the cases.

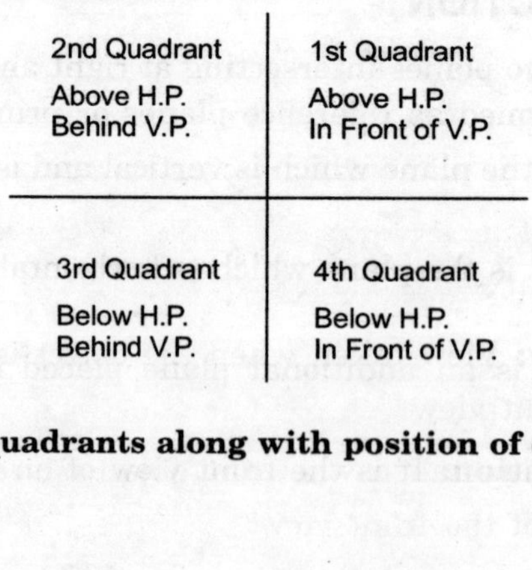

Fig. 2.8: Showing four quadrants along with position of object in each quadrant

2.6 TYPES OF ORTHOGRAPHIC PROJECTION

There are two types of orthographic projections:

(*i*) First angle projection

(*ii*) Third angle projection

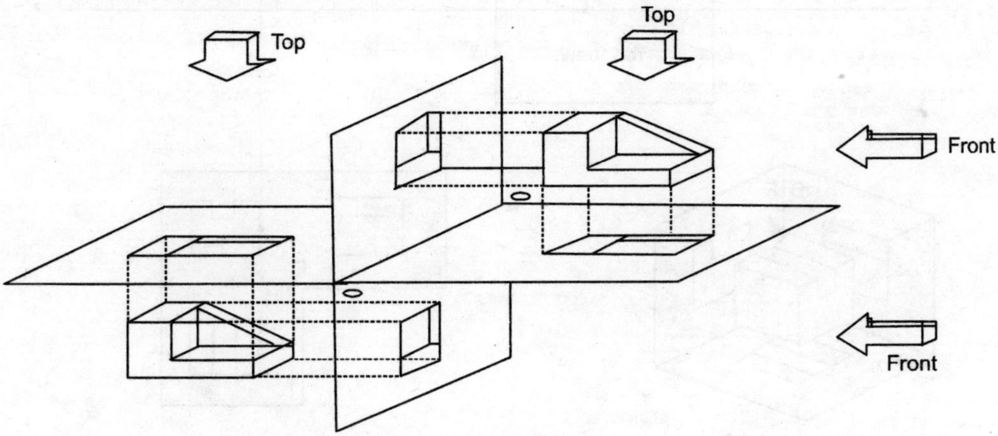

Fig. 2.9: Showing first angle and third angle projections of an object

Figure showing first angle projection (in first quadrant) and third angle projection(in third quadrant). Arrows showing the viewing direction.

2.6.1 First Angle Projection

In this the object is situated in first quadrant i.e., in front of VP and above HP. If an object is suspended in space or placed in a transparent box/cabin then 6 possible views can be obtained from six possible views. The six possible views are:

- **Front view:** View taken when the object is seen from front and is taken as the reference for drawing remaining 5 views.
- **Top view:** View taken when the object is seen from top and is placed below front view.
- **Bottom view:** View taken when the object is seen from the bottom and is placed above front view.
- **Left hand side view:** View taken when the object is seen from the left and is placed on the right side of front view.
- **Right hand side view:** View taken when the object is seen from the right and is placed at the left of front view.
- **Rear or back view:** View taken when the object is seen from the rear side and is placed on left or right of the front view.

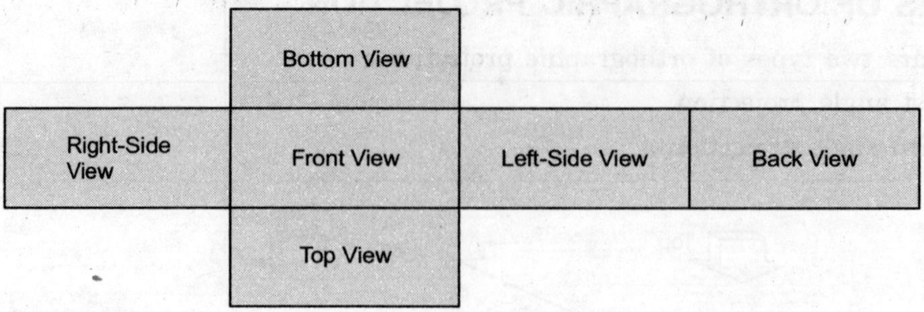

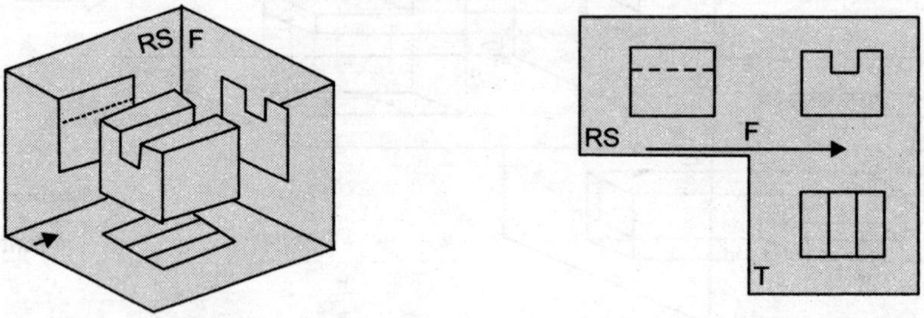

Fig. 2.10: Showing first angle projection and positions of all the possible views

2.6.2 Third Angle Projection

In this the object is situated in the third quadrant i.e., behind VP and below HP. The reference planes are considered to be transparent. If an object is suspended in space or placed in a transparent box/cabin then 6 possible views can be obtained from six possible views in third angle projection. The six possible views are:

- **Front view:** View taken when the object is seen from front and is taken as the reference for drawing remaining 5 views.
- **Top view:** View taken when the object is seen from top and is placed on top of front view.
- **Bottom view:** View taken when the object is seen from the bottom and is placed below front view.
- **Left hand side view:** View taken when the object is seen from the left and is placed on the left side of front view.
- **Right hand side view:** View taken when the object is seen from the right and is placed at the right of front view.
- **Rear or back view:** View taken when the object is seen from the rear side and is placed on left or right of the front view.

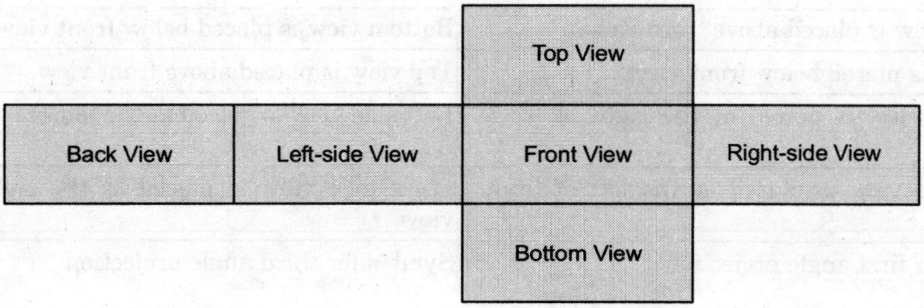

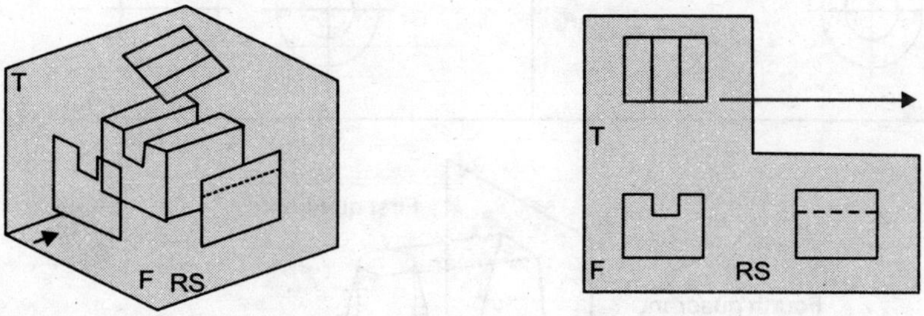

Fig. 2.11: Showing third angle projection of an object and positions of all possible views

2.7 SYMBOLS

These are the representation conventions for first angle and third angle projections according to I.S.I.

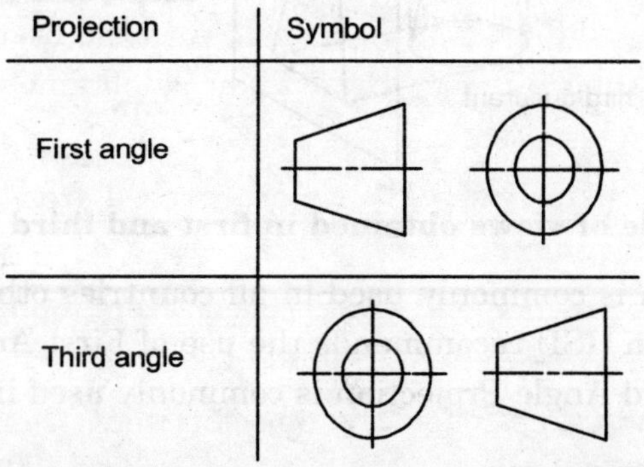

Fig. 2.11: Symbols of first and third angle projections

2.8 COMPARISON BETWEEN FIRST AND THIRD ANGLE PROJECTION

First angle projection	Third angle projection
In this method, the object is assumed to be situated in first quadrant means above HP & in front of VP.	In this method, the object is assumed to be situated in third quadrant (below HP & behind of VP).
Object is places in between observer & plane.	Planes being transparent and in between observer & object.

Bottom view is placed above front view.	Bottom view is placed below front view.
Top view is placed below front view.	Top view is placed above front view.
Left side view is placed at the right of front view	Left side view is placed at the left of front view
Right side view is placed at the left of front view	Right side view is placed at the right of front view
Symbol for first angle projection 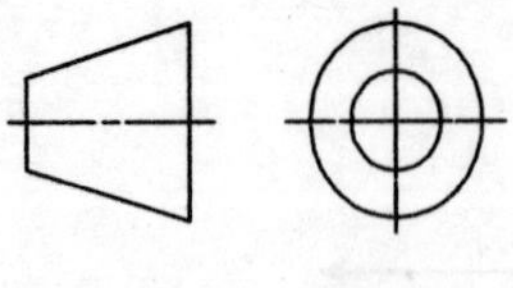	Symbol for third angle projection 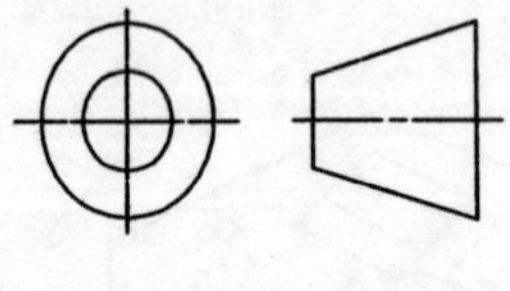

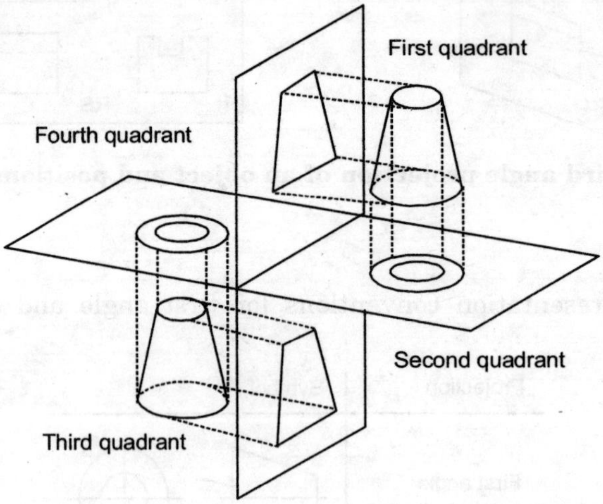

Fig. 2.12: Example of views obtained in first and third angle projections

*First Angle Projection is commonly used in all countries other than United States. The Indian Standard Institution (ISI) recommends the use of First Angle Projection method now in all the institutions. Third Angle Projection is commonly used in United States of America.

2.9 PROJECTION OF POINTS

A point may be situated in space at any position but has no magnitude. It can be represented by a small dot or a circle. Projections of a point are obtained by extending projectors perpendicular to both the planes. Projection on horizontal plane is called top view and projection on vertical plane is called front view. One of the planes is rotated to show projections on a flat surface. Vertical plane is always fixed while Horizontal plane is rotated clockwise (90°) about XY, until VP and HP come in same line. Capital letters denote the points in space, small letters denote points in top view and small letters with dashes denote front view.

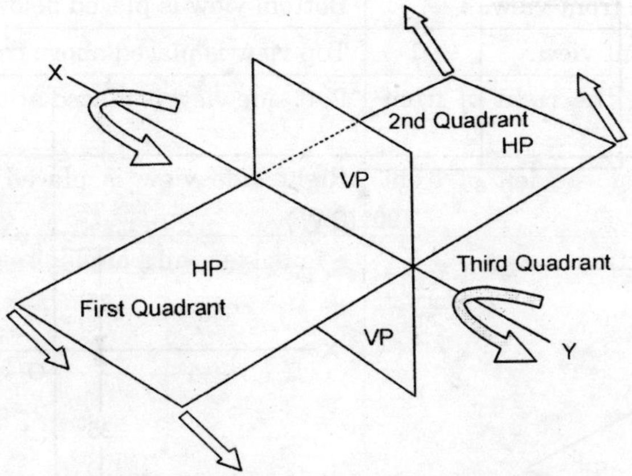

Fig. 2.13: Rotation of four quadrants

The point may be situated in space with respect to H.P. and V.P. in the following positions:

(*i*) Point is above H.P. and in front of V.P. (First Quadrant)

(*ii*) Point is above H.P. and behind V.P. (Second Quadrant)

(*iii*) Point is below H.P. and behind V.P. (Third Quadrant)

(*iv*) Point is below H.P. and in front of V.P. (Fourth Quadrant)

(*v*) Point is on H.P. and in front of V.P. (In Plane)

(*vi*) Point is above H.P. and on V.P. (In Plane)

(*vii*) Point is on H.P. and behind V.P. (In Plane)

(*viii*) Point is below H.P. and is on V.P. (In Plane)

(*ix*) Point is on H.P. and V.P. both. (In Reference Line)

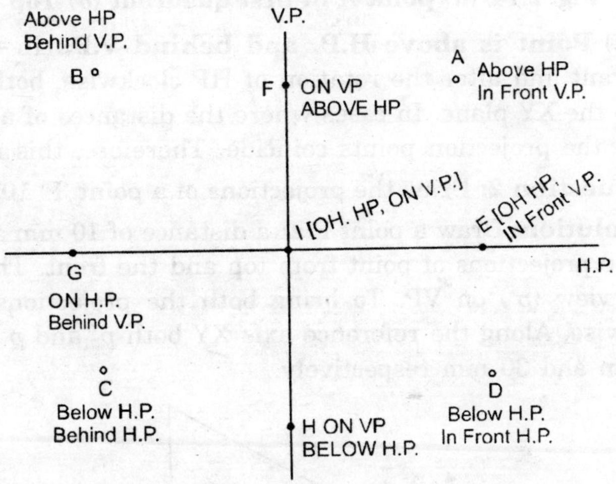

Projections of points in different positions

(*i*) **Point is above H.P. and in front of V.P:** Top view comes below XY (in H.P.) line and front view above XY (in V.P.). The distance of point above H.P. will come above XY (in V.P.) and the distance of point in front of V.P. will come below XY (in H.P.)

Question 1: Draw the projections of a point 'A' 25 mm above HP and 35 mm in front of VP.

Solution: Draw a point A at a distance of 25 mm from HP and 35 mm from VP. Look for the projections of point from top and the front. The top view (*a*) is seen on the HP and front view (*a'*) on VP. To bring both the projections on same surface HP is rotated 90° clockwise. Along the reference axis XY *a'* is plotted above XY at a distance of 25 mm and a is plotted below XY at a distance of 35 mm in the same line.

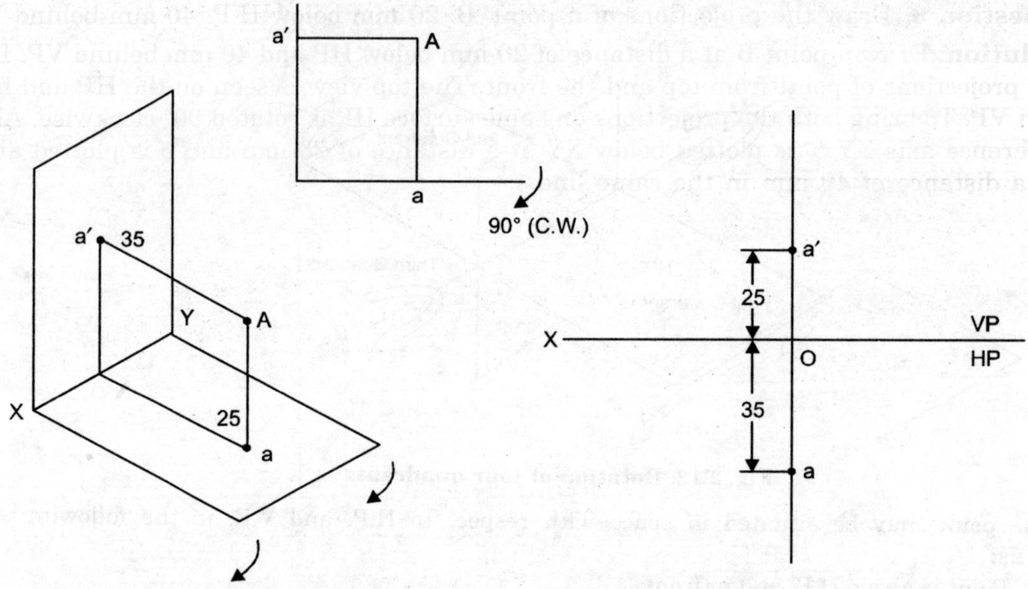

Fig. 2.14: (*a*) point A in first quadrant (*b*) Top view and front view of point A

(*ii*) **Point is above H.P. and behind V.P.:** In this case the point lies in the second quadrant and after the rotation of HP clockwise, both top view and front view are plotted above the XY plane. In cases where the distances of a point above H.P. and behind V.P. are same, the projection points coincide. Therefore, this method of projection is rarely used.

Question 2: Draw the projections of a point 'P' 10mm above HP and 30 mm behind V.P.

Solution: Draw a point P at a distance of 10 mm above HP and 30 mm behind VP. Look for the projections of point from top and the front. The top view (*p*) is seen on the HP and front view (*p*′) on VP. To bring both the projections on same surface HP is rotated 90° clockwise. Along the reference axis XY both *p*′ and *p* are plotted above XY at a distance of 10 mm and 30 mm respectively.

Fig. 2.15: (*a*) point P in second quadrant (*b*) Top view and front view of point P

(*iii*) **Point is below H.P. and behind V.P.:** The point lies in third quadrant and top view lies on HP and front view on VP. HP is rotated clockwise to bring both the projections on the same surface.

Question 3: Draw the projections of a point 'B' 20 mm below H.P. 40 mm behind V.P.

Solution: Draw a point B at a distance of 20 mm below HP and 40 mm behind VP. Look for the projections of point from top and the front. The top view is seen on the HP and front view on VP. To bring both the projections on same surface HP is rotated 90° clockwise. Along the reference axis XY b' is plotted below XY at a distance of 20 mm and b is plotted above XY at a distance of 40 mm in the same line.

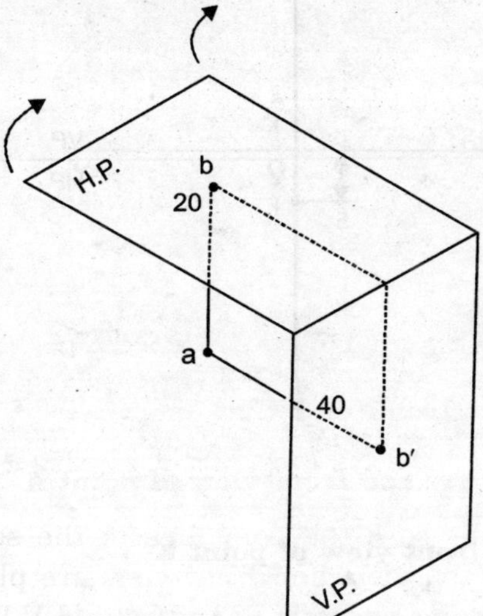

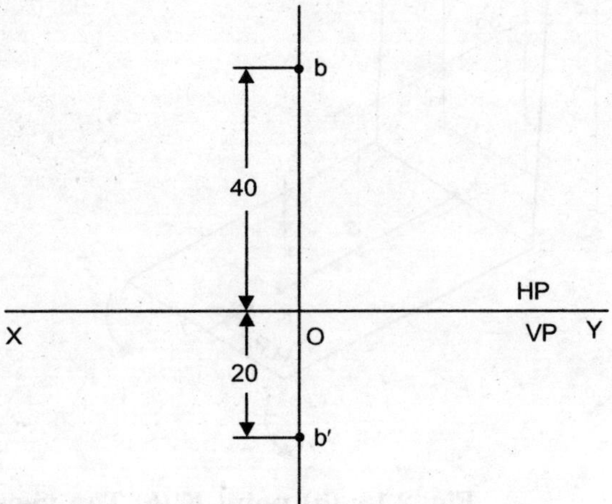

Fig. 2.16: (*a*) Point B in third quadrant (*b*) Top view and front view of point B

(*iv*) **Point is below H.P. and in front of V.P.:** The point lies in the fourth quadrant and the top view and front view both lie below the XY plane after the clockwise rotation of HP.

Question 4: Draw the projections of a point 'C' 45 mm below HP and 35 mm in front of VP.

Solution: Draw a point C at a distance of 45 mm below HP and 35 mm in front VP. Look for the projections of point from top and the front. The top view is seen on the HP and front view on VP. To bring both the projections on same surface HP is rotated 90° clockwise. Along the reference axis XY c' and c are plotted below XY at a distance of 45 mm and 35 mm respectively.

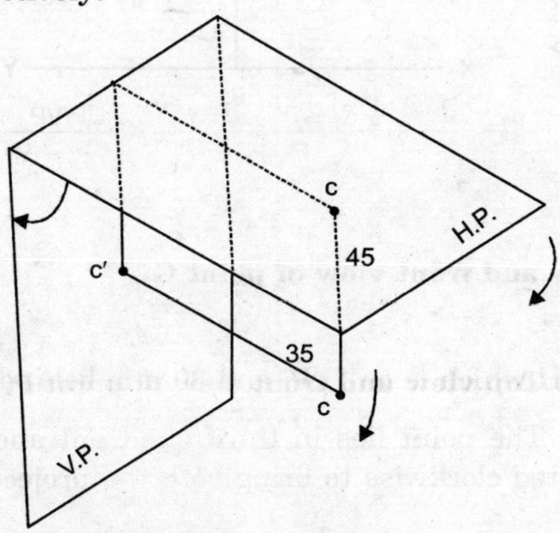

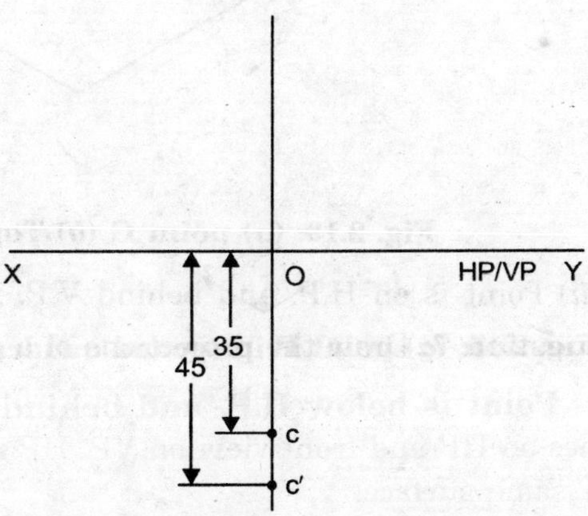

Fig. 2.17: (*a*)point C in fourth quadrant (*b*) Top view and front view of point C

(*v*) Point is on H.P. and in front of V.P.

Question 5: The point 'K' is on HP/lies on HP and 5mm in front of VP

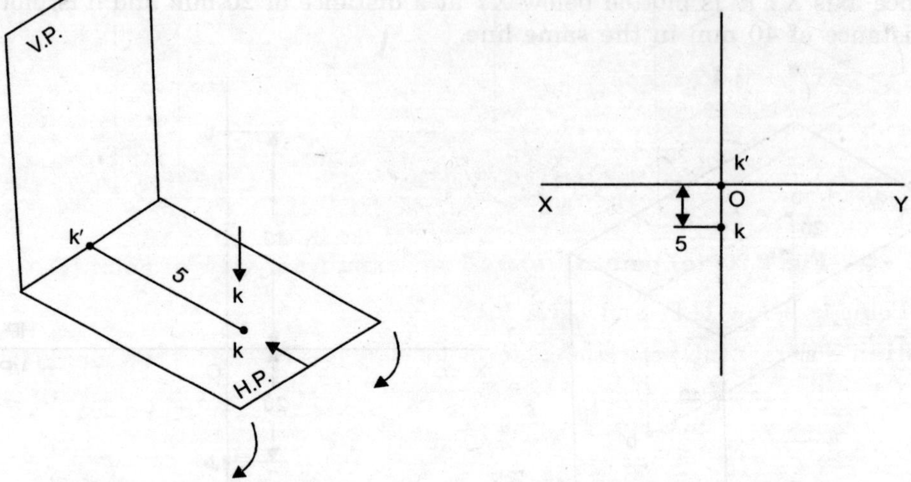

Fig. 2.18: (*a*) point K (*b*) Top view and front view of point K

(*vi*) Point is above H.P. and on V.P.

Question 6: Draw the projections of a point G which is on V.P. and 20 mm above H.P.

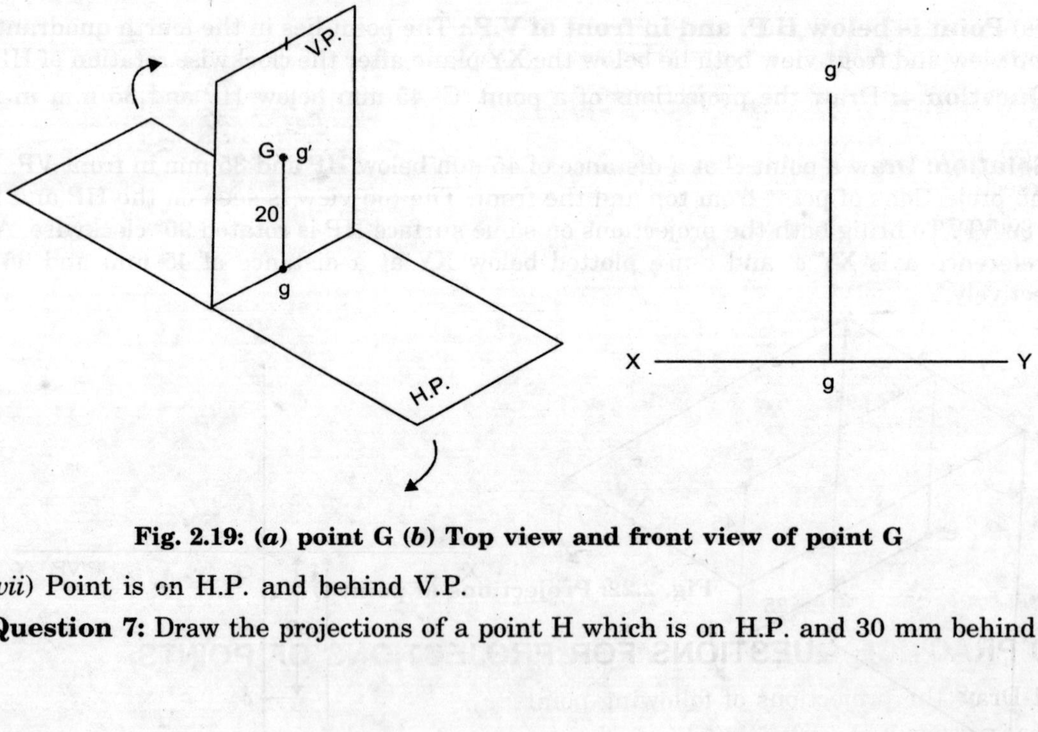

Fig. 2.19: (*a*) point G (*b*) Top view and front view of point G

(*vii*) Point is on H.P. and behind V.P.

Question 7: Draw the projections of a point H which is on H.P. and 30 mm behind V.P.

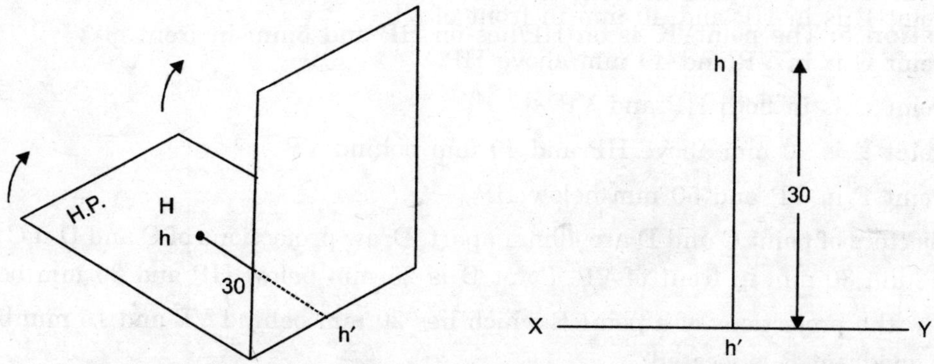

Fig. 2.20: (*a*) point H (*b*) Top view and front view of point H

(*viii*) Point is below H.P. and is on V.P.

Question 8: Draw the projections of a point which is R which is 30 mm below H.P. and on V.P.

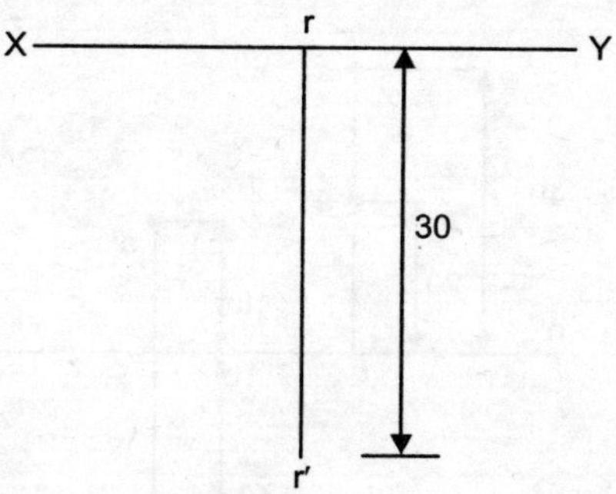

Fig. 2.21: Projections of point R

(*ix*) **Point is on H.P. and V.P. both:** There is only one possible projection at the intersection of two reference planes when a point lies on both VP and HP. Both the views will coincide in this case.

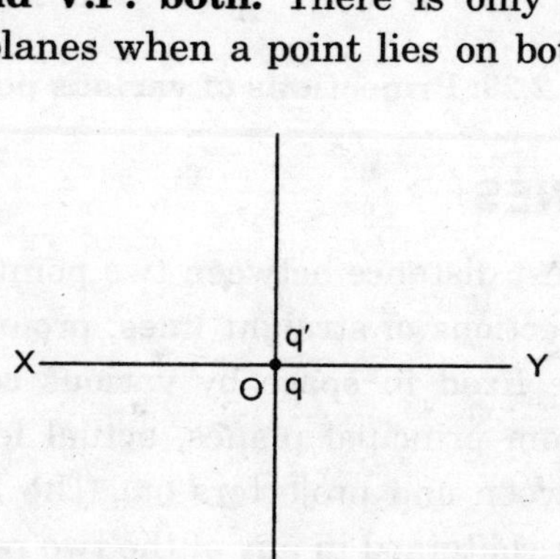

Fig. 2.22: Projections of point Q

2.10 PRACTICE QUESTIONS FOR PROJECTIONS OF POINTS

1-Draw the projections of following points:

(*i*) Point A is in HP and 40 mm behind VP

(*ii*) Point B is in HP and 40 mm in front of VP

(*iii*) Point C is in VP and 40 mm above HP

(*iv*) Point D is in both HP and VP

(*v*) Point E is 20 mm above HP and 40 mm behind VP

(*vi*) Point F in VP and 50 mm below HP

2-Projectors of point C and D are 40mm apart. Draw projections of C and D if C is 15 mm above HP and 30 mm in front of VP. Point D is 25 mm below HP and 30 mm behind VP.

3-Draw the projections of a point R which lies 20 mm behind VP and 15 mm below HP. In which quadrant R is located?

4-There is a point F which is 20 mm above HP and its shortest distance from XY is 45 mm. The point lies in first quadrant. Show its plan and elevation.

5-Projections of various points are given in figure 2.23. State their position with respect to H.P. and V.P.

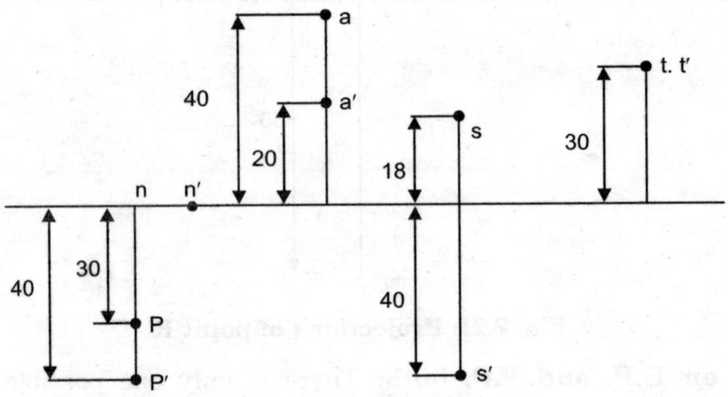

Fig. 2.23: Projections of various points

2.11 PROJECTION OF LINES

A straight line is the shortest distance between two points. It has length with negligible thickness. Hence, to draw projections of straight lines, projections of end points of the line are joined. Its position can be fixed in space by various combinations of inputs such as distances of extreme points from principal planes, actual length of line, inclinations with reference planes, distance between end projectors etc. The line may be drawn parallel to, perpendicular to, inclined to, or contained in any of the two reference planes or both of them.

Inclination of a straight line to a reference plane is the angle which line makes with the plane. The point of intersection of any line with reference planes, when produced are termed as *traces*. The point of intersection of straight line with vertical plane is *vertical trace* and the point of intersection of straight line with horizontal plane is *horizontal trace*. Actual length of the line is called the *true length*.

(*i*) **Straight line parallel to both the reference planes:** When a straight line is parallel to both the planes its projections on both the planes are parallel along reference axis (XY) and are true lengths. There is no inclination with either vertical plane or horizontal plane.

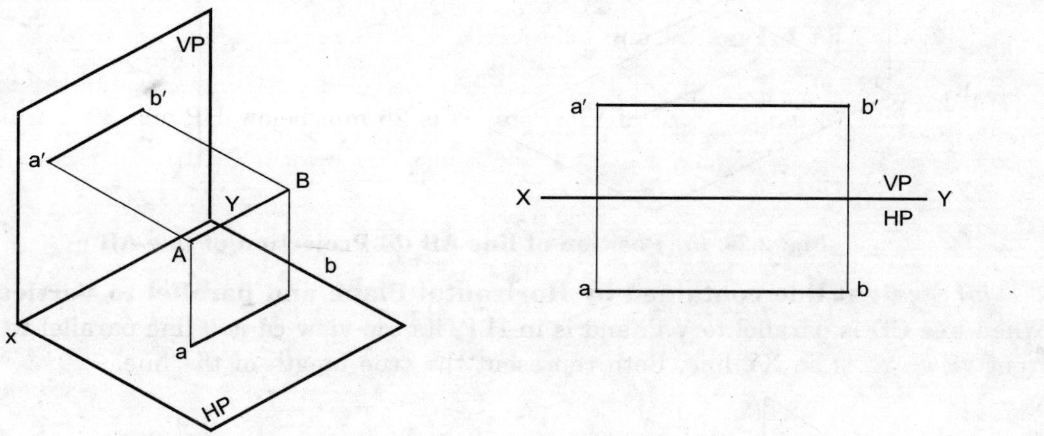

Fig. 2.24: (*a*) Position of line AB (*b*) Projection of line AB

(*ii*) **Straight line perpendicular to HP and parallel to VP.:** When the line is perpendicular to a reference plane, the projection on plane to which it is perpendicular is a point and projection of line on the other reference plane is the true length perpendicular to XY plane. In this case *a′b′* is the true length of the line.

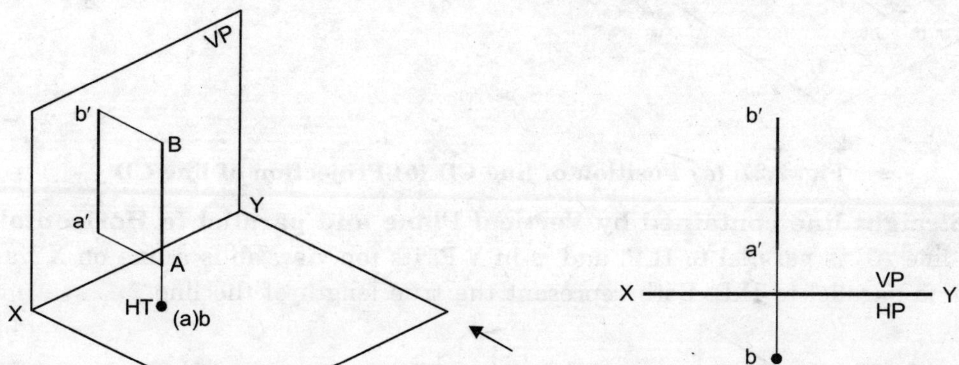

Fig. 2.25: (*a*) Position of line AB (*b*) Projection of line AB

(*iii*) **Straight line perpendicular to VP and parallel to HP.:** The top view gives the true length of line AB and front view will be a point *b′* (*a′*). While looking from front *a′* and *b′* will coincide, *a′* lying behind *b′*.

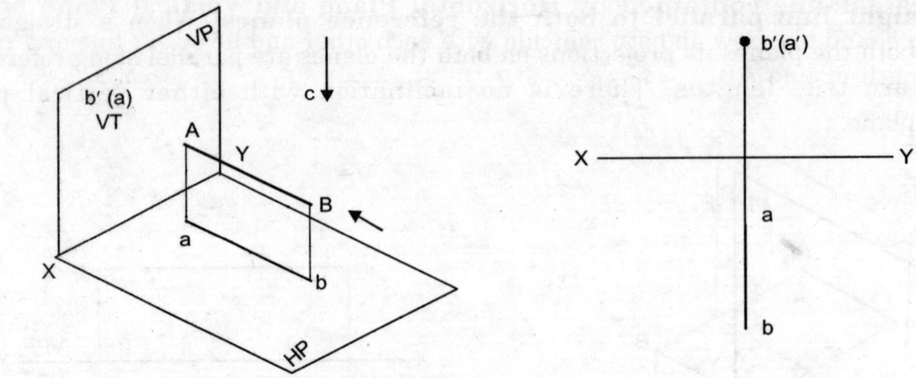

Fig. 2.26: (*a*) Position of line AB (*b*) Projection of line AB

(*iv*) **Straight line contained by Horizontal Plane and parallel to Vertical Plane:** When line CD is parallel to V.P. and is in H.P. its top view cd is a line parallel to V.P. and front view *c′d′* is on XY line. Both represent the true length of the line.

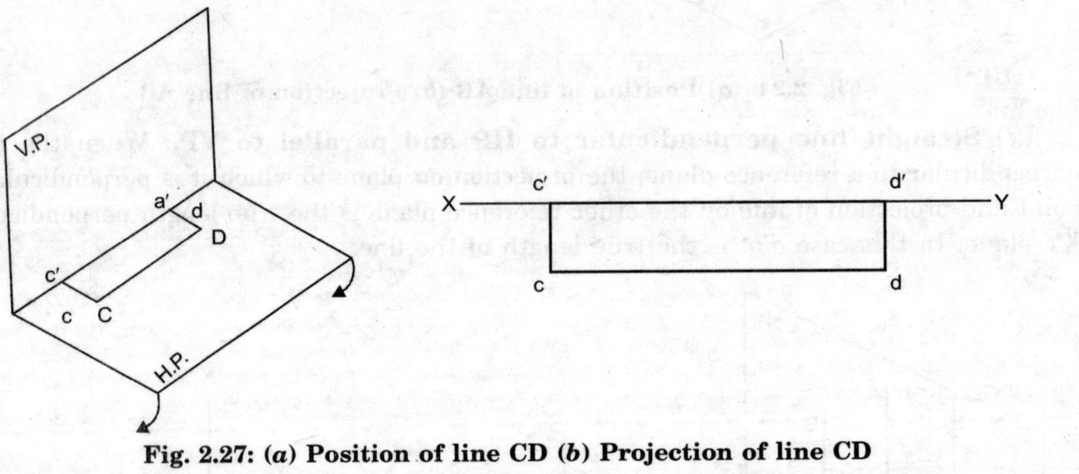

Fig. 2.27: (*a*) Position of line CD (*b*) Projection of line CD

(*v*) **Straight line contained by Vertical Plane and parallel to Horizontal Plane:** When a line AB is parallel to H.P. and is in V.P. its top view ab is a line on XY and front view *a′b′* is parallel to H.P. Both represent the true length of the line.

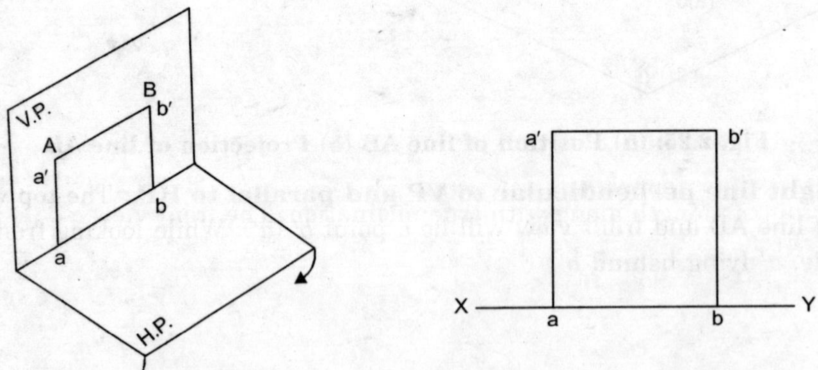

Fig. 2.28: (*a*) Position of line AB (*b*) Projection of line AB

(*vi*) **Straight line contained by Horizontal Plane and Vertical Plane both:** The front view *a'b'* and top view ab both coincide with each other and lie on XY line and represent the true length of line AB.

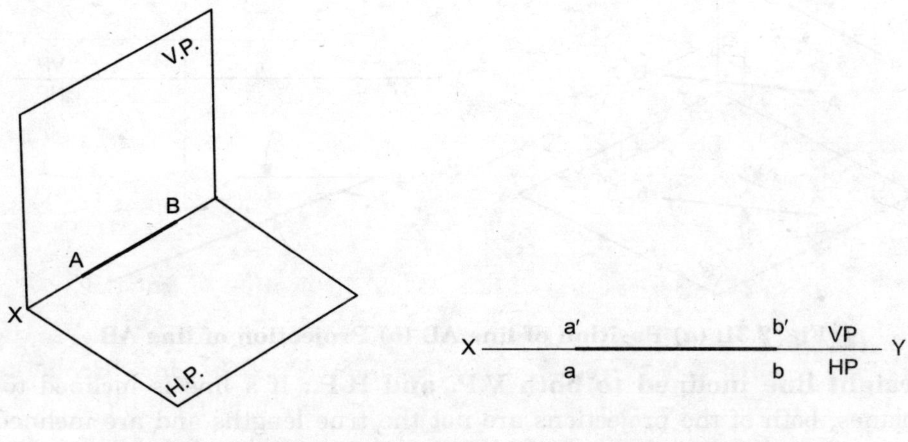

Fig. 2.29: (*a*) Position of line AB (*b*) Projection of line AB

(*vii*) **Straight line inclined to Horizontal Plane and parallel to Vertical Plane:** The line AB is inclined at angle θ to the horizontal plane. The front view *a'b'* is equal to the line AB and gives its true length along with true inclination. The top view ab is a straight line shorter than AB.

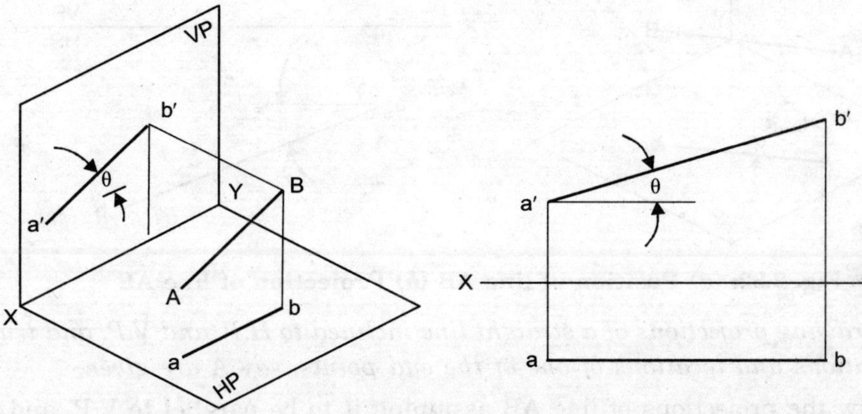

Fig. 2.30: (*a*) Position of line AB (*b*) Projection of line AB

(*viii*) **Straight line inclined to Vertical Plane and parallel to Horizontal Plane:** The line AB is inclined at angle φ to the vertical plane. The top view ab is equal to the line AB and gives its true length along with true inclination. The front view *a'b'* is a straight line shorter than AB.

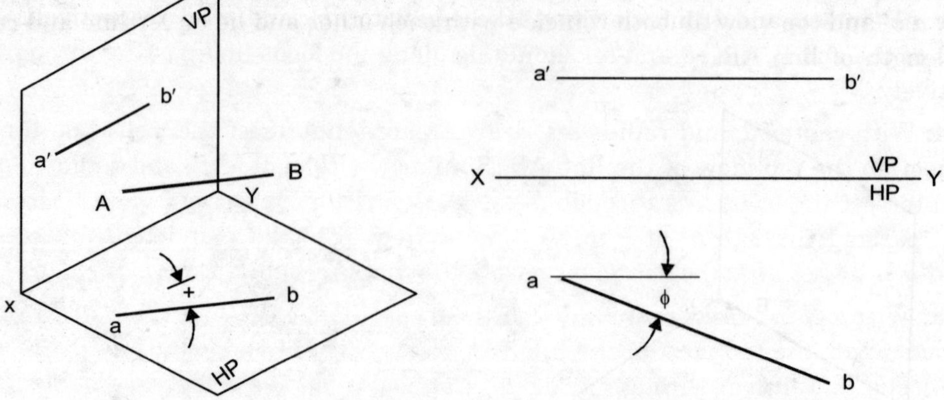

Fig. 2.31: (*a*) Position of line AB (*b*) Projection of line AB

(*ix*) **Straight line inclined to both V.P. and H.P.:** If a line is inclined to both the reference planes, both of the projections are not the true lengths and are inclined at some angles α, β greater than the actual inclinations with respective reference planes.

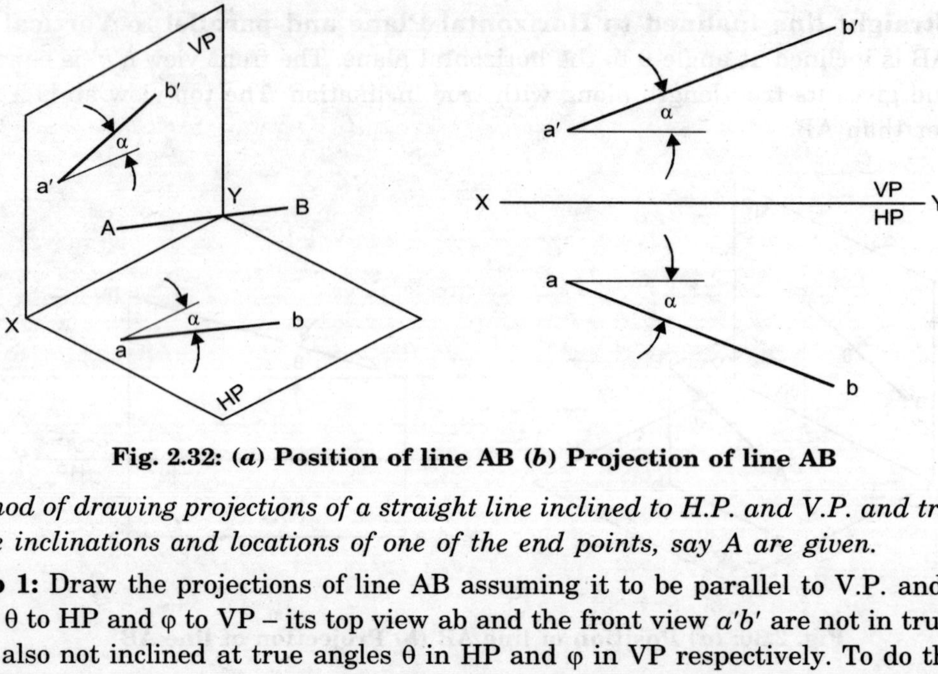

Fig. 2.32: (*a*) Position of line AB (*b*) Projection of line AB

Method of drawing projections of a straight line inclined to H.P. and V.P. and true length and true inclinations and locations of one of the end points, say A are given.

Step 1: Draw the projections of line AB assuming it to be parallel to V.P. and inclined at angle θ to HP and φ to VP – its top view ab and the front view *a'b'* are not in true lengths and are also not inclined at true angles θ in HP and φ in VP respectively. To do this rotate the line AB about the end A, keeping θ constant. This rotation of the line will bring the end B to the new position B_1. Project AB_1 on VP and HP. $a'b_1'$ is in true length inclined at *q* to the XY line.

Step 2: Rotate the line AB about the end A keeping φ the inclination of AB with VP constant, till it becomes parallel to HP. AB_2 is the new position of the line AB. Project AB_2 on HP and VP. Since AB_2 is parallel to HP, ab_2, the projection of AB_2 on HP is in true length inclined to XY line. The line is rotated back to its original position AB.

Step 3: The conclusions that can be drawn from the above two steps are (*i*) the length of the line AB in top and front views will be equal to ab_1 and $a'b_2'$ respectively and the projections of the end B, (i.e., *b* and *b'*) should lie along the locus line passing through b_2 and b_1' respectively.

Step 4: With center a, and radius ab_2 draw an arc to intersect the locus line through b_2 at *b*. Connect ab the top view of the line AB. Similarly with center *a'*, and radius $a'b_2'$ draw an arc to intersect the locus line through b_1' at *b'*. Connect *a'b'* the front view of the line AB. Draw $a'b_1'$ passing through *a'* at θ to XY line and equal to the true length of AB. $a'b_1'$ is projected down to get ab_1, the top view parallel to the XY line.

Step 5: With centrea and radius ab_1, draw an arc to cut the locus line drawn through b_2 at *b*. Connect ab, the top view of the line AB. With centre *a'* and radius $a'b_2'$, draw an arc to cut the locus line drawn through b_1' at *b'*. Connect a '*b*', the front view of the line AB.

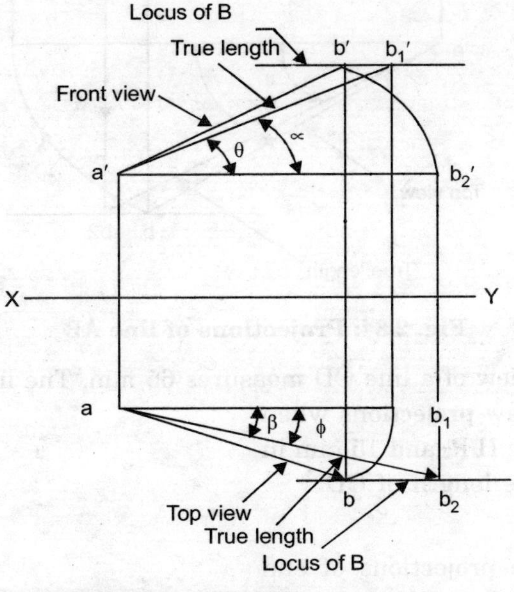

Fig. 2.33: Projection of line AB

Method of finding positions of a straight line (true length and true inclinations) when projections inclined to V.P. and H.P. are given.

If positions of end points of line are not given, one end may be assumed to lie either in H.P. or V.P. or in both H.P. and V.P.

Assuming that the position of end point is given steps to find true lengths are:

Step 1: Draw the top view *ab* and the front view *a'b'*.

Step 2: With center a and radius abrotate the top view to the new position ab_1 and make it parallel to the XY line. As ab_1 is parallel to the XY line, its corresponding front view will give the true length and at true inclination.

Step 3: With center *a'* and radius a '*b*' rotate the front view to the new position $a'b_2'$ parallel to the XY. Since $a'b_2'$ is parallel to the XY, its corresponding top view will give true length and true inclination.

Step 4: With a as center and ab_1 as radius, draw an arc to intersect at *b*. Join *ab*.

Step 5: With *a′* as center and ab_2' as radius, draw an arc to intersect at *b′*. Join *a′b′*.

Step 6: Connect ab_2 which will be in true length and true inclination which the line AB makes with VP. Connect a 'b_1'. As a check, connect *bb′*. It should be in a straight line.

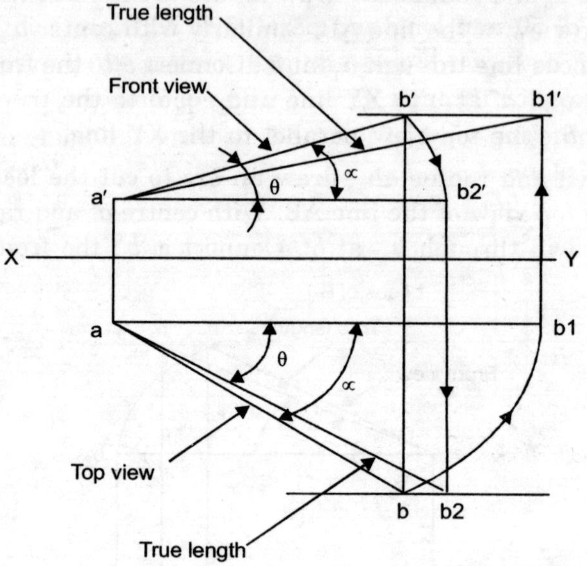

Fig. 2.34: Projections of line AB

Problem 1. The top view of a line CD measures 65 mm. The line is parallel to V.P. and inclined at 35° to H.P. Draw projections when end point C is 10mm above H.P. and 15 mm in front of V.P. Find the true length of CD.

Solution.

- **Step 1:** Draw the projections of end point C – *c'* 10mm above XY line and *c* which is 15 mm below XY line.
- **Step 2:** Top view of line is parallel to V.P. So draw a straight line parallel to XY line such that *ab* = 65 mm.
- **Step 3:** Draw a line at 35° with H.P. through *c′* and project *b* vertically upwards such that it intersects the line at *b′*.
- **Step 4:** *cd* and *c′d′* are the projections of line CD. Measure *c′d′* which gives true length of the line

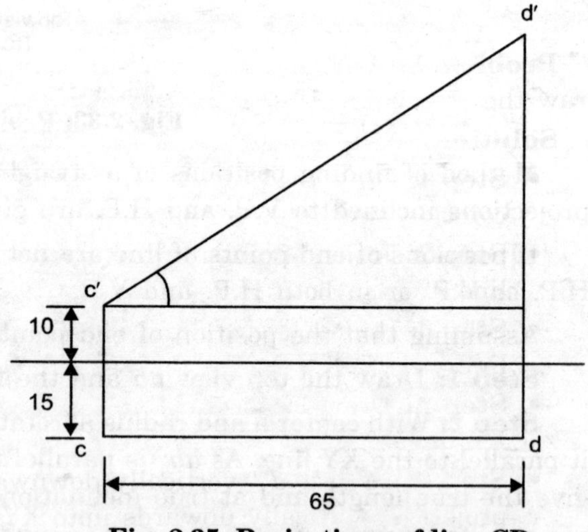

Fig. 2.35: Projections of line CD

Problem 2. The elevation of a line CD of length 70 mm measures 50 mm. The line is parallel to H.P. Draw the projections and determine the inclination of line with V.P. if one end is 10 mm above H.P. and 12 mm in front of V.P. (111)

Solution.

- **Step 1:** Draw *c′* and *c*.
- **Step 2:** Draw *c ‘d’* 50 mm long and parallel to XY line.
- **Step 3:** Project *d′* downwards and taking distance of 70 mm and *c* as center cut the distance from projector through *d′* at *d* using a compass.
- **Step 4:** Join *cd*, the top view of line and measure the angle of inclination φ of line CD with vertical plane.

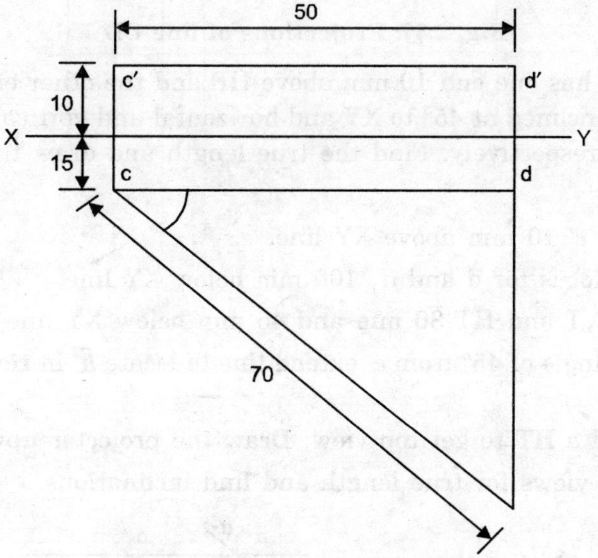

Fig. 2.36: Projections of line CD

Problem 3. A line CD 75 mm long makes an angle of 30° with H.P. and 45° with V.P. Draw the projections if point C lies in H.P. and V.P. both.

Solution.

- **Step 1:** Draw the projections of point C, which lies in both H.P and V.P. *c* and *c′* will coincide and lie on XY line.
- **Step 2:** From *c′* draw a line at an angle of 30° with XY line and cut 75 mm from this line at c_1'.
- **Step 3:** From *c*, draw a line at an angle of 45° with XY line and cut 75 mm from this line at c_2.
- **Step 4:** Draw locus of d_1' and d_2 by drawing lines through these point parallel to XY line.
- **Step 4:** Project d_1' vertically downwards upto XY and mark this point as d_1 and project d_2 vertically upwards upto XY at d_2'.
- **Step 5:** With *c′* as the center and $c'd_2'$ as radius draw an arc cutting the locus of d_1 at *d′*. Join *c′d′*. Repeat the same process and join *cd* which is the required top view of the line.

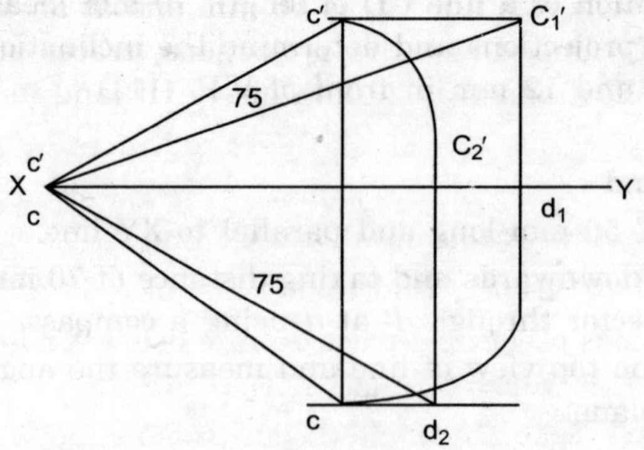

Fig. 2.37: Projections of line CD

Problem 4. A line has one end 10 mm above HP and the other end 100 mm in front of V.P. The front view is inclined at 45° to XY and horizontal and vertical trace are 45 mm and 35 mm below XY line respectively. Find the true length and draw the projections

Solution.

- **Step 1:** Locate c' 10 mm above XY line.
- **Step 2:** Draw locus for d and d_1 100 mm below XY line.
- **Step 3:** Draw VT and HT 30 mm and 45 mm below XY line respectively.
- **Step 4:** At an angle of 45° from c' extend line to locate h' in reverse direction. Locate HT, VT and v.
- **Step 5:** Extend v HT to get top view. Draw the projector upwards and locate b'.
- **Step 6:** Rotate views for true length and find inclinations.

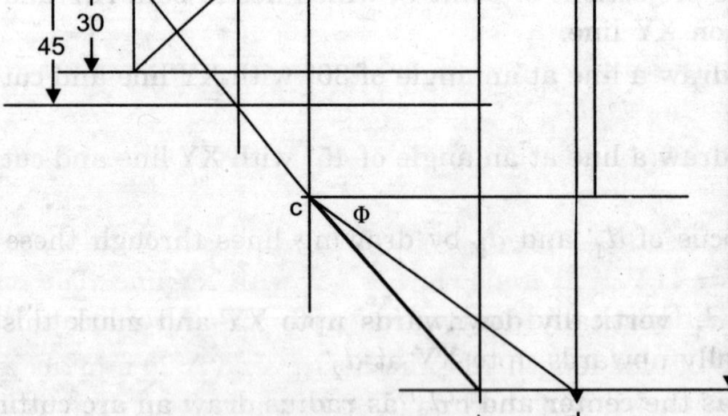

Fig. 2.38: Projections of line CD

Problem 4. The front view of a line CD is 60 mm and at an angle 45° to XY line. The top view is at 30° with XY line. Draw the projections, traces of line and also find the inclinations with HP and VP if end C is 20 mm above HP and its vertical trace is 15 mm below HP.

Solution.

- **Step 1:** Draw front view *c′* 15 mm above XY line.
- **Step 2:** Take angel 45° from *c′* and taking distance 60 mm locate point *d′*.
- **Step 3:** Draw the locus of vertical trace 15 mm below XY line and extend front view to this locus and locate vertical trace.
- **Step 4:** From point V take 30° angle downwards as top view. Locate d by drawing projector from *d′*.
- **Step 5:** Use the method of rotation of views and find the true lengths and angle of inclinations.

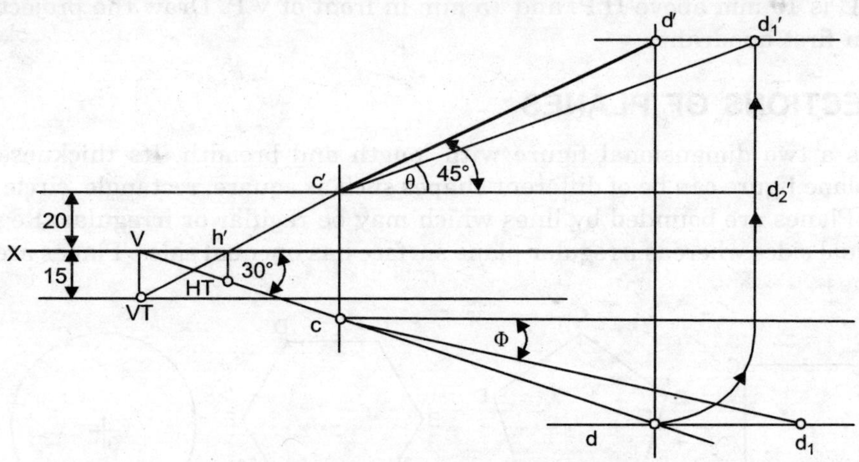

Fig. 2.39: Projections of line CD

2.12 PRACTICE QUESTIONS FOR PROJECTIONS OF LINES

1. Define straight lines and trace of a line.
2. A line AB is 75 mm long and it is 30° & 40° Inclined to H.P. and V.P. respectively. End point A is 12 mm above H.P. and 10 mm in front of V.P. Draw projections given that line is in 1st quadrant.
3. A line AB 60 mm long is parallel to V.P. & H.P. both. The end A is 30 mm above H.P. and 40 mm in front of V.P. Draw the projection of line AB.
4. The end A of a line AB is 10 mm in front of V.P. and 20 mm above H.P. The line is inclined at 30° to H.P. and front view is 45° with XY line. The top view is 60 mm long. Find the true length and inclination with V.P. Also, locate the HT and VT.
5. A straight line AB of true length 100 mm has its end A 20 mm above HP and 30 mm in front of VP. The top view of the line is 80 mm and front view is 70 mm. Draw the projections (TV and FV) of the line AB and obtain the true inclinations of the line AB with HP and VP.

6. Front view of line AB is 500 inclined to XY and measures 55 mm long while its top view is 600 inclined to XY line. If end A is 10 mm above H.P. and 15 mm in front of V.P. draw its projections. Find True length and inclinations of line with H.P. & V.P.
7. A line AB 50 mm long is inclined at 30° to H.P. and 45° to V.P. End A is 10mm away from both the planes H.P. and V.P. Draw the projections of line if it lies in the second quadrant.
8. A line AB 55 mm long has its end A, 15 mm away from both V.P. and H.P. The end point B is 25 mm away from H.P. and V.P. Draw its projections and determine the traces.
9. The front view of a line 70 mm long, measures 50 mm. The line is parallel to H.P. and one of kits ends is in V.P. and 15 mm above H.P. Draw the projections and determine its inclination with V.P.
10. A line is 75 mm long. Its front view is 50 mm long and top view is 60 mm long. End point C is 10 mm above H.P. and 15 mm in front of V.P. Draw the projections if end D is in first quadrant.

2.13 PROJECTIONS OF PLANES

A plane is a two dimensional figure with length and breadth. Its thickness is always neglected. A plane figure can be of different shapes such as square, rectangle, circle, pentagon, hexagon, etc. Planes are bounded by lines which may be regular or irregular. Regular plane surface has equal sides whereas irregular plane surface has unequal sides. Planes are generated by moving lines.

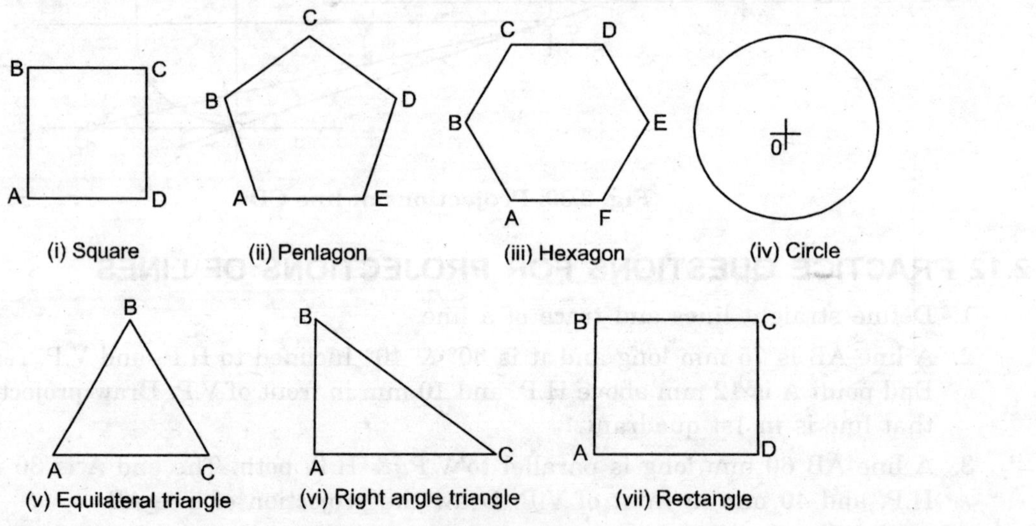

Fig. 2.40: Types of planes

2.14 TYPES OF PLANES

There are two types of planes used in engineering drawings:

1. Perpendicular planes
2. Oblique planes

1. **Perpendicular planes:** The planes perpendicular to one or both the reference planes i.e., H.P. and V.P. are termed as perpendicular planes. These can be:

- Planes which are perpendicular to H.P. and parallel to V.P.
- Planes which are perpendicular to V.P. and parallel to H.P.
- Planes which are perpendicular to V.P. and H.P. both.
- Planes which are perpendicular to V.P. and inclined to H.P. These are termed as auxiliary inclined planes (A.I.P.).
- Planes which are perpendicular to H.P. and inclined to V.P. These are termed as auxiliary vertical planes (A.V.P.).

Traces of Planes: The extension of a given plane shape to the reference plane (HP or VP) to which it is perpendicular or inclined. The plane meets the H.P. or V.P. as a line and this line is called *trace of a plane*. The line where the plane meets H.P. is called the *horizontal trace* (H.T.) and the line where the plane meets V.P. is called the *vertical trace* (V.T.)

2.15 PROJECTIONS OF PERPENDICULAR PLANES

(*i*) **Plane perpendicular to H.P. and parallel to V.P.:** A rectangle ABCD is perpendicular to H.P. and parallel to V.P. Its front view will be a rectangle *a′ b′ c′ d′* showing the true shape and size of the plane. Top view will be a line ***ad* – *bd*** which is parallel to XY. It has only horizontal trace and no vertical trace.

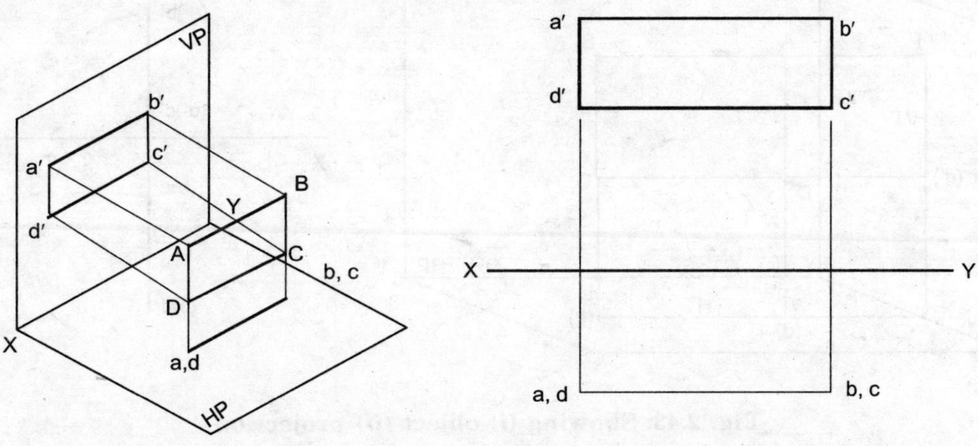

Fig. 2.41: Showing (*i*) object ABCD (*ii*)projection

(*ii*) **Plane perpendicular to V.P. and parallel to H.P.:** Figure 2.42 shows a rectangle ABCD in spaceperpendicular to V.P. and parallel to H.P. It has only vertical trace and no horizontal trace. The top view abcd gives the actual shape and size of the object where as the front view is a straight line parallel to XY.

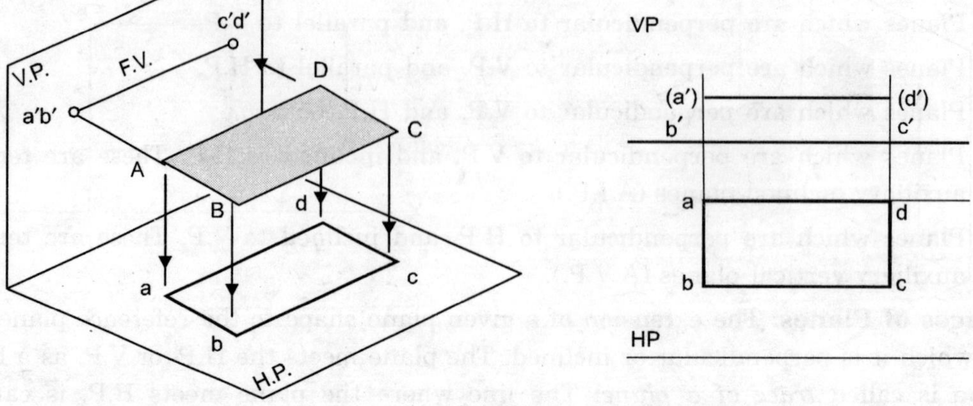

Fig. 2.42: Showing *(i)* object *(ii)*projection

(*iii*) **Plane perpendicular to H.P. and V.P.:** Figure showing a square lamina perpendicular to both VP and HP. It has both vertical and horizontal traces. The front view (straight line) $a'b' - c'd'$ and the top view $ad - cb$ (straight line) coincide with V.T. and H.T. respectively.

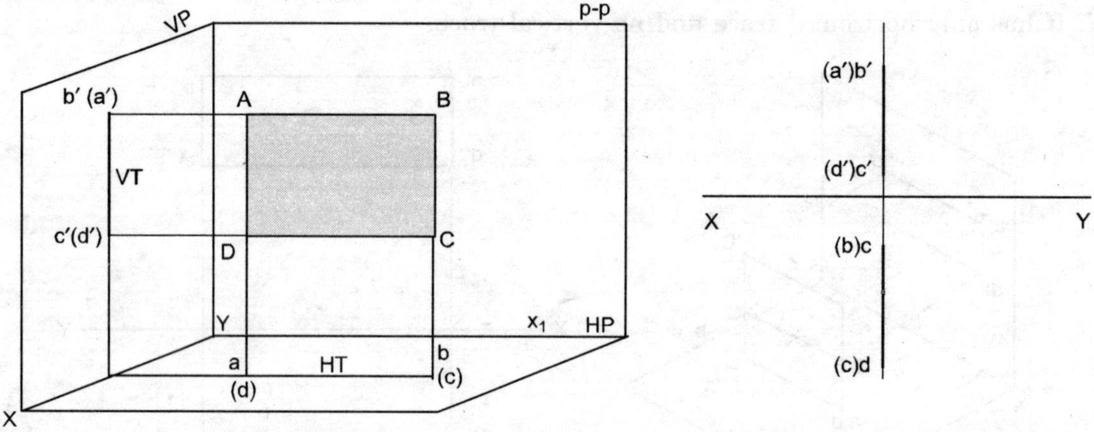

Fig. 2.43: Showing *(i)* object *(ii)* projection

(*iv*) **Plane perpendicular to H.P. and inclined to V.P.:** Figure showing a rectangular lamina ABCD perpendicular to H.P. and inclined to V.P at angle φ. It has horizontal trace inclined at angle φ to XY and vertical trace perpendicular to XY line. The front view is shortened and less than the actual dimensions.

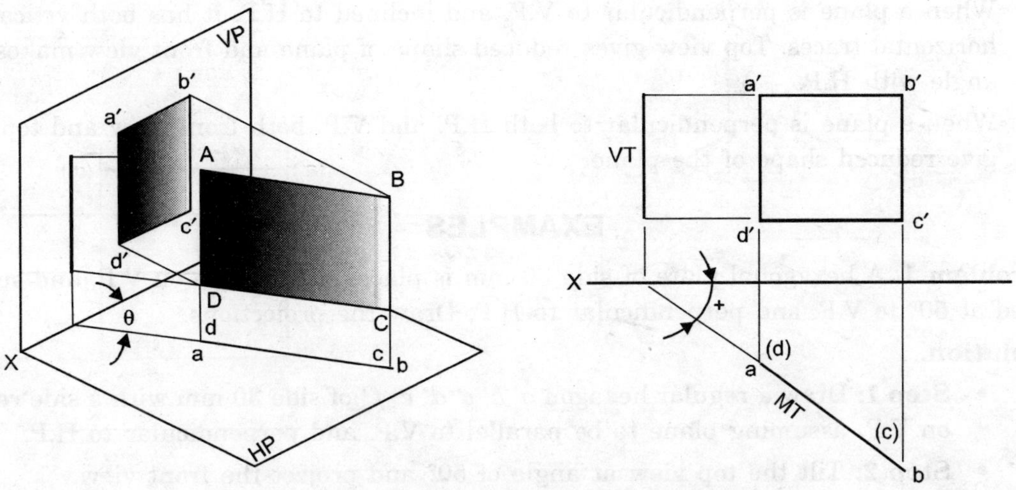

Fig. 2.44: Showing (*i*) object (*ii*) projection

(*v*) **Plane perpendicular to V.P. and inclined to H.P.:** Figure showing a rectangular lamina ABCD perpendicular to V.P. and inclined to H.P at angle θ. It has vertical trace inclined at angle θ to XY and horizontal trace perpendicular to XY line. The top view is shortened and less than the actual size.

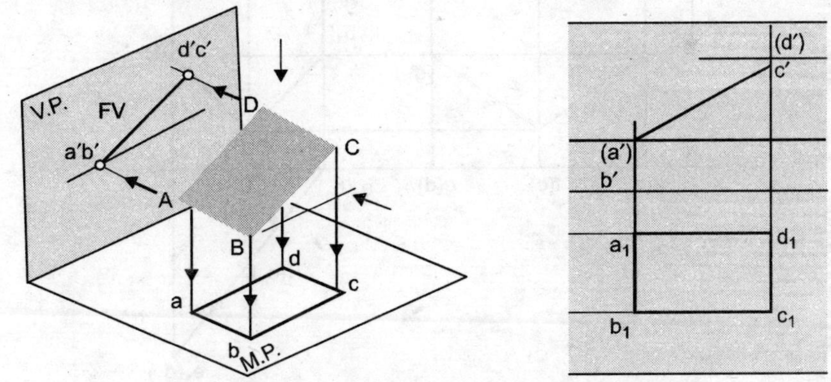

Fig. 2.45: Showing (*i*) object (*ii*) projection

POINTS TO REMEMBER

- When a plane is perpendicular to H.P. and parallel to V.P., it has only one trace i.e., horizontal trace and when a plane is perpendicular to V.P. and parallel to H.P., it has only one trace which is vertical trace.
- In cases when plane is perpendicular to both the planes H.P. and V.P., it has both vertical and horizontal traces.
- When a plane is perpendicular to H.P. and inclined to V.P. it has both vertical and horizontal traces. Front view gives reduced shape of plane and top view makes true angle with V.P.

- When a plane is perpendicular to V.P. and inclined to H.P. it has both vrtical and horizontal traces. Top view gives reduced shape of plane and front view makes true angle with H.P.
- When a plane is perpendicular to both H.P. and V.P. both front view and top view give reduced shape of the plane.

EXAMPLES

Problem 1. A hexagonal plate of side 30 mm is places with a side on V.P. and surface inclined at 50° to V.P. and perpendicular to H.P. Draw the projections.

Solution.

- **Step 1:** Draw a regular hexagon *a' b' c' d' e' f'* of side 30 mm with a side resting on V.P. assuming plane to be parallel to V.P. and perpendicular to H.P.
- **Step 2:** Tilt the top view at angle of 50° and project the front view.

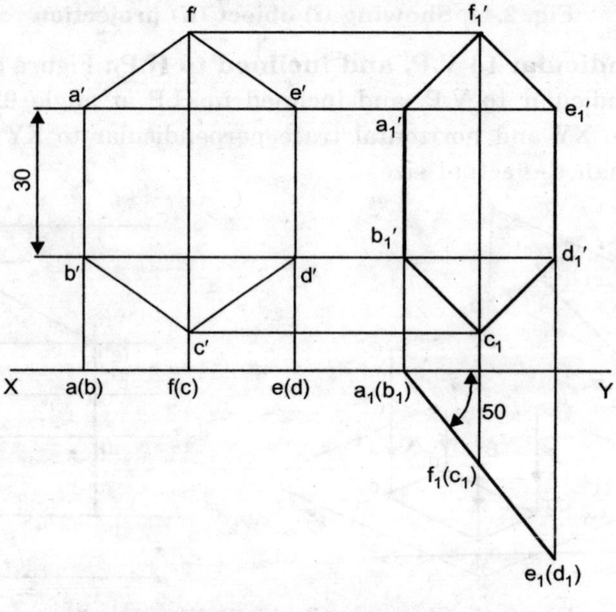

Fig. 2.46: Projection of hexagonal plate

Problem 2. Draw the projections of a circle of 5 cm diameter having its plane vertical and inclined at 30° to the V.P. Its center is 3 cm above H.P. and 2 cm in front of V.P.

Solution.

Steps:

- Draw a circle of diameter 5 cm with center 3 cm above the XY line.
- Divide it into 12 equal parts and project top view on XY line
- Reproduce a new top view separately, making an angle 30° to the V.P.
- Project the new front view upwards with the help of new top view and initial top view. The new front view is an ellipse.

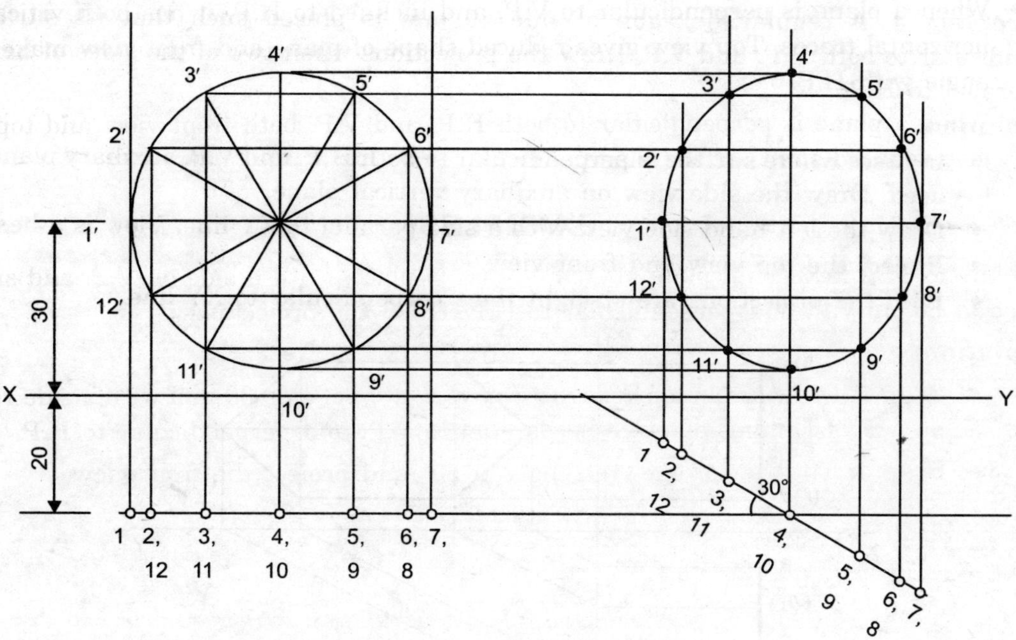

Fig. 2.47: Projection of circle

3. Draw the projections of a rhombus with diagonals 100 mm and 40 mm long. The smaller diagonal is parallel to both the refernce planes and larger diagonal is inclined at 35° to the H.P.
   - **Step 1:** Draw top view of rhombus in true shape abcd in a way that ac is parallel to XY line. Project the corresponding top view which is a straight line.
   - **Step 2:** Tilt and incline *a′c′* at angle 35° to the H.P. Project the second top view $a_1b_1c_1d_1$.
   - Reproduce another top view such that smaller diagonal b_1d_1 is parallel to both the planes and project the final front view accordingly.

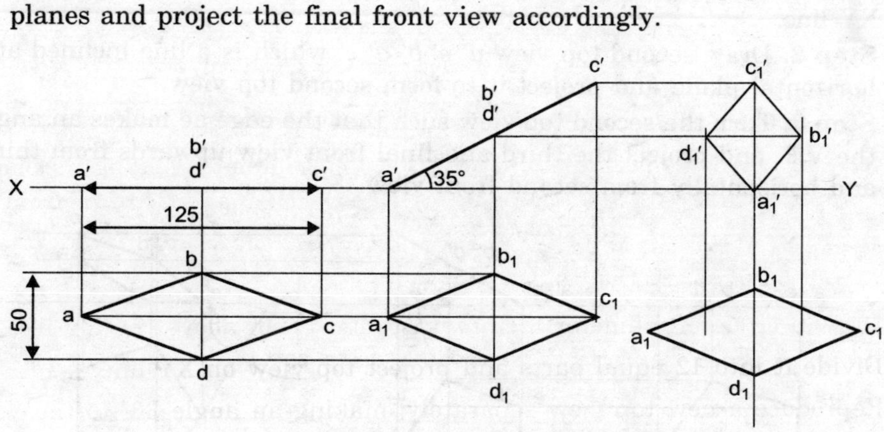

Fig. 2.48: Projection of a rhombus

Problem 4. A regular hexagon of side 40 mm is placed such that its surface is perpendicular to both H.P. and V.P. Draw the projections when two of the sides of hexagon are parallel to H.P.

Solution.

- In cases where surface is perpendicular to both H.P. and V.P. auxiliary planes are used. Draw the side view on auxiliary vertical plane.
- Draw the left hand side view with a side parallel to XY line. View is a hexagon.
- Project the top view and front view.
- Both the projections are straight lines perpendicular to XY line.

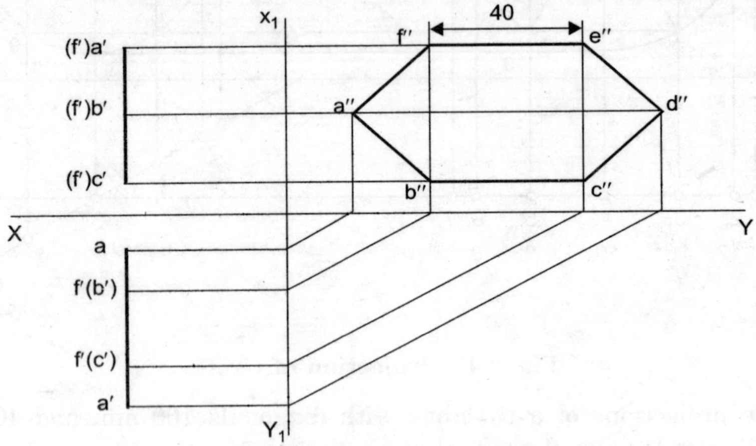

Fig. 2.49: Projection of regular hexagon

Problem 5. A pentagonal plane of 25 mm side is resting in H.P. on one of its sides which makes an angle of 60° with V.P. and the surface of plane makes an angle of 30° with H.P. Draw the projections.

Solution.

- **Step 1:** Assume that the pentagon is lying on H.P. with an edge perpendicular to V.P. Draw top view (pentagon) *abcde* and front view *a' b' c' d' e'* (line) on XY line.
- **Step 2:** Draw second top view *a' e' b' d' c'* which is a line inclined at 30° to the horizontal plane and project it to form second top view.
- **Step 3:** Turn the second top view such that the edge ae makes an angle 60° with the V.P. and project the third and final front view upwards from third top view and horizontally from second front view.

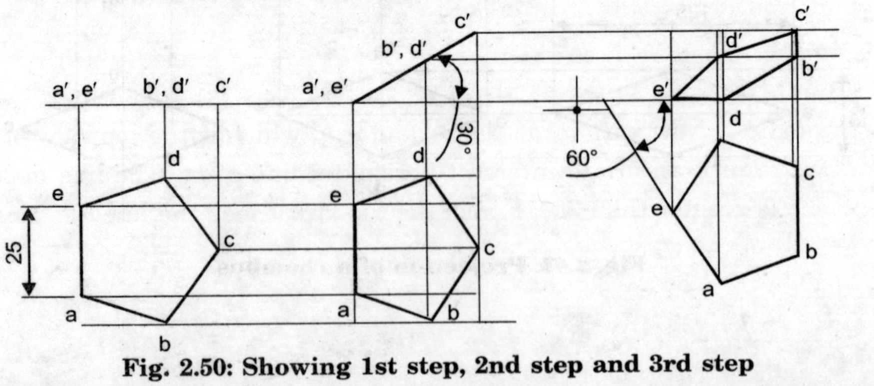

Fig. 2.50: Showing 1st step, 2nd step and 3rd step

Problem 6. A rectangular plate 30 mm * 50 mm is resting on HP on one small side which is at angle 30° to V.P., while the surface of plane makes an angle of 45° with H.P. Draw the projections.

Solution.

- **Step 1:** Draw the top view *abcd* (rectangular) and front view *a'd'* – *c'd'* (line) such that shorter side is perpendiculat to V.P.
- **Step 2:** Draw the second top view $a_1b_1c_1d_1$ as a square of side 30 mm (smaller side) and project the 2nd front view at an angle 45° by drawing vertical projections.
- **Step 3:** Tilt $a_1b_1c_1d_1$ such that edge a_1b_1 makes an angle 30° with V.P. Project the front view a_1' b_1' c_1' d_1'.

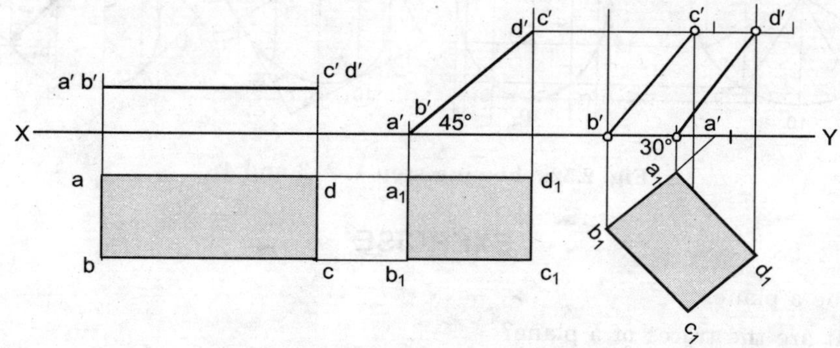

Fig. 2.51: Showing 1st step, 2nd step and 3rd step

Problem 7. A circular plate having diameter 60mm is resting on H.P. on point A of its rim (circumference). The plane is inclined at 45° to the H.P. and (*i*) the top view of diameter makes an angle of 30° with V.P. (*ii*) Diameter makes an angle 30° with the V.P.

Solution.

- **Step 1:** Draw the top view and front view of circle and bisect circle in 12 equal parts.
- **Step 2:** Tilt the front view at 45° to H.P. and project second top view.
- **Step 3:** Tilt second top view in such a way that diameter is inclined at 30° with V.P. and project the front view.
- **Step 4:** As given in question, diameter makes an angle 30° with the V.P. Draw a line a_1b_1 equal to the diameter *ab* at 30° to the V.P and draw a horizontal line through b_2. Now with a_1 as center and a_1b_1 which is the top view of diameter as radius, mark an arc to intersect horizontal line at b_1. The line makes an angle which is greater than 30°. Finish the top view along the line a_1b_1 and project the final front view.

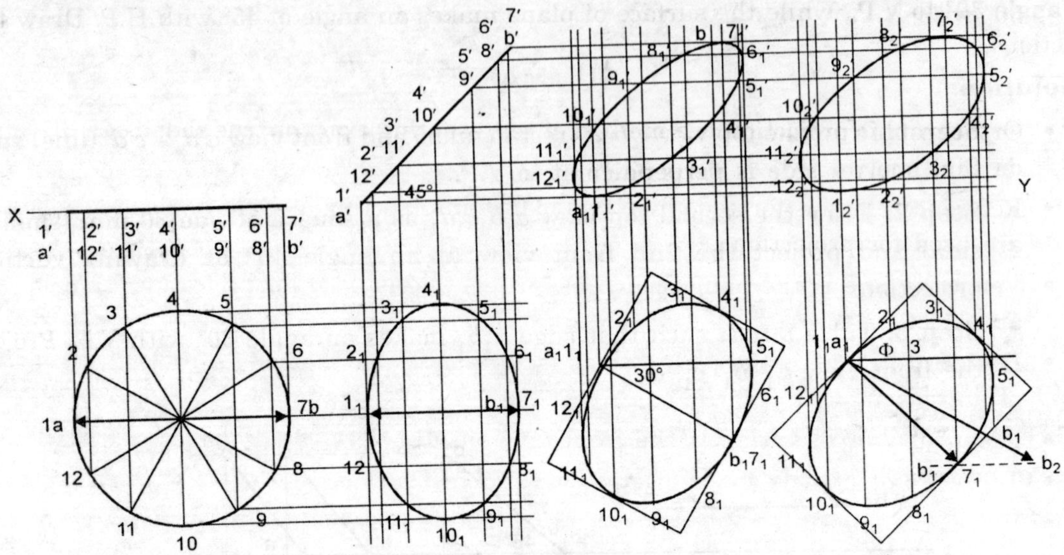

Fig. 2.52: showing step 1, 2, 3 and 4

EXERCISE

1. Define a plane.
2. What are the traces of a plane?
3. What are auxiliary planes?
4. A Rectangle 40 mm and 70 mm sides is resting on HP on one small side which is 40° inclined to VP, while the surface of the plane makes 45° inclination with HP. Draw it's projections.
5. A regular pentagon of 30 mm sides is resting on HP on one of the side while its opposite vertex (corner) is 30 mm above HP. Draw projections when side in HP is 30° inclined to VP.
6. A square lamina of side 80 mm rests on a corner on H.P. and it is inclined with H.P. such that its plan is a rhombus with a diagonal of 40 mm. The long diagonal is inclined with the V.P. at 45°. Determine its inclination with H.P. and draw its projections.
7. A circular plate of 30 mm side is resting in V.P. on a point of its rim. The diameter of plate is inclined at 30° to V.P. and diameter is parallel to both the planes. Draw the projections.
8. A rhombus having diagonals 50 mm and 25 mm is resting in V.P. on one of its corners. The rhombus is inclined with V.P. such that its front view is a square of 25 mm long diagonals. Draw projections.
9. Draw the projections and traces of a square of side 30 mm with one of its sides perependicular to V.P. and makes an angle 35° to H.P.

10. Draw the projections of a pentagon with side 35 mm having one side in H.P. The plane is inclined at 30° to H.P. and the side with which it rests makes an angle of 50° with H.P.

SUMMARY

- Orthographic projection is a method of representing exact or true shape of a 3D object on 2D sheet in two or more views.
- Reference plane is formed when two planes intersecting at right angles to each other are used for projection.
- Vertical Plane is the plane which is vertical and is used for projecting front view of an object.
- Horizontal Plane is the plane which is horizontal and is used for projecting top view of an object.
- Front view or elevation is the front view of an object and is projected on VP.
- Top view or plan is the projection of an object on horizontal.
- Vertical plane is always fixed while Horizontal plane is rotated clockwise (90°) about XY, until VP and HP come in same line.
- *Inclination* of a straight line to a reference plane is the angle which line makes with the plane.
- The point of intersection of any line with reference planes, when produced are termed as *traces*. The point of intersection of straight line with vertical plane is *vertical trace* and the point of intersection of straight line with horizontal plane is *horizontal trace*.
- Actual length of the line is called the *true length*.
- A plane is a two dimensional figure with length and breadth. Its thickness is always neglected.
- The planes perpendicular to one or both the reference planes i.e., H.P. and V.P. are termed as perpendicular planes.

CHAPTER

3

AUXILIARY PLANES AND VIEWS AND PROJECTIONS OF SOLIDS

Learning Objectives

- Describe auxiliary reference planes and auxiliary views
- Understand how points and lines are projected on auxiliary reference planes
- Learn about types of solids
- Understand projections of solids in various positions
- Learn about the visibility of various surfaces of the solid in different views and in different positions
- Learn the concept of floor plans

3.1 INTRODUCTION

Many objects have such shapes that their principal planes and lines of sight are not parallel to the regular principal projection planes. The projections on principal planes are either insufficient to describe the complete geometry or difficult to draw. There are infinite possible auxiliary views of an object.

Auxiliary views can help in finding:

- True lengths and inclinations of a line.
- Point view of a line.
- Edge view of a plane.
- True shape of a plane.
- Projections of solids
- True shapes and inclinations of solids etc.

In practice, auxiliary plane is assumed in position which best facilitates auxiliary views leaving principal views clear.

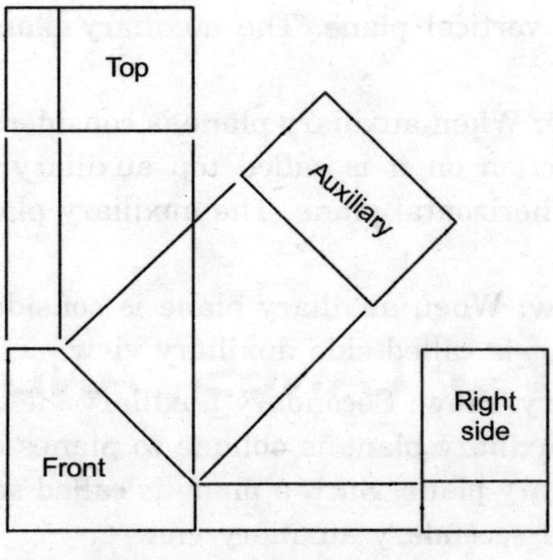

Fig. 3.1: Auxiliary view

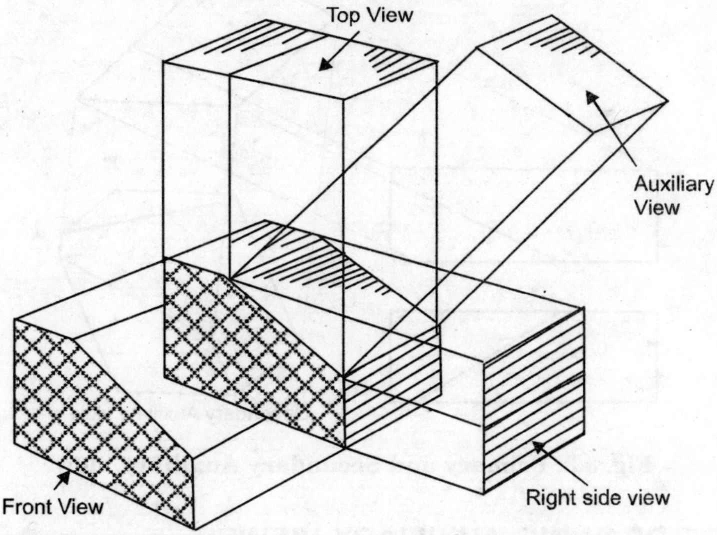

Fig. 3.2: An example of different views of an object

3.2 TYPES OF AUXILIARY VIEWS

(*i*) Primary auxiliary views

(*ii*) Secondary auxiliary view

When true shapes of inclined surfaces are not shown in their principal planes and inclined surface has important features primary auxiliary views are used.

(*i*) **Primary Auxiliary View:** If auxiliary plane is perpendicular to principal plane of projection, primary auxiliary view is obtained.

These are of three types:

- **Auxiliary Front view:** When auxiliary plane is considered perpendicular to principal horizontal plane, projection on it is called front auxiliary view. It may be inclined or parallel to principal vertical plane. The auxiliary plane is assumed hinged to the principal plane.

- **Auxiliary Top view:** When auxiliary plane is considered perpendicular to principal vertical plane, projection on it is called top auxiliary view. It may be inclined or parallel to principal horizontal plane. The auxiliary plane is assumed hinged to the principal plane.
- **Auxiliary Side view:** When auxiliary plane is considered perpendicular to profile plane, projection on it is called side auxiliary view.

(*ii*) **Secondary Auxiliary View:** Secondary auxiliary view is drawn after drawing the primary auxiliary view. If auxiliary plane is oblique to planes of projection, it is considered perpendicular to some auxiliary plane. Such a plane is called secondary auxiliary plane and view projected on it is called secondary auxiliary view.

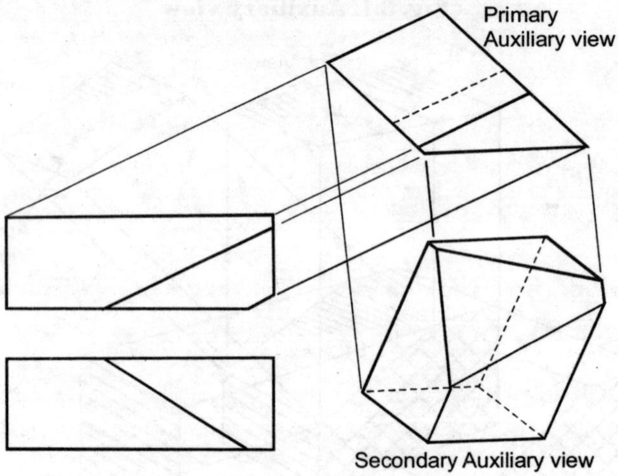

Fig. 3.3: Primary and Secondary Auxiliary view

3.3 METHOD OF DRAWING AUXILIARY VIEWS

- Draw the top view and front view of the object.
- Draw an auxiliary plane line X_1Y_1 parallel to inclined surface of front view.
- Draw projections perpendicular to XY line from various points of inclined surface.
- Use compass to mark the point on projections and complete the auxiliary view by joining points in correct sequence.

Problem 1: A hexagonal base pyramid of 30 mm side and 60 mm long axis has an edge of its base on H.P. and makes an angle of 35° with H.P. and is parallel to V.P. Draw its projections using auxiliary plane method.

Solution: Draw the top view and front view of the object.

- Draw an auxiliary plane X_1Y_1 inclined at 35° with the axis.
- Project all the points to form top view on X_1Y_1. Keep distance of all the points from X_1Y_1 equal to distance of initial top view from XY line.

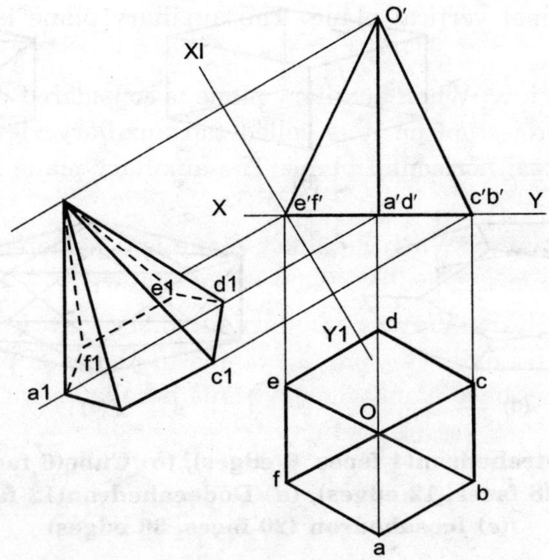

Fig. 3.4: Projections of a hexagonal base pyramid

3.4 PROJECTIONS OF SOLIDS

A solid is a 3 dimensional object having length, breadth and thickness bounded by a surface or surfaces which may be plane or curved. To describe shape of a solid Vertical Plane (V.P.), Horizontal Plane (H.P.) and Profile Plane (P.P.) is drawn.

3.4.1 Classification of Solids

Solids are mainly classified as:

(*i*) Polyhedra

(*ii*) Solids of revolution

(*i*) **Polyhedra:** A solid bounded by plane surfaces is called a polyhedral. The polyhedra can be classified as:

(*a*) **Regular polyhedra:** Regular polyhedral is the solid bounded only by plane surfaces (faces). All the faces are similar, equal and regular. All the angles between the faces are also equal. They can be further sub-classified as:

- **Tetrahedron:** The solid has four equal equilateral triangular faces.
- **Cube:** The solid has six equal square faces.
- **Octahedron:** The solid has eight equal equilateral triangular faces
- **Dodecahedron:** The solid has twelve equal and regular pentagonal faces.
- **Icosahedron:** The solid has twenty equal triangular faces.

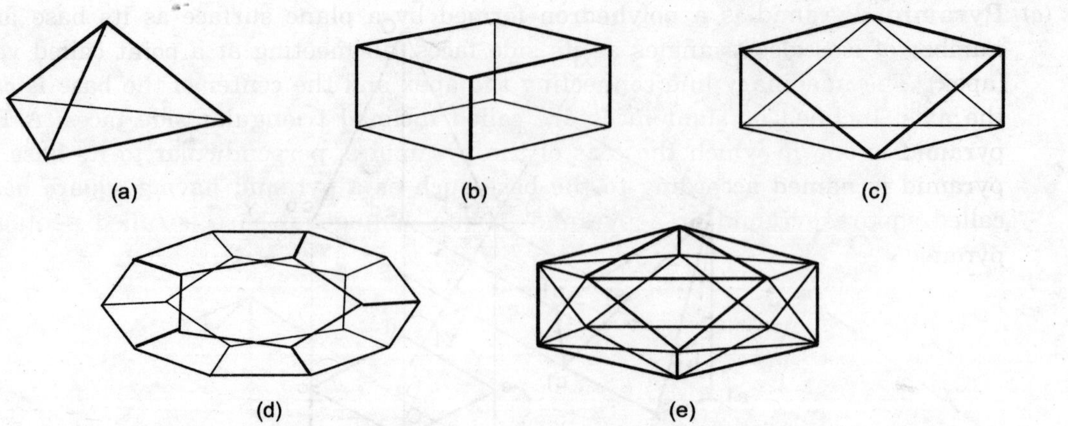

Fig. 3.5: (*a*) Tetrahedron(4 faces, 6 edges), (*b*) Cube(6 faces, 12 edges), (*c*) Octahedron(8 faces, 12 edges), (*d*) Dodecahedron(12 faces, 30 edges), (*e*) Icosahedron (20 faces, 30 edges)

(*b*) **Prism:** A polyhedron formed by two equal, similar and parallel end bases and connected by side faces which are either rectangles or parallelograms. The imaginary line joining the centers of bases is called the axis. The prism is named according to the base such as a prism having square bases is called square prism or a prism having pentagonal bases is called pentagonal prism. A prism is said to be right regular prism when its axis is perpendicular to bases and all the faces are equal rectangles.

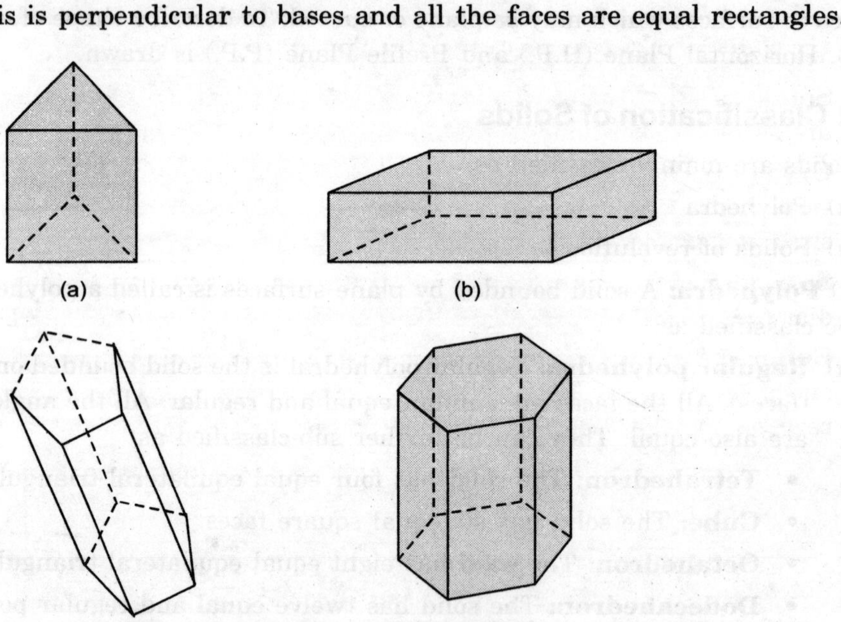

Fig. 3.6: (*a*) Triangular prism, (*b*) Rectangular prism, (*c*) Pentagonal prism, (*d*) Hexagonal prism

(*c*) **Pyramid:** Pyramid is a polyhedron formed by a plane surface as its base and a number of isosceles triangles as its side faces, all meeting at a point called vertex (apex). The imaginary line connecting the apex and the center of the base is called the axis. Inclined or slant faces are called inclined triangular side faces. A Right pyramid is one in which the axis of the pyramid is perpendicular to its base. The pyramid is named according to the base such as a pyramid having square base is called square pyramid or a pyramid having pentagonal base is called pentagonal pyramid.

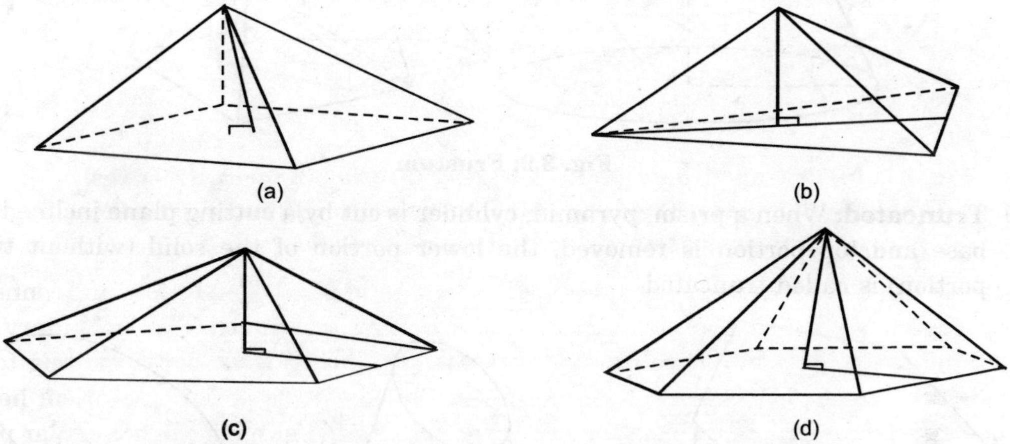

Fig. 3.7: (*a*) Square pyramid, (*b*) Triangular pyramid, (*c*) Pentagonal pyramid, (*d*) Hexagonal pyramid

(*ii*) **Solids of Revolution:** Solids of revolution are generated when some of the plane figures are revolved about one of their sides. These are classified as:

(*a*) **Cylinder:** The solid which is generated when a rectangle is revolved about one of its sides is called cylinder. It has two equal circular bases. The fixed line about which circle revolves is called the axis and is perpendicular to the bases.

(*b*) **Sphere:** Sphere is a solid which is generated when a semi-circle is revolved about its diameter which remains fixed. The mid-point of diameter is the center of sphere.

(*c*) **Cone:** The cone is a solid which is generated when a right triangle is revolved about one of its perpendicular sides which remains fixed. The fixed line about which triangle revolves is called the axis and circle described by the other side is called the base.

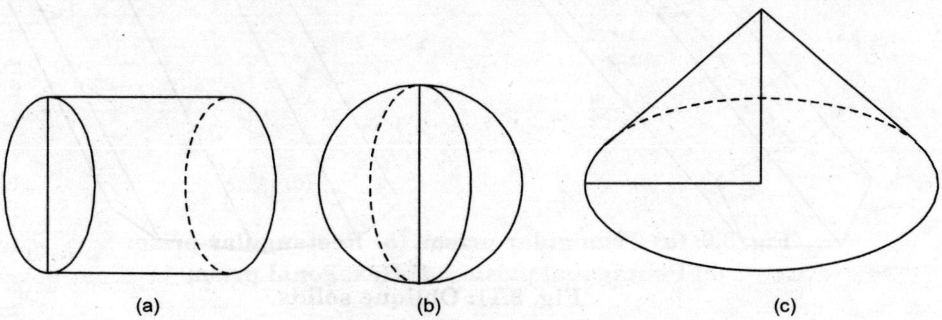

Fig. 3.8: (*a*) Cylinder, (*b*) sphere, (*c*) cone

(*iv*) **Frustum:** When a prism, pyramid, cylinder is cut by a cutting plane parallel to the base and top portion is removed, the lower portion of the solid (without the top portion) is called frustum.

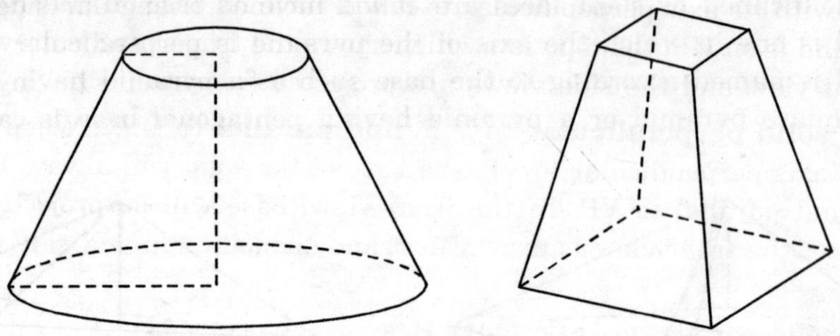

Fig. 3.9: Frustum

(*v*) **Truncated:** When a prism, pyramid, cylinder is cut by a cutting plane inclined to the base and top portion is removed, the lower portion of the solid (without the top portion) is called truncated.

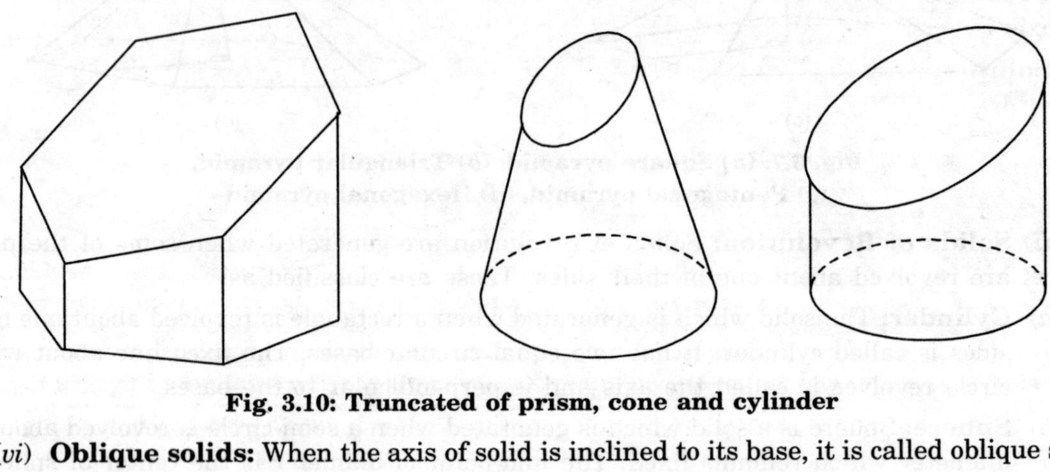

Fig. 3.10: Truncated of prism, cone and cylinder

(*vi*) **Oblique solids:** When the axis of solid is inclined to its base, it is called oblique solid.

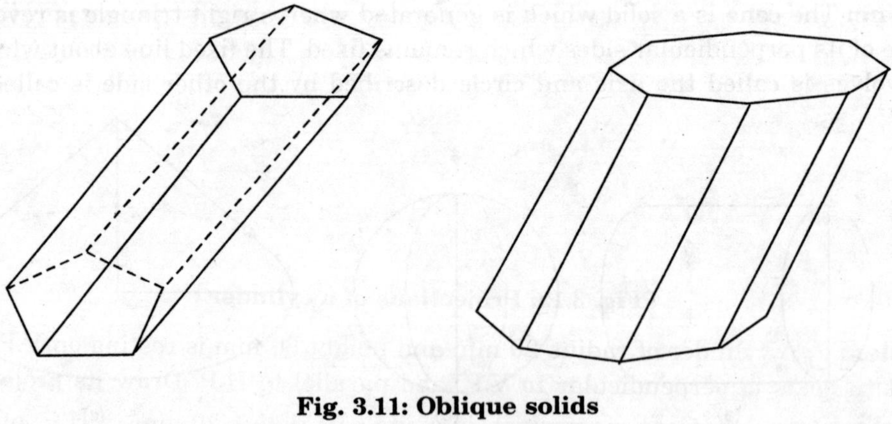

Fig. 3.11: Oblique solids

3.4.2 Projections of Solids

(*i*) **Axis of solid perpendicular to H.P. and parallel to V.P.:** A solid when placed on H.P. with its axis perpendicular to it will have its base on H.P. So, when the base of the solid is on HP, the top view which shows the true shape is drawn first and then the front view and the side views are projected from it.

(*ii*) **Axis of solid perpendicular to V.P. and parallel to H.P.:** When a solid is placed with its axis perpendicular to VP, the base of the solid will always be perpendicular to HP and parallel to VP. In the front view, base will be projected in true shape Therefore, the front view is drawn first and then the top and side views are drawn from it.

(*iii*) Axis of solid parallel to H.P. and V.P.

(*iv*) Axis of solid parallel to H.P. and inclined to V.P.

(*v*) Axis of solid parallel to V.P. and inclined to H.P.

(*vi*) Axis of solid inclined to H.P. and V.P.

Problem 1: A cylinder of radius 20 mm and height 60 mm is resting on H.P. on its base. Its axis is perpendicular to H.P. and parallels to V.P. Draw its projections.

Solution: Draw its top view which is a circle of radius 20 mm in H.P. and let center be o and divide it in 8 equal parts. Project the front view on XY to obtain a rectangle.

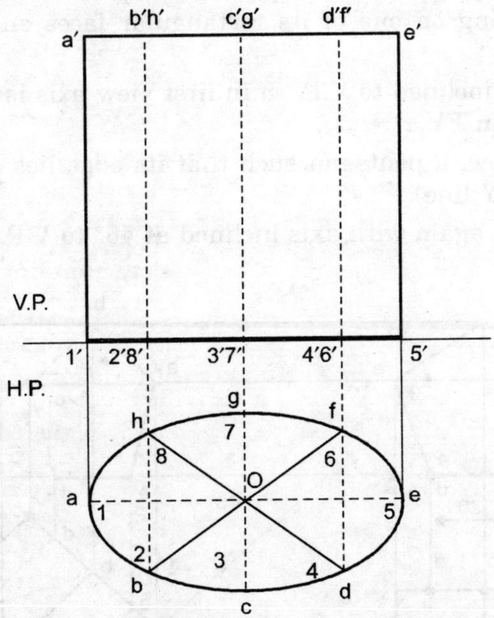

Fig. 3.12: Projections of a cylinder

Problem 2: A cylinder of radius 20 mm and height 60 mm is resting on V.P. on its base such that the axis is perpendicular to V.P. and parallel to H.P. Draw its projections.

Solution: Draw its front view which is a circle of radius 20 mm in H.P. and let center be and divide it in 8 equal parts. Project the top view on XY to obtain a rectangle.

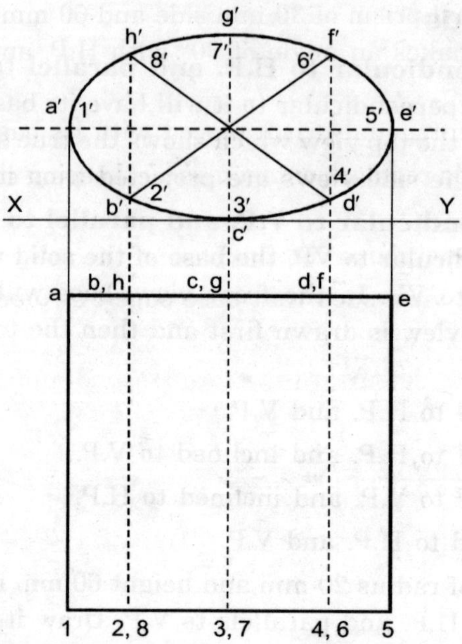

Fig. 3.13: Projections of a cylinder (problem 2)

Problem 3: Draw the projections of a pentagonal prism whose base is 30 mm and axis is 60 mm long. It is resting on one of its rectangular faces on H.P. and axis inclined at 45° to V.P.

Solution: The axis is inclined to V.P. so in first view axis is kept perpendicular to V.P. and true shape is drawn in FV.

- Draw the front view, a pentagon such that its edge lies on XY line and project the top view (below XY line).
- Make the top view again with axis inclined at 45° to V.P. and project the final front view.

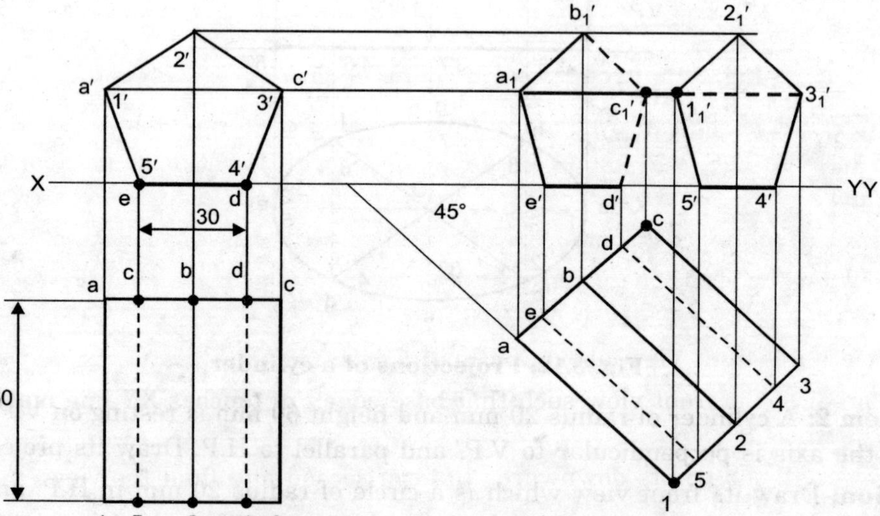

Fig. 3.14: Projections of pentagonal prism

Problem 4: A pentagonal prism of 30 mm side and 60 mm axis is resting with one of its edge of base on H.P. and makes an angle of 30° with H.P. and is parallel to V.P. Draw its projections.

Solution: Assuming pentagonal prism placed with its base on H.P.

- Draw the top view of a regular pentagon *abcde*, keeping right hand side *cd* perpendicular to XY.
- Project the front view above XY line.
- Reproduce the front view such that base edge *c′d′* touches XY line making an angle 30° with H.P.
- Project all the points downwards to form second and final top view.

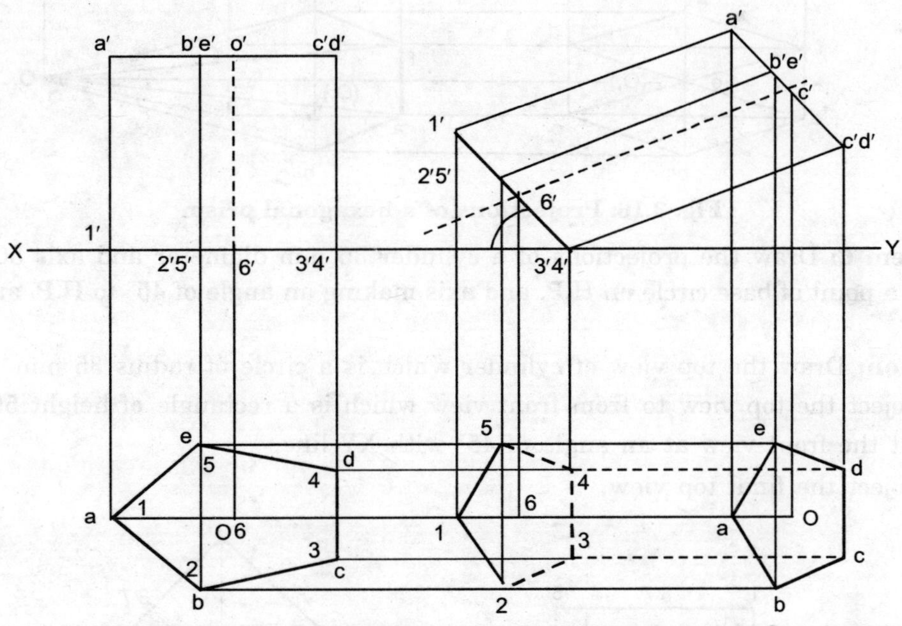

Fig. 3.15: Projections of pentagonal prism (problem 4)

Problem 5: A hexagonal pyramid of 30 mm side and 60 mm axis is resting with one of its edge of base on H.P. and makes an angle of 35° with H.P. and is parallel to V.P. Draw its projections.

Solution: Assuming hexagonal prism placed with its base on H.P.

- Draw the top view of a regular hexagon *abcdef*, keeping right hand side *cd* perpendicular to XY.
- Project the front view above XY line.
- Reproduce the front view such that base edge *c′d′* touches XY line making an angle 35° with H.P.
- Project all the points downwards to form second and final top view.

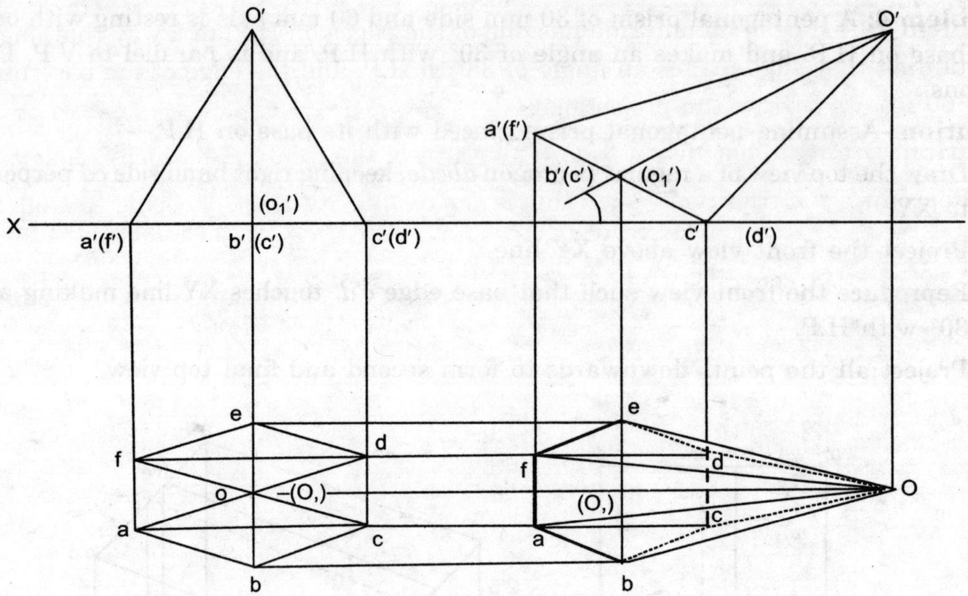

Fig. 3.16: Projections of a hexagonal prism

Problem 6: Draw the projections of a cylinder 35 mm diameter and axis 50 mm long resting on a point of base circle on H.P. and axis making an angle of 45° to H.P. and parallel to V.P.

Solution: Draw the top view of cylinder which is a circle of radius 35 mm

- Project the top view to from front view which is a rectangle of height 50 mm.
- Tilt the front view at an angle of 45° with XY line.
- Project the final top view.

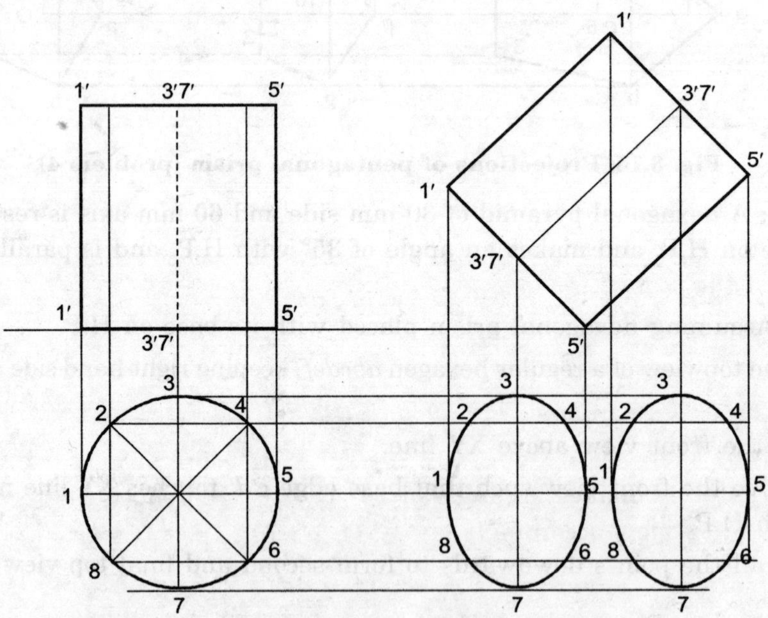

Fig. 3.17: Projections of a cylinder (problem 6)

Problem 7: A cube of 35 mm side has one of the edges on H.P. in a way that one square face containing that edge makes an angle of 35° to H.P. and the edge of the base makes an angle of 60° to V.P. Draw the projections.

Solution: Draw the top view a square and project the front view which is also a square.

- Reproduce the front view such that it makes 35° with H.P. keeping 2′3′ on XY line.
- Project the next top view.
- Reproduce second top view such that edge 23 makes an angle of 60° with the V.P. Project the final front view.

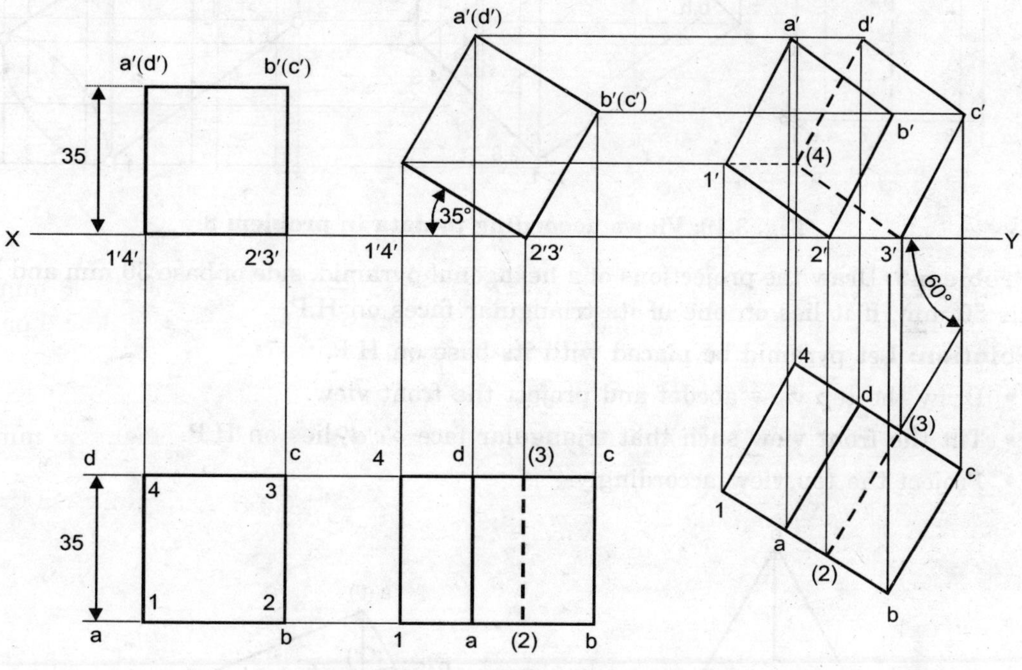

Fig. 3.18: Projections of a cube

Problem 8: Draw the projections of a right circular cylinder of 25 mm radius and 60 mm axis resting on one point of base circle on V.P. and its axis makes an angle of 50° with V.P. The front view of the axis is inclined at 30° to H.P.

Solution: Draw the front view and top view in their initial positions. The top view is a rectangle of height 60 mm and front view is a circle of diameter 50 mm.

- Tilt the top view at an angle of 50° with XY line. Project the second front view about XY line.
- As the front view makes an angle of 30° to H.P., reproduce the second front view. Project the top view accordingly.

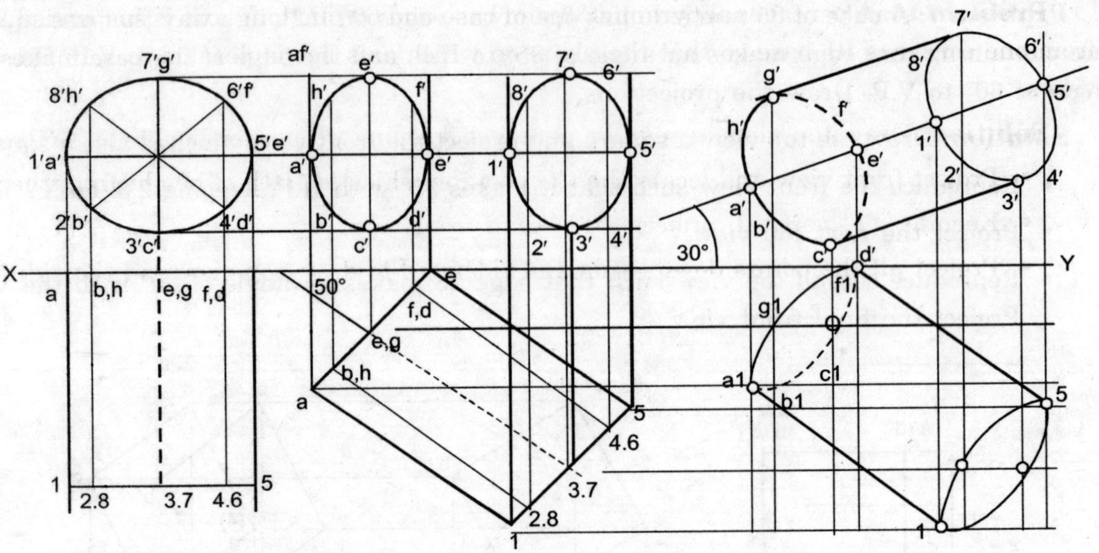

Fig. 3.19: Views according to data in problem 8

Problem 9: Draw the projections of a hexagonal pyramid, side of base 30 mm and height of axis 50 mm, if it lies on one of its triangular faces on H.P.

Solution: Let pyramid be placed with its base on H.P.

- Draw the top view abcdef and project the front view.
- Tilt the front view such that triangular face *o′c′d′* lies on H.P.
- Project the top view accordingly.

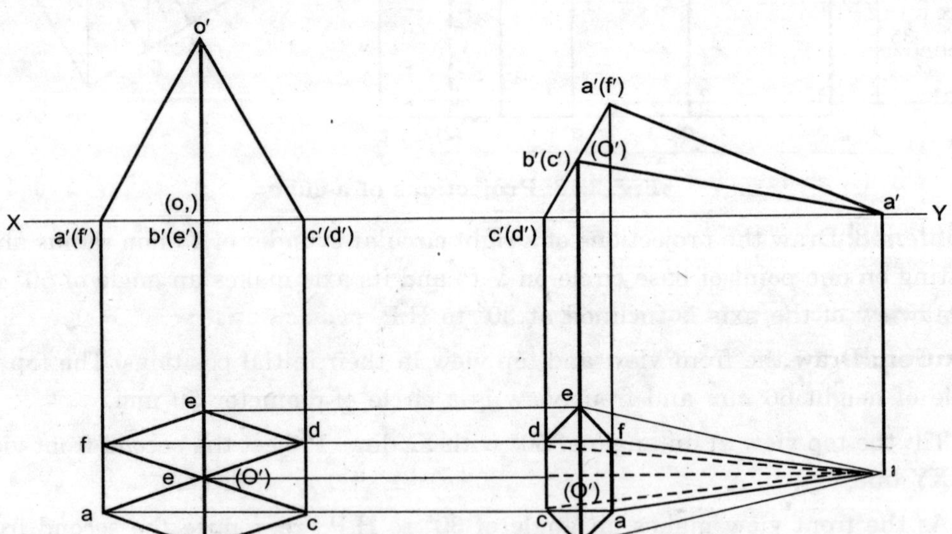

Fig. 3.20: Views according to data in problem 9

Problem 10: A pentagonal pyramid 35 mm base and 60 mm long axis is suspended freely from one corner of its base so that the plane on which axis lies is parallel to V.P. Draw the views.

Solution: Draw top view, a regular pentagon keeping *ab* perpendicular to XY line.

- Project front view and locate the CG position which is 1/4th of height from the base.
- Keeping *g'd'* vertical, project the second front view.
- Project all the points downwards and sideward to draw top view and the side view.

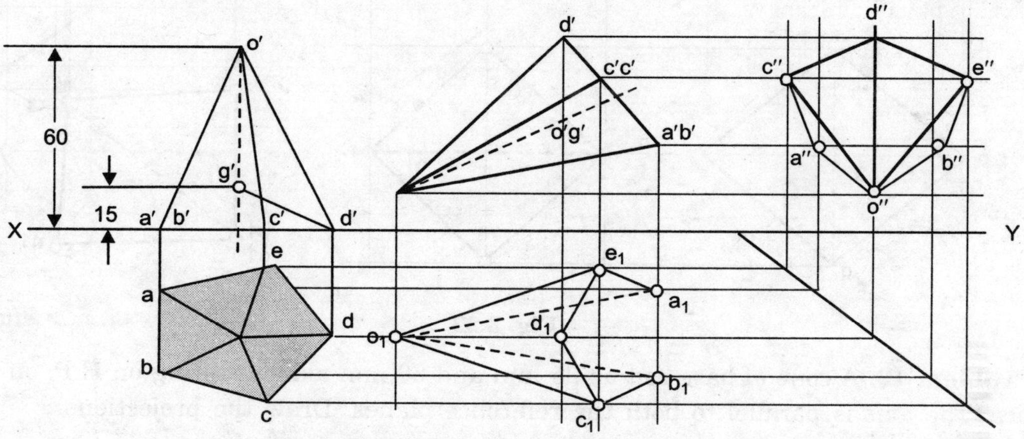

Fig. 3.21: Projections of a pentagonal pyramid

Problem 11 : A square pyramid of 40 mm base side and axis of 60 mm length rests on it apex on H.P. Draw the projections if one slant edge is vertical and triangualr face through it is perpendicular to V.P.

Solution: Draw the top view, square which shows the true length.

- Draw the projections to draw front view. The pyramid is resting on apex, so the view will be an inverted triangle.
- Reproduce second front view keeping *o*'p' vertical. Draw the projections to form top view.
- Reproduce second top view keeping p_1s_1 perpendicular to V.P.
- Project the final front view.

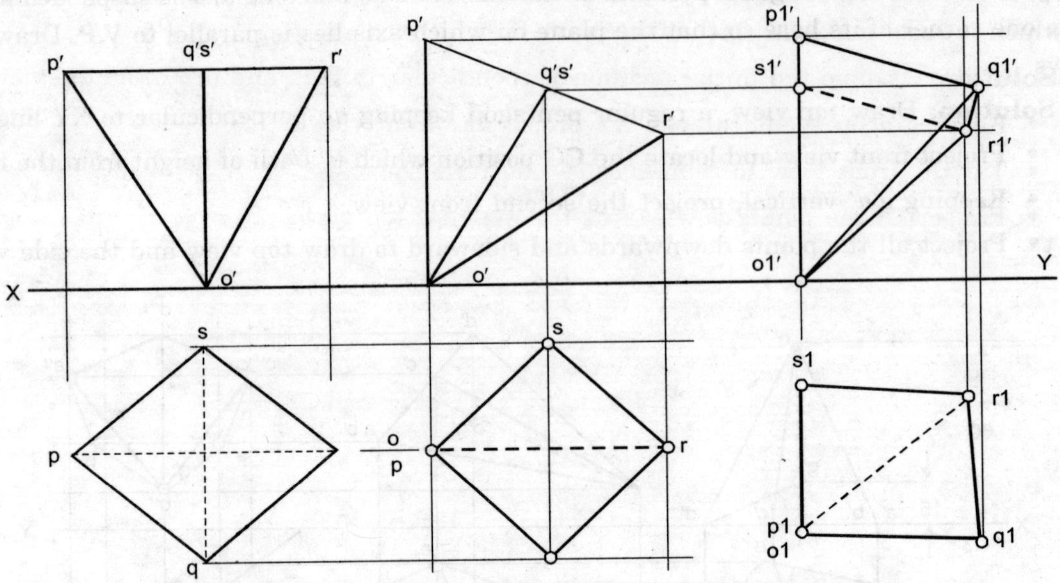

Fig. 3.22

Problem 12: A cone of base radius 25 mm and 60 mm axis is resting on H.P. on one of its face. The axis is parallel to both the reference planes. Draw the projections.

Solution: Axis is considered perependicular to profile plane so side view is drawn first.

- Draw a circle representing the side view keeping a point of circle on XY line.
- Divide the circle in 8 equal parts.
- Project the front view horizontally from circle and then project down the top view from front view and the side view.

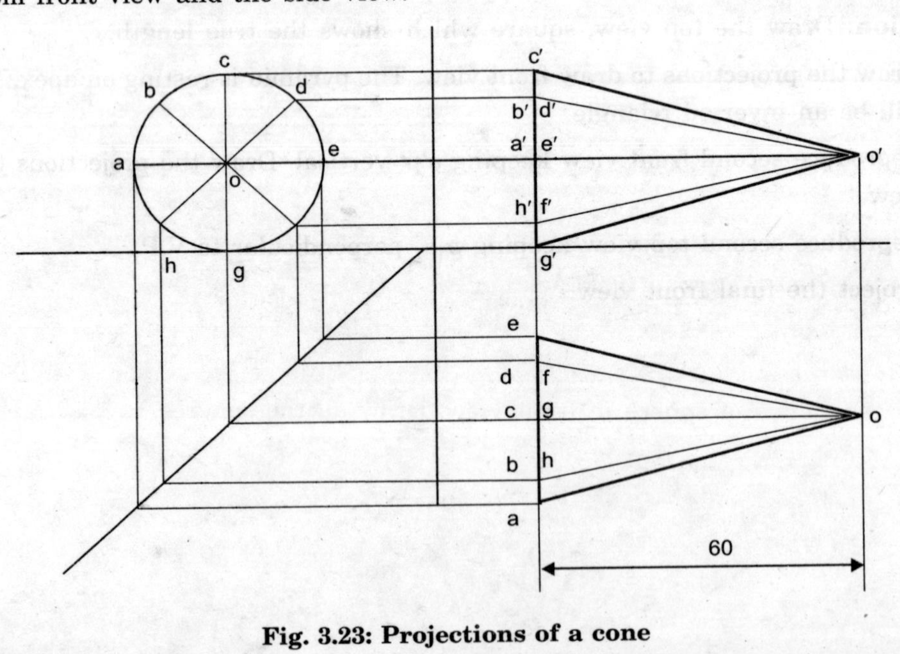

Fig. 3.23: Projections of a cone

Problem 13: Draw the projections of a pentagonal pyramid of 40 mm base and 60 mm axis if it rests on one of its base on H.P. such that the two base edges containing the corner on which it rests make 35° inclination with H.P.

Solution: Assume the initial position perpendicular to H.P. and draw the top view.

- Keeping ab perpendicular to XY project the front view.
- Tilt the front view to reproduce second front view.
- Project all the points downwards to form new top view.

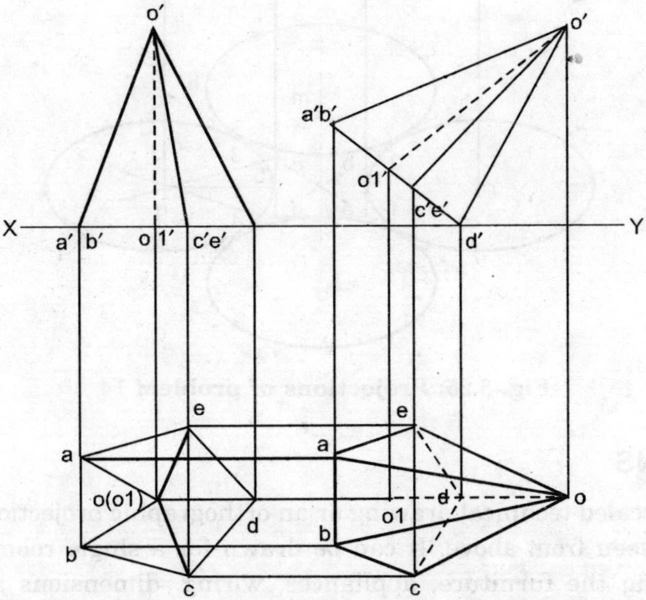

Fig. 3.24: Projections of a pentagonal pyramid

Problem 14: A square prism of 30 mm side and 60 mm axis is resting on its base on the H.P. with two faces perpendicular to V.P. Draw the projections and find the radius if 4 spheres of equal size rest on H.P. such that each sphere touches the face of prism and other two spheres.

Solution: Draw the top and front views of prism and join diagonals of square in top view.

- Find the bisectors of angles *dch* and *fdc* which intersect at a point *q*.
- From point *q* draw a perpendicular *qs* which is the radius of spheres.
- Draw four spheres touching prism and other two spheres.
- Bisect the angle *c*'3' Y such that it intersects the projector from *q* at *q*'. This point is the radius of sphere in front view. Draw all the spheres.

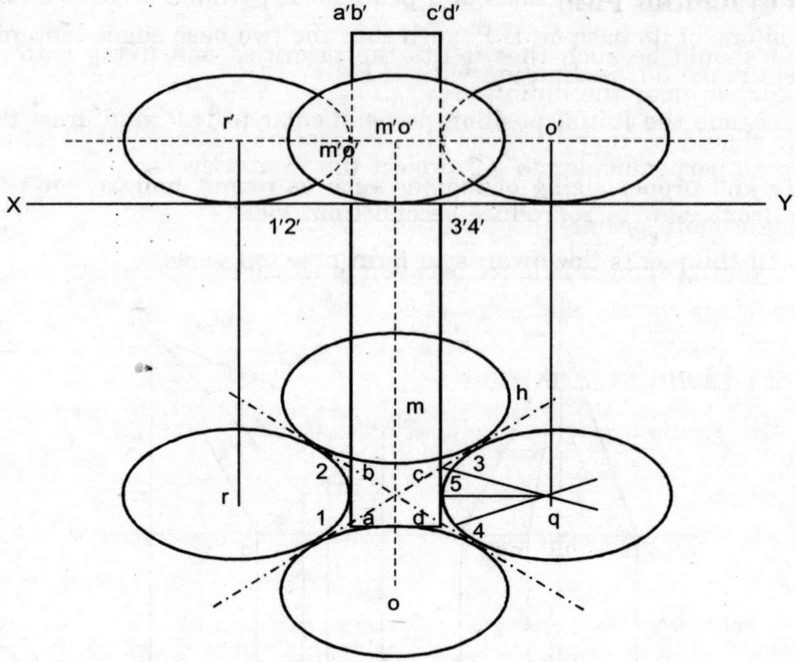

Fig. 3.25: Projections of problem 14

3.5 FLOOR PLANS

A floor plan is a scaled technical drawing or an orthographic projection of layout of a room or a building when seen from above. It can be drawn for a single room, floor or a building with details including the furniture, appliances, wiring, dimensions and other necessary information.

Elevation is an orthographic projection of an object when it is viewed from the front.

Elevation Plan and elevation views are commonly used on job sites.

3.5.1 Types of Drawings used in Floor Plans

Assembly drawings are used to represent items that consist of more than one component. Component drawings provide detailed information about the individual units and give information about the component dimensions, tolerances etc.

Concept drawings/sketches are freehand quick and simple drawings to explore initial ideas for designs. The working drawings provide dimensioned, graphical information that is useful for a contractor to construct the works, or by suppliers to fabricate components of the works or to assemble or install components. Design drawings are used to develop and communicate ideas about a developing design in the early stages. Detail drawings are large scale drawings that provide a detailed description of the geometric form of a part of an object such as a building, bridge, machine, plant etc.

3.5.2 Features of A Floor Plan

(*i*) Floor plan should be such that it fits the priorities and living patterns. Example keeping kitchen near the dining area.

(*ii*) Flexibility should be there in floor plan so that any change could be made easily.

(*iii*) Versatility and proper sizing of regions such as rooms, hallway etc is necessary to accommodate more people and furniture.

(*iv*) An ideal layout keeping a balance between practical considerations and architectural designs to ensure safety and comfort.

3.5.3 Method of Drawing Floor Plan

Flowchart showing step-by-step method of drawing floor plan:

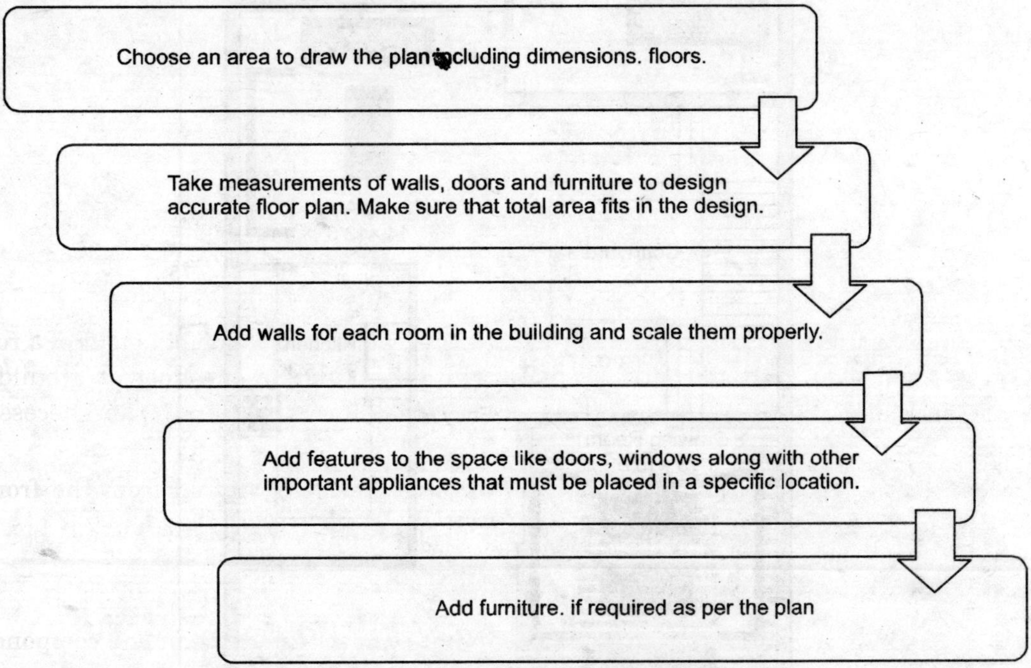

Fig. 3.26: Method of drawing floor plan

Figure of a ground floor plan showing the dimensions and location of each room, position of doors, windows, bed, sink, closets, location of furniture etc.

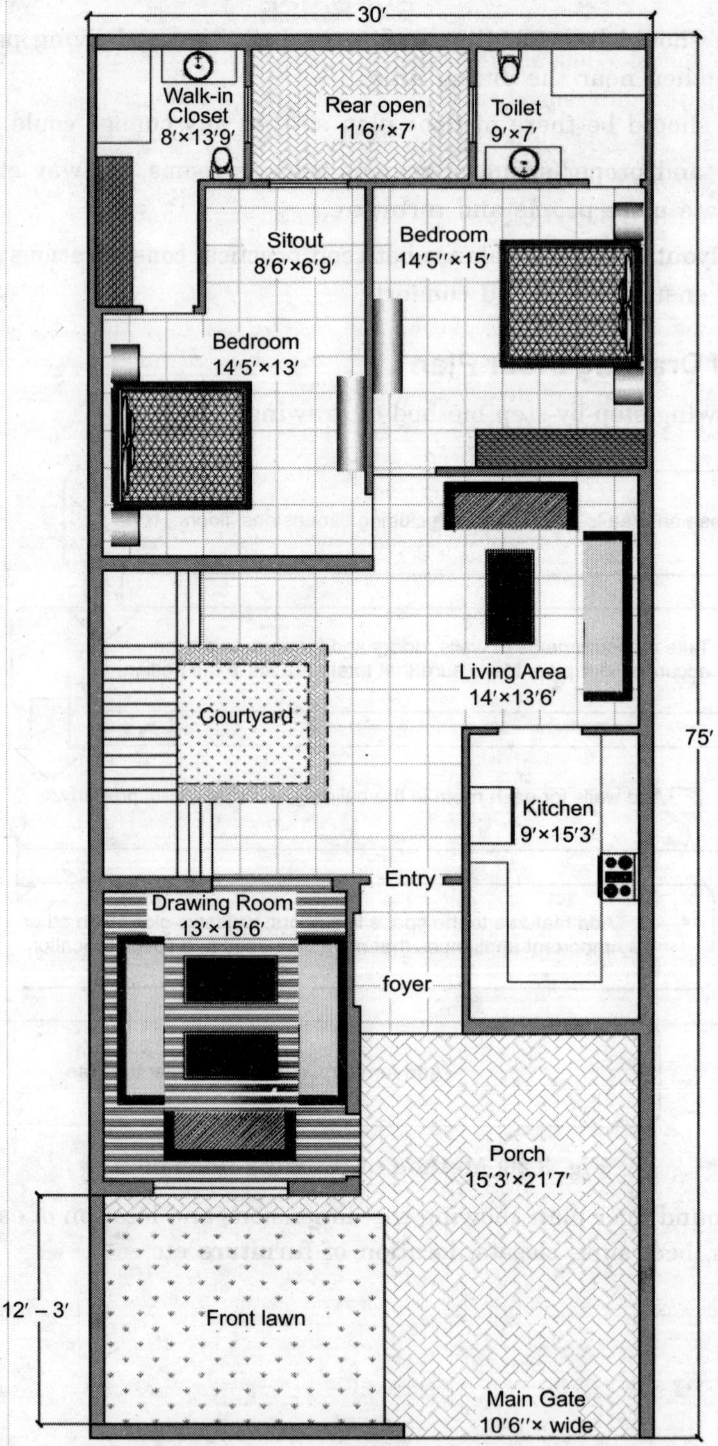

Fig. 3.27: An example of floor plan

EXERCISE

1. Define solids. What are the types of solids?
2. What is the difference between a prism and a pyramid?
3. What are the types of auxiliary views?
4. What are floor plans?
5. What is the difference between frustum and truncated?
6. Draw the projections of a pentagonal pyramid with base 25 mm edge and axis 45 mm long is resting on horizontal plane with one of its base edge inclined at 40° to V.P.
7. A hexagonal prism of side 30 mm and axis 60 mm is resting on V.P. on its base such that the axis is perpendicular to V.P. and parallel to H.P. Draw its projections.
8. A cylinder of 30 mm diameter and 50 mm axis is resting on H.P. on a point on its base circle such that the axis makes an inclination of 35° with V.P. and top view of the axis makes an angle of 50° with V.P. Draw the projections.
9. A cone of 30 mm base diameter and 70 mm axis is resting on H.P. on one of its generators and the axis makes an angle of 50° with V.P. Draw the projections.
10. Draw the projections of a square base pyramid of 35 mm side and 65 mm axis is freely suspended from one of its corners of base and the axis makes an angle of 45° with the V.P.
11. A triangular pyramid of base 35 mm and height 65 mm is resting on a side of base on H.P. The axis of pyramid makes an angle of 40° to H.P. and 55° to V.P. Draw its projections.
12. Draw the projections of a pentagonal prism, side of base 30 mm, axis 50 mm long which is resting on its base edges on horizontal plane with its axis inclined at 50° to H.P.
13. A hexagonal prism of 25 mm side and 65 mm axis is resting in H.P. on one of its face. The axis is parallel to both the reference planes. Draw the projections.
14. A triangular pyramid of base 25 mm and height 60 mm is resting on a side of base on H.P. The axis of pyramid makes an angle of 30° to H.P. and 45° to V.P. Draw its projections.
15. Three equal sphere of 40 mm diameter are placed centrally in a hemispherical bowl of 120 mm diameter, which is resting on its curved surface on H.P. Each sphere touches the other two spheres. Draw the projections of bowl and sphere when the line joining the centers of two spheres is perpendicular to V.P. Find the height of centers of sphere above H.P. ignoring the thickness of bowl.
16. Fill in the blanks:

    (*i*) An object having three dimensions is called a ....

    (*ii*) The solid which is generated when a rectangle is revolved about one of its sides is called a ....

    (*iii*) There are .... possible auxiliary views of an object.

(*iv*) A solid having four equal equilateral triangle faces is called ....

(*v*) The shortest distance from a point to a plane is seen in the ....view

(*vi*) A .... is an aerial view is a two-dimensional scale drawing of the planning, size and direction of rooms, doors, walls and windows.

(*vii*) A .... is one in which the axis of the pyramid is perpendicular to its base.

SUMMARY

- Auxiliary views can help in finding true lengths and inclinations and point view of a line, edge view of a plane, true shape of a plane, projections of solids, true shapes and inclinations of solids etc.
- If auxiliary plane is perpendicular to principal plane of projection, primary auxiliary view is obtained.
- A floor plan is a scaled technical drawing or an orthographic projection of layout of a room or a building when seen from above.
- Elevation is an orthographic projection when viewed from the front.
- Assembly drawings are used to represent items that consist of more than one component.
- Sketches are freehand quick and simple drawings to explore initial ideas for designs.
- Working drawings provide dimensioned, graphical information.
- A floor plan should meet the needs of the family which may include their preferences, storage requirements; specific features for particular rooms etc. such as keeping bedrooms and libraries away from noise.

CHAPTER

4

SECTIONAL VIEWS

Learning Objectives

- Describe the concept and need for sectional views
- Describe cutting plane and types of sections
- Create 2D drawings, construct and Interpret views and sectional views correctly
- Describe development of surfaces
- Draw the development of surfaces for simple solids such as prisms, pyramids, cylinders, and cones

4.1 INTRODUCTION

The interiors of an object can be described by using hidden lines in exterior views. In case of simple drawings, dotted lines in orthographic projection can be used to show the interior portions but in case of complicated drawings, large numbers of dotted lines create confusion. To avoid this, sectioning is done. A section view is drawn cutting the object by an imaginary section or cutting plane so as to reveal the interiors. The section surfaces are shown by section lines, evenly spaced and inclined at 45° to the horizontal.

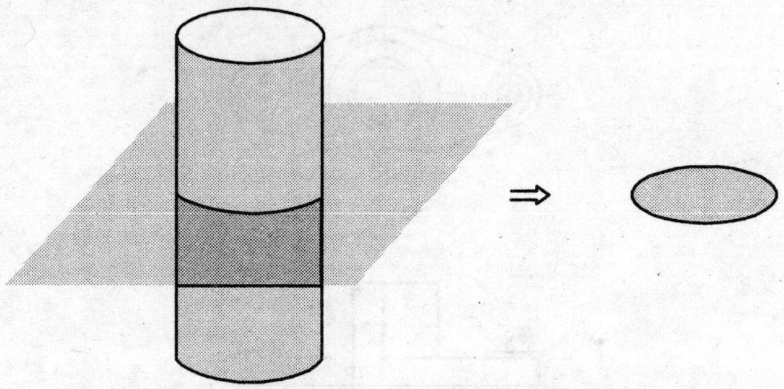

Fig. 4.1: Sectional view of a cylinder

The section obtained when the cutting plane cuts an object along the length is called longitudinal section. If plane cuts entirely across the object, then section is called full section.

*Boundary of section lines should be firm and there should be no firm lines inside the boundary.

4.2 TYPES OF SECTIONAL VIEWS

(*i*) Full section view is the one where the section plane goes completely through the object.

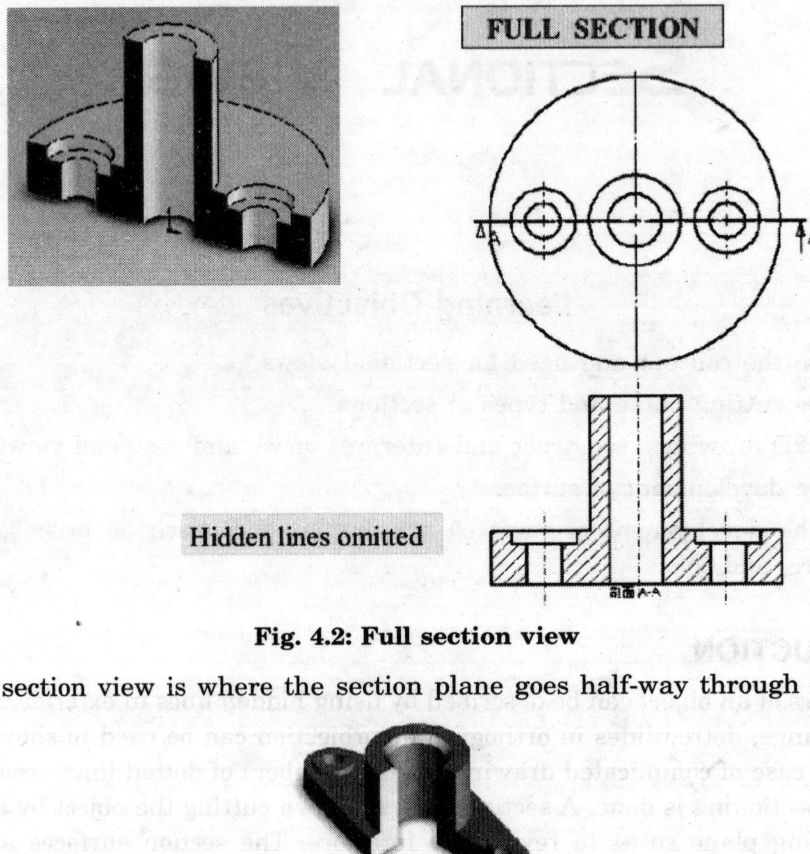

Fig. 4.2: Full section view

(*ii*) Half section view is where the section plane goes half-way through the object.

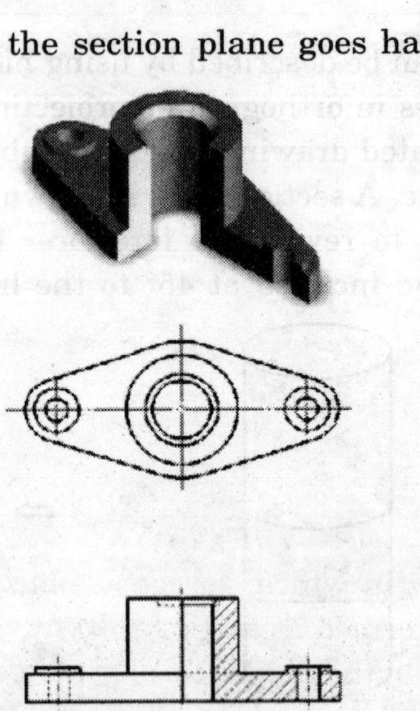

Fig. 4.3: Half section view

(*iii*) **Offset section, where the sectional plane is bent through the features that are not aligned.**

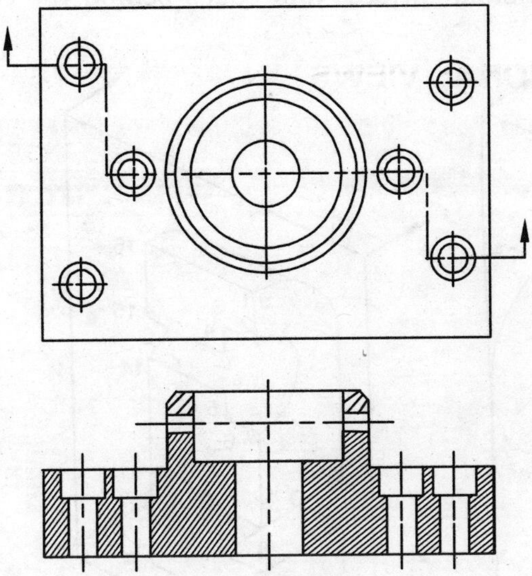

Fig. 4.4: Offset section view

(*iv*) Broken-out section is the one where the section goes through part of the object.

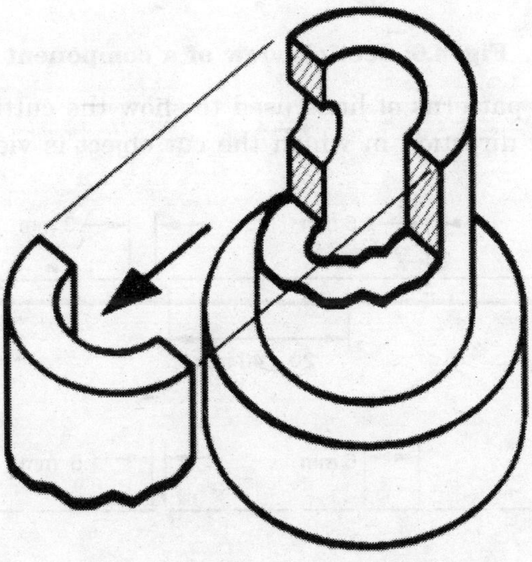

Fig. 4.5: Broken out section

Sectional view is the view in which section is obtained. If the elevation represents the section, then the elevation is termed as sectional front view or sectional elevation. And if the plan represents the section, then the plan is termed as sectional top view or sectional plan. The cutting plane is an imaginary plane and it is assumed to be used to cut or section the solid.

A section view is created by passing an imaginary cutting plane vertically through the centre of the object. In this case the right half of the object is removed in order to reveal the interior features. The line of sight for the section view is perpendicular to the cut surfaces, which means they are drawn true size and shape in the section view. No hidden lines are drawn and all visible surfaces and edges behind the cutting plane are drawn as object lines.

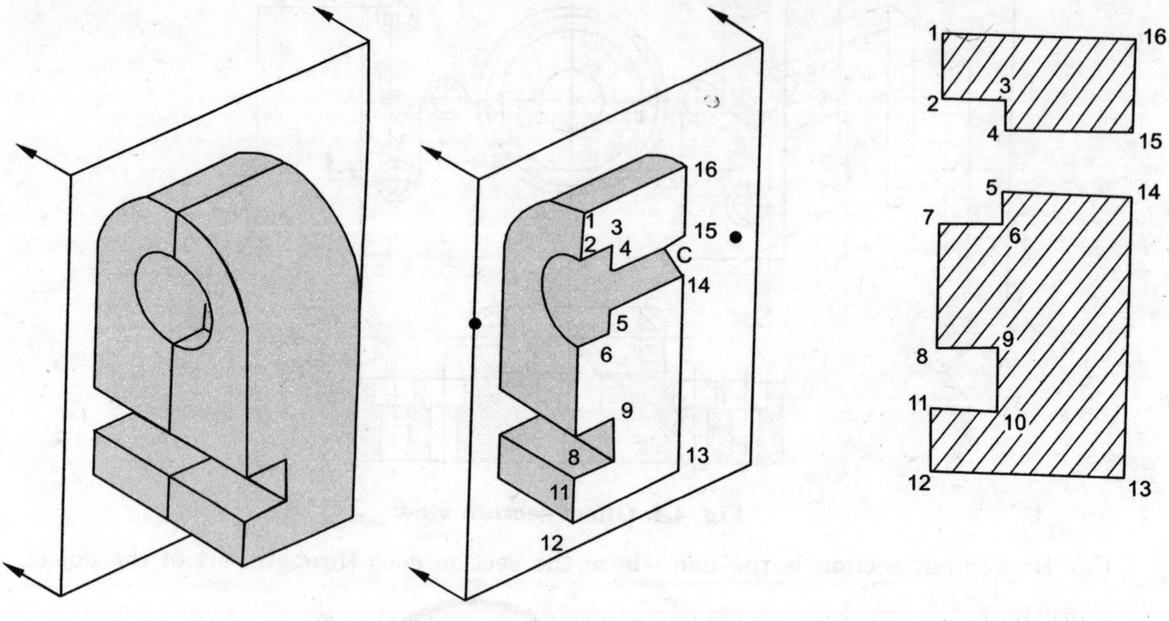

Fig. 4.6: Section view of a component

Figure 4.7 shows two patterns of lines used to show the cutting-plane line. Direction of Arrowheads indicates the direction in which the cut object is viewed.

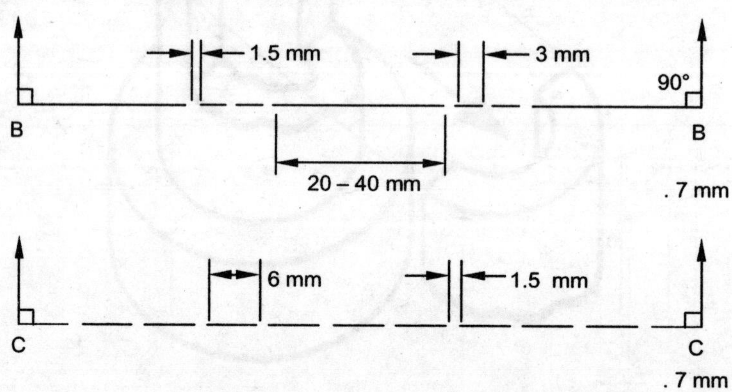

Fig. 4.7: Representation of cutting plane lines

The possible positions of section with respect to reference plane are:

(*i*) **When section plane is parallel to H.P. and perpendicular to V.P.:** When section plane is parallel to H.P. and perpendicular to V.P. it is vertical trace and will be a straight line parallel to XY line and will be above XY line.

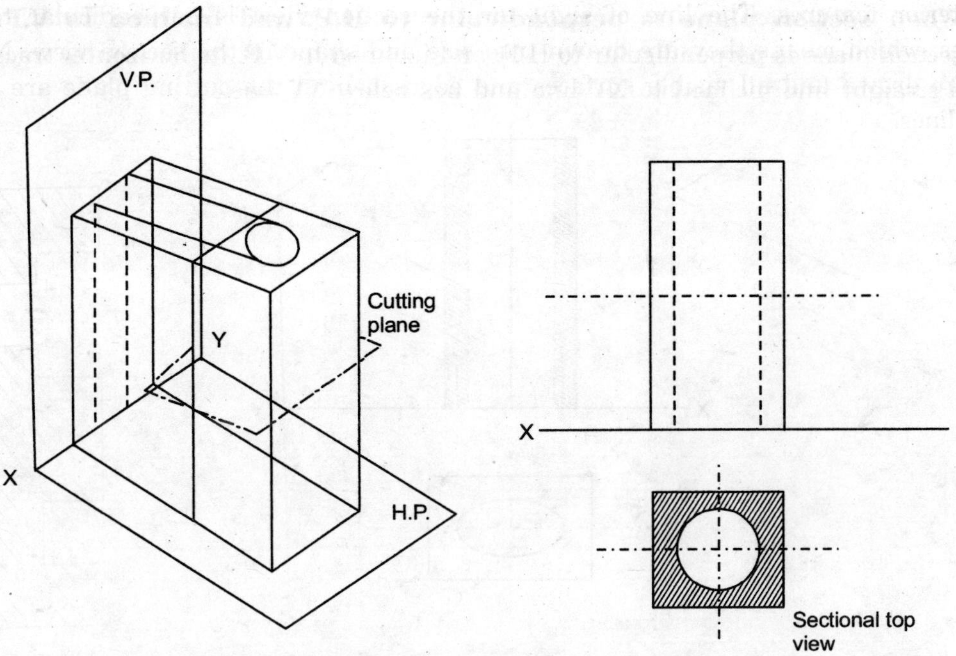

Fig. 4.8: Projections of a rectangular block when section plane is parallel to H.P. and perpendicular to V.P.

(*ii*) **When section plane is parallel to V.P. and perpendicular to H.P.:** When section plane is parallel to H.P. and perpendicular to V.P. it is horizontal trace and is a straight line parallel to XY line and lies below XY line.

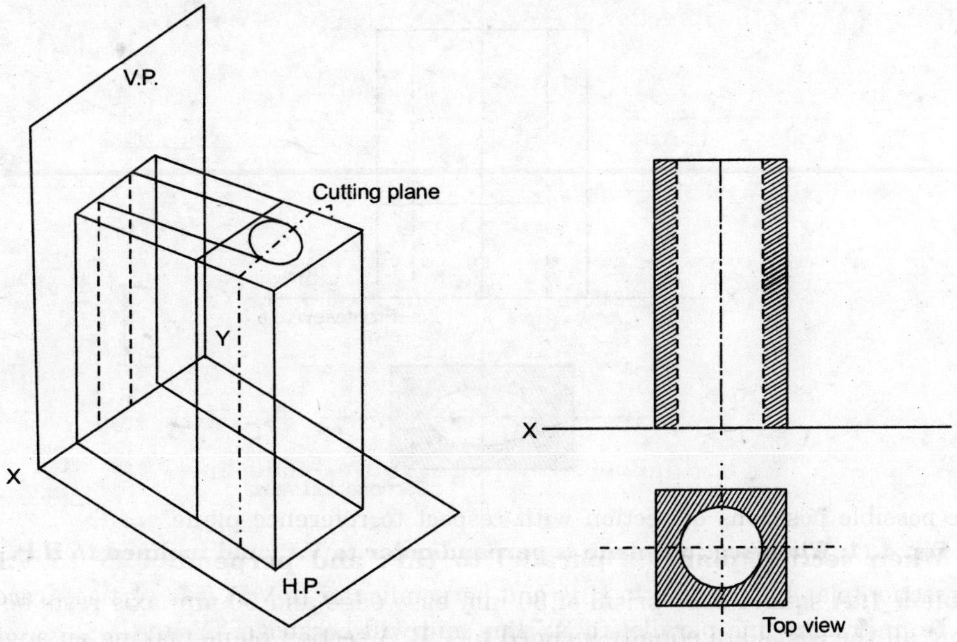

Fig. 4.9: Projections of a rectangular block when section plane is parallel to V.P. and perpendicular to H.P.

(*iii*) **When section plane is perpendicular to H.P. and inclined to V.P.:** When section plane is perpendicular to H.P. and inclined to V.P. its horizontal trace will be a straight line inclined to XY line and lies below XY line.

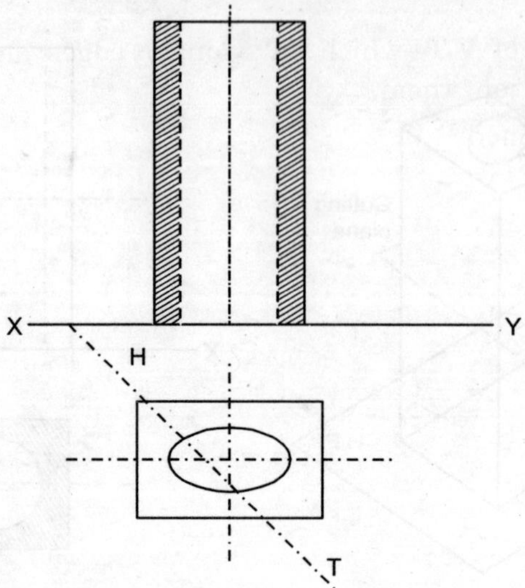

Fig. 4.10: When section plane is perpendicular to H.P. and inclined to V.P.

(*iv*) **When section plane is perpendicular to V.P. and inclined to H.P.:** When section plane is perpendicular to V.P. and inclined to H.P. its vertical trace will be a straight line inclined to XY line and lies above XY line.

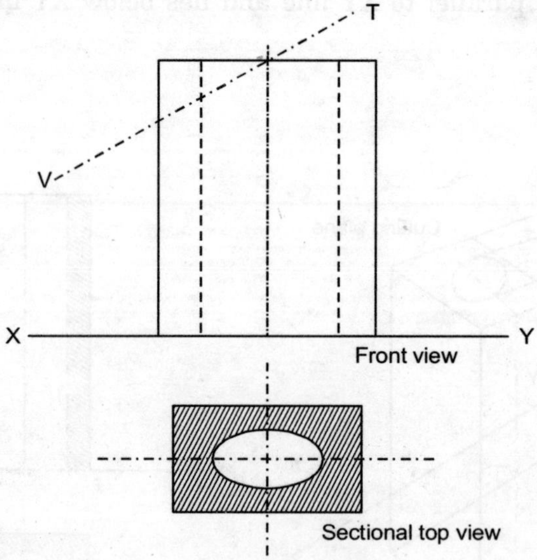

Fig. 4.11: When section plane is perpendicular to V.P. and inclined to H.P.

Problem 1: A square base prism of 30 mm base edge and 50 mm axis rests on H.P. in a way that all the edges are equally inclined to V.P. A section plane making an angle of 55°

with H.P. and perpendicular to V.P. passes through a point 40 mm from base. Draw front view, sectional top view and true shape of the section.

Solution: Draw the projections of prism and draw cutting plane as per the given data as shown in the figure.

- Project the points of V.T. which cuts longer edges and upper ends at p', q' and r' respectively and join them.
- True shape is obtained.

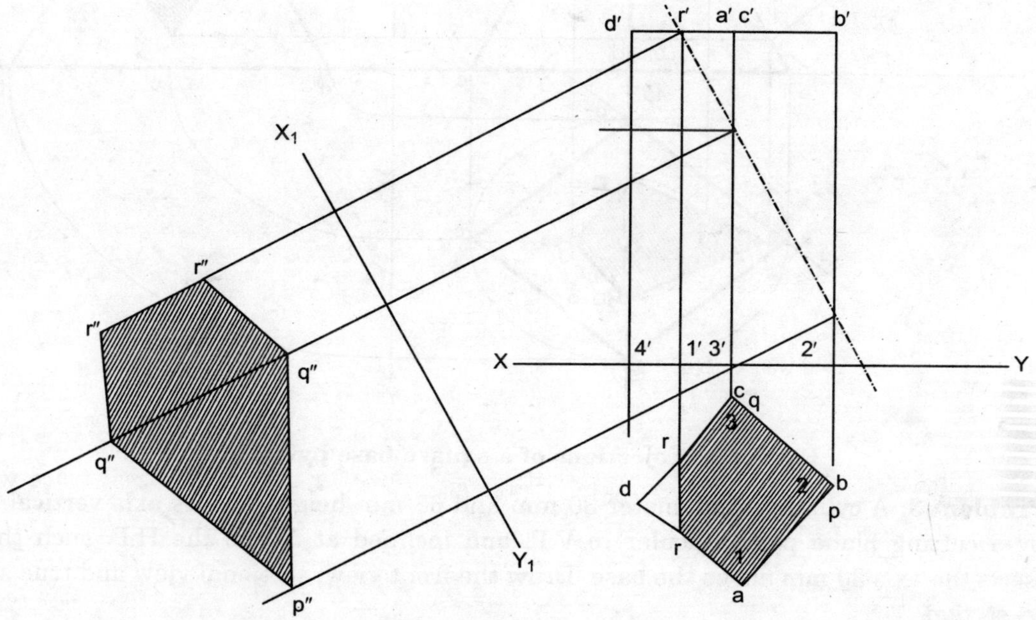

Fig. 4.12: Projections of a prism

Problem 2: A square base pyramid with base side of 30 mm and axis 60 mm long is resting on H.P. on its base in such a way that its edges are equally inclined to V.P. A section plane perpendicular to V.P. inclined at 45° to H.P., bisects the axis. Draw the sectional view and true shape of the section.

Solution: Draw the projections of pyramid in required position and draw cutting plane as per the given data as shown in the figure 4.13.

- Project the points of V.T. which cuts slant edges at different points on corresponding lines in top view to obtain sectional view.
- True shape is obtained.

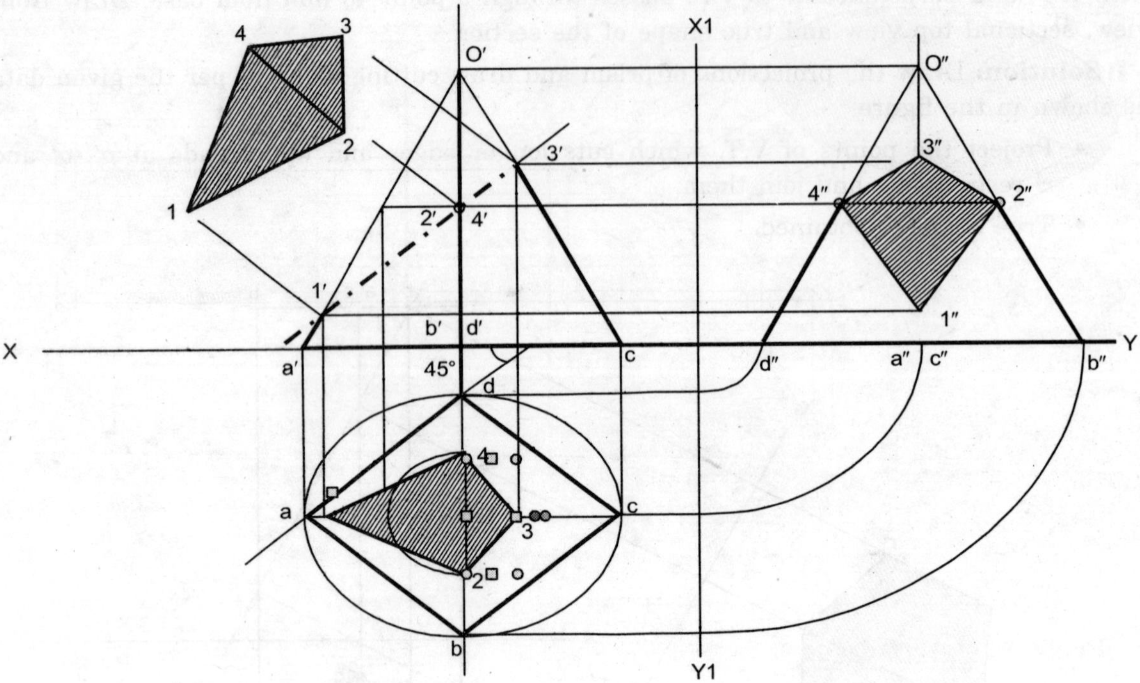

Fig. 4.13: Projections of a square base pyramid

Problem 3: A cylinder of diameter 30 mm and 65 mm height has its axis vertical. It is cut by a cutting plane perpendicular to V.P. and inclined at 50° to the H.P. such that it intersects the axis 30 mm above the base. Draw the front view, sectional view and true shape of the section.

Solution: Draw the projections of cylinder in required position and draw cutting plane as per the given data as shown in the figure 4.14.

- Project the points of V.T. which cuts generators at different points on corresponding lines in top view to obtain sectional view.
- Draw end view and project cutting points on front view horizontally. Join all the points to obtain sectional end view.
- True shape is obtained.

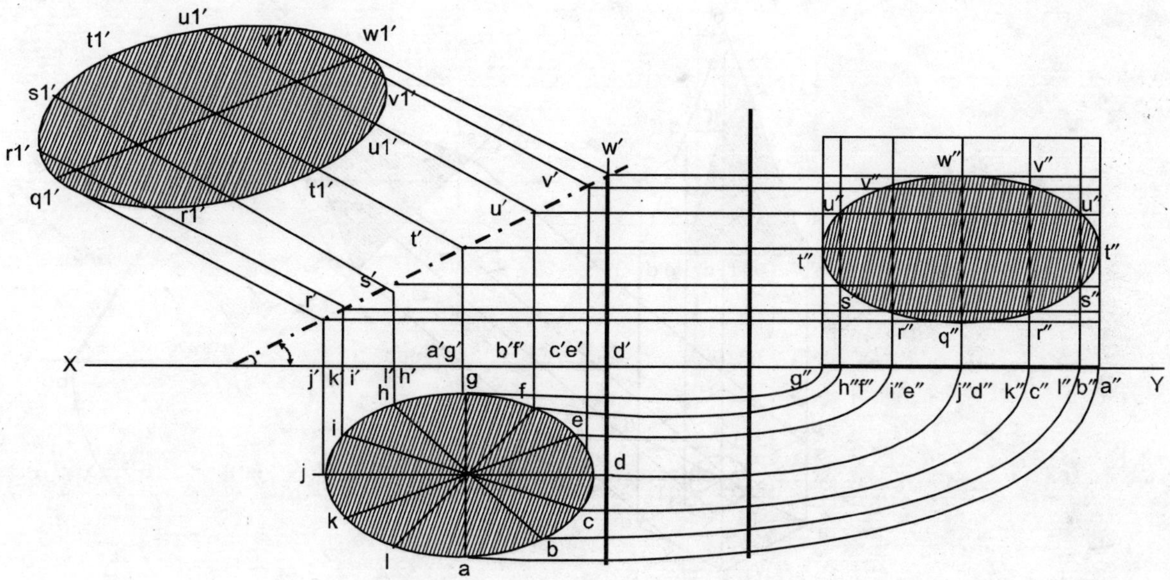

Fig. 4.14: Projections of a cylinder

Problem 4: A hexagonal pyramid of base radius 25 mm and 70 mm axis is resting on its base in H.P such that one of the edges of base is perpendicular to V.P. It is cut by a section plane perpendicular to H.P. inclined at 45° to the H.P. and passing through pyramid at a distance of 15 mm from the axis. Draw front view, sectional side view and true shape of section.

Solution: Draw the projections of pyramid in required position and draw cutting plane as per the given data as shown in the figure 4.15.

- Project the points of V.T. which cuts generators at different points on corresponding lines in top view to obtain sectional view.
- Project cutting points in front view.
- For point q draws an arc with o as centre and oq as radius to obtain q_1. Project the point to obtain q_1'.
- Join all the points to obtain sectional end view.
- True shape is obtained.

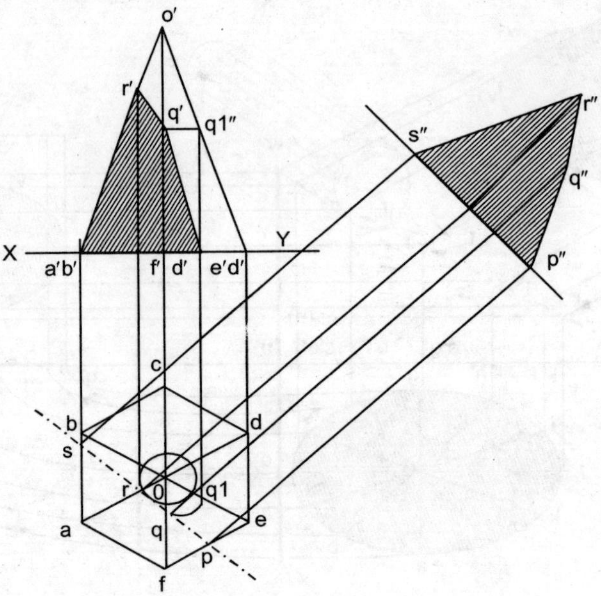

Fig. 4.15: Projections of a hexagonal pyramid

4.3 DEVELOPMENT OF SURFACES OF RIGHT REGULAR SOLIDS

A geometrical surface is generated by the motion of a geometric line, straight or curved. Moving line which generates the surface is called 'generatrix' of surface. The line (lines) which govern the movement of generatrix is called directrix,

Ruled surface: A ruled surface is one that may be generated by sweeping a straight line (generatrix), along a straight or curved path.

Plane surface: A plane is a ruled surface generated by a line, one point of which moves along a straight path while the generatrix remains parallel to its original position. Solids bounded by plane surfaces are tetrahedron, prism, and pyramid.

Single-curved surface: A single-curved surface can be unrolled to coincide with a plane. These are composed of consecutive pairs of straight-line elements in the same plane. Solids bounded by single-curved surface are cone, cylinder.

Double curved surface: A surface generated by revolving a curved line about a straight line in the plane of the curve is called a double-curved surface of revolution. Solids bounded by double-curved surface are sphere, paraboloid, spheroid, torus, and hyperboloid.

Warped surface: A warped surface is a ruled surface that is not developable. Warped surfaces cannot he unrolled or unfolded to lie flat. Solids bounded by warped surface are conoid, cylindroid, helicoid, hyperbolic paraboloid.

A layout of the complete surface of a 3 dimensional object on a plane surface is called a development or pattern. Development consists of drawing successive surfaces of object in their true shape and size with joined common edges. Development of right cylinder is a rectangle of width equal to circumference of cylinder.

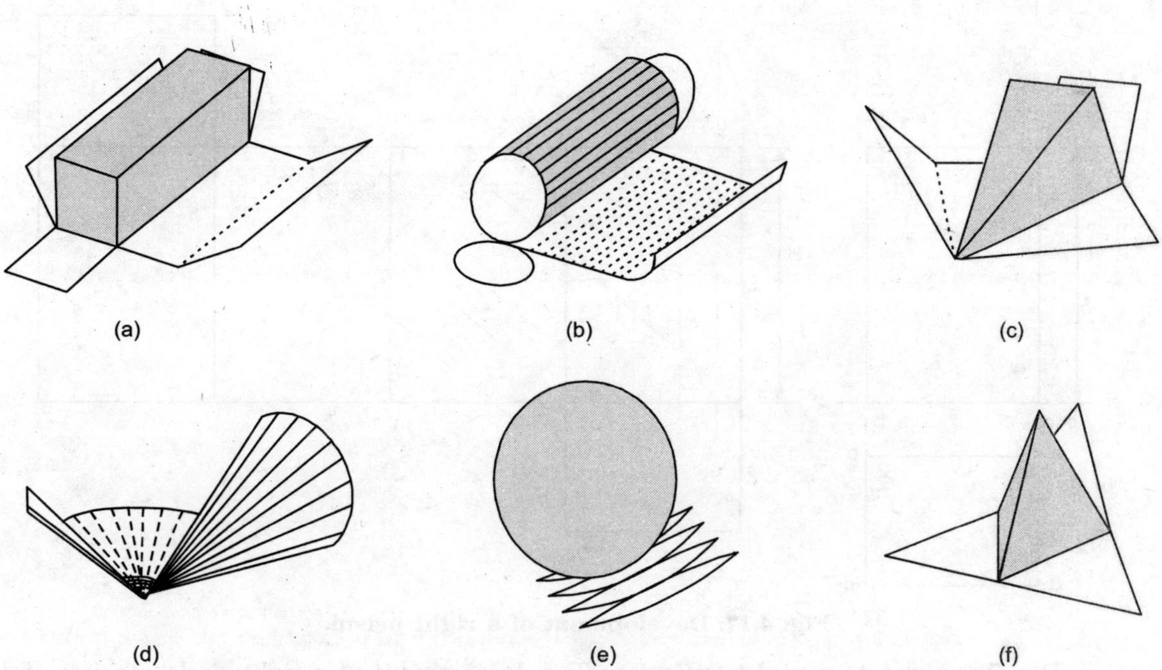

Fig. 4.16: (*a*) Prism and (*b*) Cylinder (Parallel development); (*c*) Pyramid and (*d*) Cone (Radial line development); (*e*) Sphere (Approximation development); (*f*) Tehrahedron (Triangulation development)

4.4 METHODS OF PATTERN DEVELOPMENT

1. **Parallel line method:** Parallel line development method uses parallel lines to construct the expanded pattern of 3-dimensional shape. In this the surface is divided into a series of parallel lines to determine the shape of a pattern. These lines are called stretch out lines. For example prisms and cylinders

(*i*) **Development of a right prism:** To develop a right prism, following steps are followed:

- The right prism consists of rectangular faces in contact. The sides of rectangle are equal to length of base edge and height of the prism.
- Draw the front view and top view of prism in required position.
- Stretch out horizontal line equal to perimeter of the base and divide it in four equal parts.
- Erect perpendiculars at each division and mark their heights equal to the height of prism.
- Join the points and get the development of the right prism.

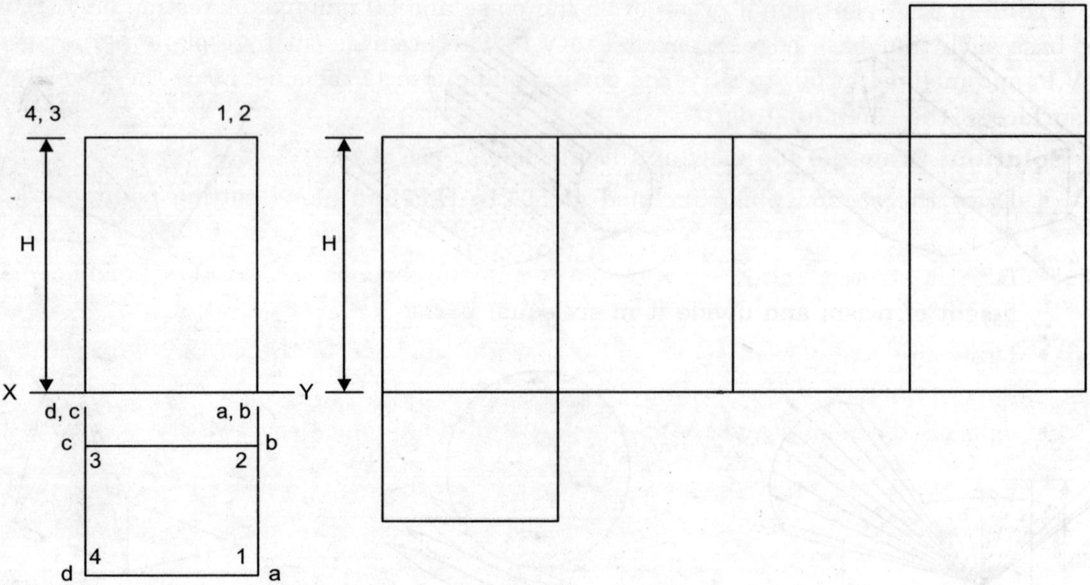

Fig. 4.17: Development of a right prism

(*ii*) **Development of a right cylinder:** The development of a right circle consists of a rectangle with one side equal to the circumference of base circle and other side equal to the height of the cylinder. The steps followed in development of a right cylinder are:

- Draw the front view and top view of prism in required position and divide the circle in 12 equal parts.
- Stretch out horizontal line equal to perimeter of circle and height equal to height of cylinder.
- Show the generators by dividing circumference in 12 equal parts.

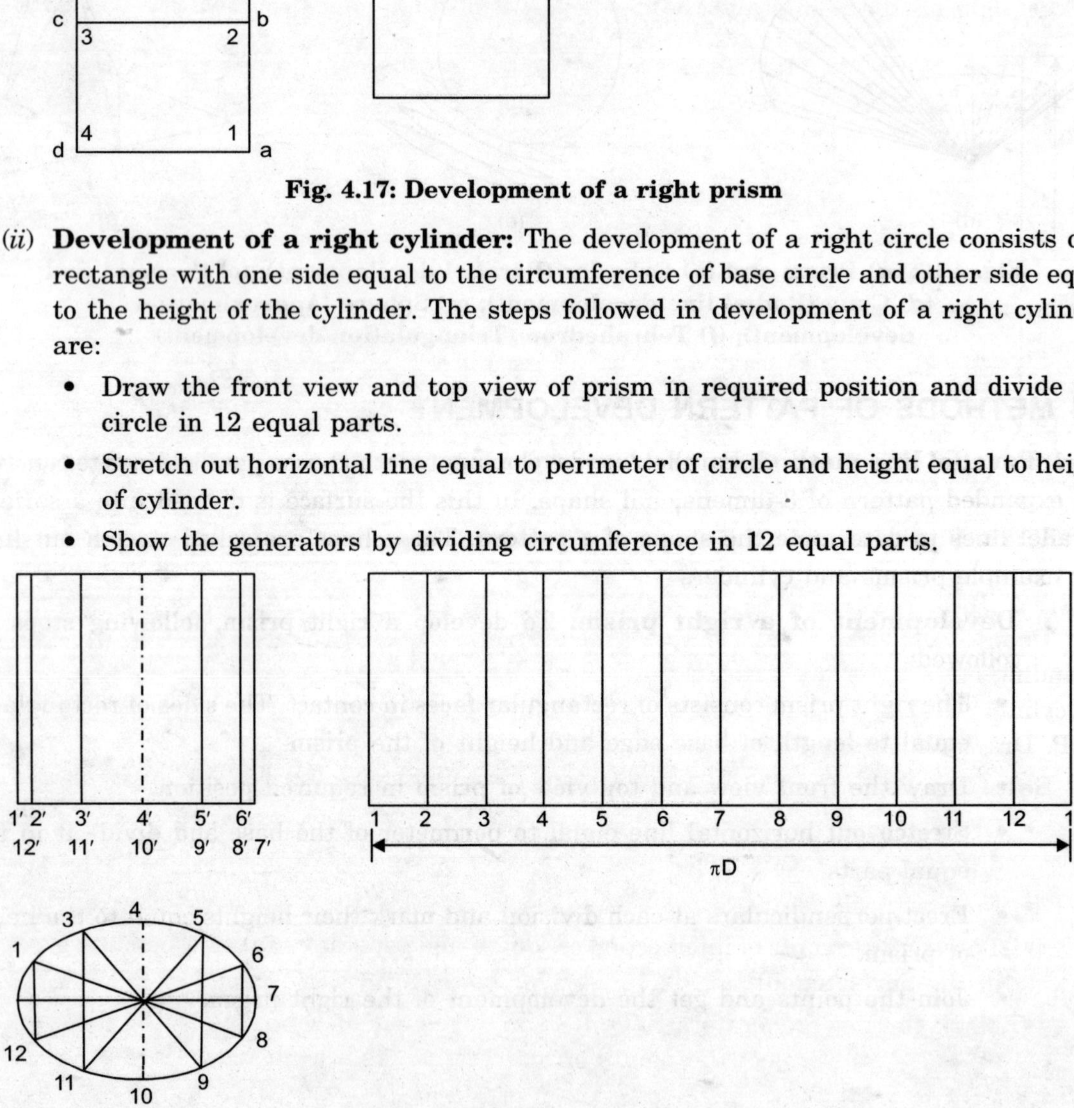

Fig. 4.18: Development of a right cylinder

Problem 1: A pentagonal prism of 25 mm base and 50 mm axis is resting on H.P. with the base such that base edge is parallel to V.P. The prism is cut by a plane perpendicular to V.P. and inclined at 50° to H.P. The cutting plane bisects the axis. Draw the development of surface of the remaining part.

Solution: Draw the top view and front view of prism.

- Draw the section plane inclined at 50° to H.P and name cutting points as l', m', n', o'.
- Draw a stretch line 1 – 1 = 6 × *ab* (equal to perimeter of base) of height equal to height of prism and divide it in six equal parts.
- Draw projectors from l', m', n', o' parallel to base to the corresponding lines on development of prism surface.
- Join all the points and the development of remainder prism will be formed.

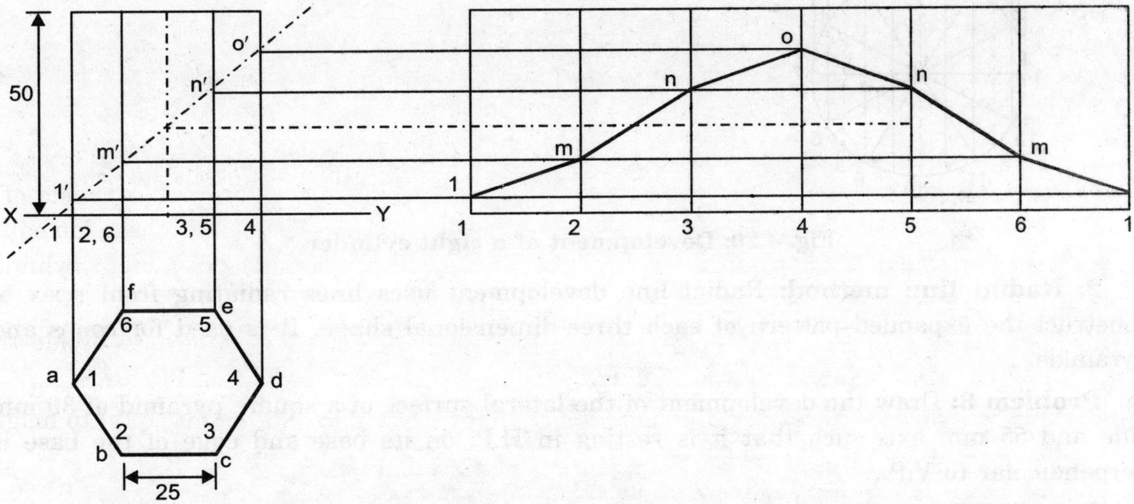

Fig. 4.19: Development of a pentagonal prism

Problem 2: A right circular cylinder with 20 mm base radius and 50 mm long axis is standing on H.P. on the base. There is a square hole of 20 mm whose axis bisects the axis of cylinder and is perpendicular to V.P. The faces of square hole are equally inclined with H.P. Draw the lateral surface of cylinder.

Solution: Draw the top and front view of the cylinder and draw its development.

- Draw a square of 20 mm side such that its axis bisects the axis of and cuts the different generators at various points.
- Project these points in top view, dotted lines are used to show hole cut.
- Transfer the points of intersection of square and generators to corresponding generators in the development.
- Mark the new generators P, Q, R, S in development according to the distance from top view and project the points of intersection from front view.

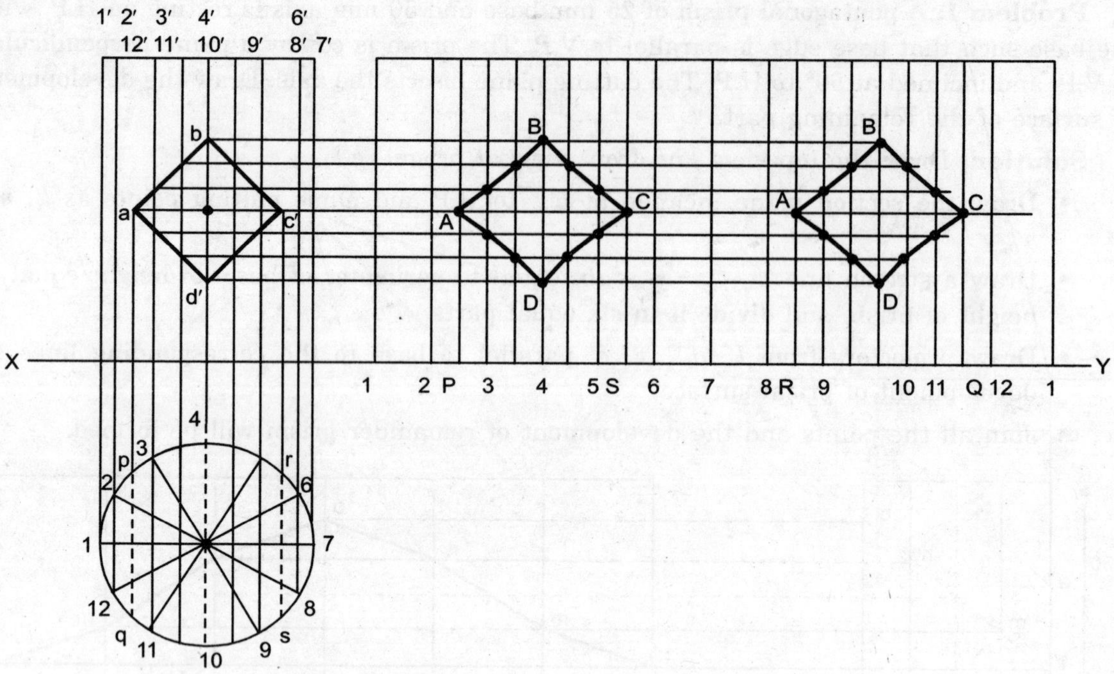

Fig. 4.20: Development of a right cylinder

2. Radial line method: Radial line development uses lines radiating from apex to construct the expanded pattern of each three-dimensional shape. It is used for cones and pyramids.

Problem 3: Draw the development of the lateral surface of a square pyramid of 30 mm side and 55 mm axis such that it is resting in H.P. on its base and edge of the base is perpendicular to V.P.

Solution: Draw the top view and front view of the pyramid.

- None of the view gives true shape taking *o* as centre and *oc* as radius draw an arc to cut at c_1 from *c*. Project the point on XY line as c_1'.
- $o'c_1'$ gives the true length of slant edge. So taking $o'c_1'$ as radius and *o* as centre draw an arc.
- Mark four divisions of length 30 mm. Join them with centre which leads to formation of four triangles. Erase the arc.

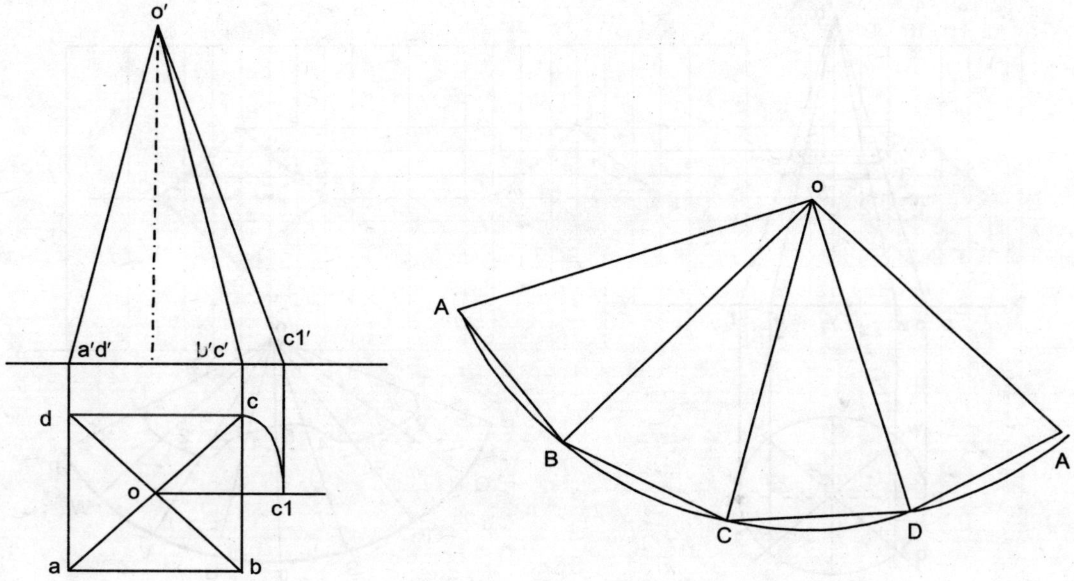

Fig. 4.21: Development of lateral surface of square pyramid

Problem 4: A cone of 25 mm diameter and axis of 55 mm is resting on its base on H.P. It is cut by a plane perpendicular to V.P. and inclined at 35° to H.P. which bisects the axis. Draw the development of lateral surface.

Solution: Draw the projections of cone.

- Divide circle in 8 equal parts.
- Taking *o* as centre and *o′p′* or *o′t′* as radius draw an arc of length equal to

$$\theta = \frac{\text{radius of right cone}}{\text{slant height}} \times 360°$$

- Cut the arc of angle θ and divide it in 8 equal parts.
- Name the points on the section plane
- Project the points on section plane horizontally on the outer generator and name them 2″, 3″, 4″ and 5″.
- Taking the distances *o′* 1″, *o′* 2″, *o′* 3″, *o′* 4″, *o′* 5″ in compass, cut the arcs on corresponding generators in development.
- Join the points on the curve.

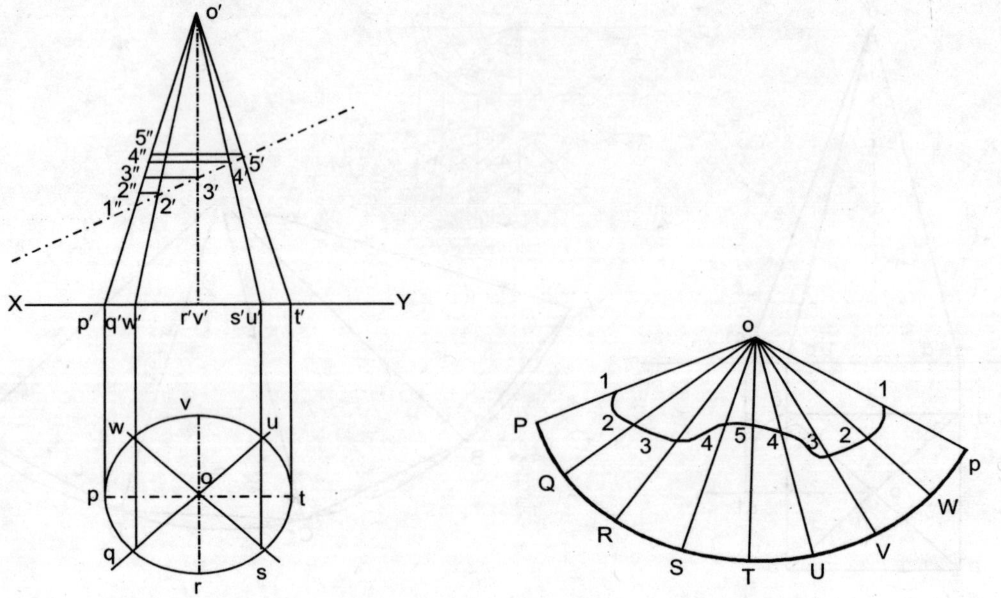

Fig. 4.22: Development of cone in given case

3. **Triangulation method:** Triangulation developments are made from polyhedrons, singlecurved surfaces, and wrapped surfaces. It is employed to obtain the development of transition pieces. This method consists of dividing a surface into a number of triangles and then transferring them to form the development.

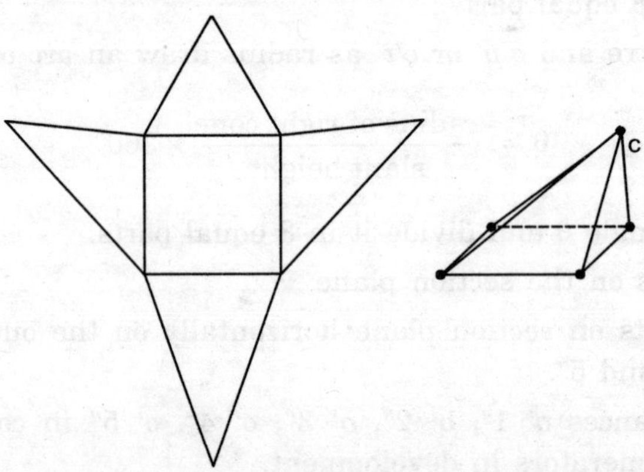

Fig. 4.23: Triangulation method

4. **Approximate method:** In approximate development, the shape obtained is only approximate. After joining, the part is stretched or distorted. There are two methods for the development of a sphere:

(*a*) **Zone method:** In zone method sphere is cut into a number of horizontal sections. Each section can be taken as frustum of a cone such that the apex of cone lies at the intersection of extended chords. Large number of zones give better approximation.

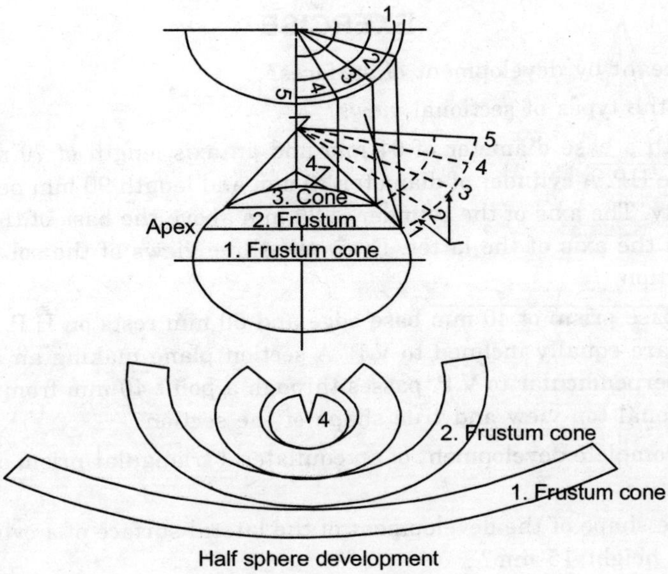

Fig. 4.24: Zone method

(*b*) **Lune method:** In this method the sphere is cut into a number of equal meridian sections which are called lunes. For complete development of sphere, each lune is drawn separately. For example 12 lunes each of length equal to half the circumference of sphere are made if the top view of sphere is divided into 12 equal parts.

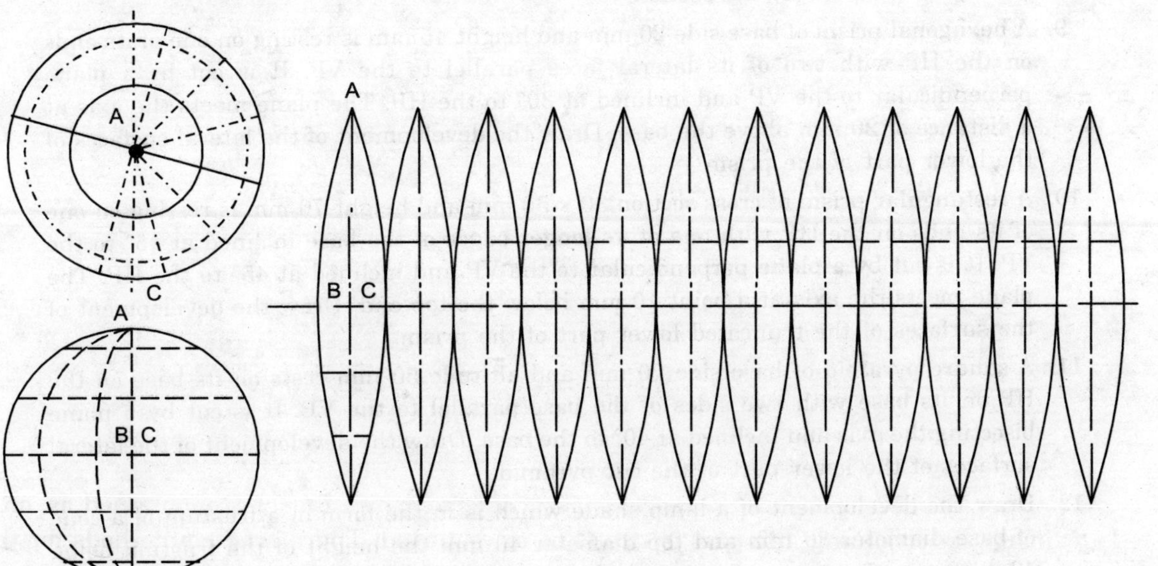

Fig. 4.25: Lune method

EXERCISE

1. What is meant by development of surfaces?
2. What are the types of sectional views?
3. A cone with a base diameter of 64 mm and an axis length of 70 mm is kept on its base on the HP. A cylinder of diameter 30 mm and length 90 mm penetrates the cone horizontally. The axis of the cylinder is 25 mm above the base of the cone and 6 mm away from the axis of the latter. Draw the three views of the solids showing curve of intersection.
4. A square base prism of 40 mm base edge and 60 mm rests on H.P. in a way that all the edges are equally inclined to V.P. A section plane making an angle of 50° with H.P. and perpendicular to V.P. passes through a point 40 mm from base. Draw front view, sectional top view and true shape of the section.
5. Show the complete development of an equilateral triangular prism of base side A and height H.
6. What is the shape of the development of the lateral surface of a cylinder of diameter 6 mm and height 15 mm?
7. A cone, having base diameter of 50 mm and height of 70 mm is resting on its base on HP. It is cut by a section plane such that the true shape of the section in front view is a rectangular hyperbola with a base of 40 mm. Find front and top views.
8. A cone base 75 mm diameter and axis 100 m long, has its base on the HP. A section plane parallel to one of the end generators and perpendicular to the FP cuts the cone intersecting the axis at a point 75 mm from the base. Draw the sectional Top View and the true shape of the section.
9. A hexagonal prism of base side 20 mm and height 45 mm is resting on one of its ends on the HP with two of its lateral faces parallel to the VP. It is cut by a plane perpendicular to the VP and inclined at 30° to the HP. The plane meets the axis at a distance of 20 mm above the base. Draw the development of the lateral surfaces of the lower part of the prism.
10. A rectangular prism of cross section 30 × 50 mm and height 70 mm is resting on one of its ends on the HP with one of its longer edges of the base inclined at 45° to the VP. It is cut by a plane perpendicular to the VP and inclined at 45° to the HP. The plane meets the axis at a point 10 mm below the top end. Draw the development of the surfaces of the truncated lower part of the prism.
11. A square pyramid of base side 20 mm and altitude 50 mm rests on its base on the HP on its base with two sides of the base parallel to the VP. It is cut by a plane bisecting the axis and inclined at 40° to the base. Draw the development of the lateral surfaces of the lower part of the cut pyramid.
12. Draw the development of a lamp shade which is in the form of a frustum of a cone of base diameter 80 mm and top diameter 40 mm the height of the frustum being 40 mm.

13. A cone of base diameter 60 mm and height 70 mm rests vertically on its base on the ground. A string is wound round the curved surface of the cone starting from left extreme point on the base and ending at the same point. Find the shortest length of the string required. Also trace the path of the string in the front and top views.

SUMMARY

- A section view is drawn cutting the object by an imaginary section or cutting plane so as to reveal the interiors.
- In full section view section plane goes completely through the object.
- A ruled surface is one that may be generated by sweeping a straight line (generatrix), along a straight or curved path.
- A plane is a ruled surface generated by a line, one point of which moves along a straight path while the generatrix remains parallel to its original position.
- A section view is created by passing an imaginary cutting plane vertically through the centre of the object.
- In third-angle projection the sectional view is placed on the side behind the sectioning viewing plane.
- Material which has been cut by the cutting plane is hatched.
- Dimensions are NOT inserted in hatched areas.
- No hidden lines are drawn and all visible surfaces and edges behind the cutting plane are drawn as object lines.
- Solids bounded by single-curved surface are cone, cylinder.
- A warped surface is a ruled surface that is not developable.
- A layout of the complete surface of a 3 dimensional object on a plane surface is called a development or pattern.
- Development of right cylinder is a rectangle of width equal to circumference of cylinder.

CHAPTER

5

Isometric Projections

Learning Objectives

- Describe the objective and nature of isometric views
- Construct isometric objects
- Understand the concept of sketching
- Learning the conversion of isometric view to orthographic and vice-versa

5.1 INTRODUCTION

In engineering field, it is normal to draw two or more than two orthographic projections to give the shape and size of the object. Sometime, less experienced technician often required hard exercise of the brain to visualize the orthographic projections. To make it easy to understand the shape of the object, it needs an additional view known as Isometric view. The isometric view represents the overall appearance of the object and makes it easy to understand some details which may be difficult to understand in the orthographic projections.

5.2 ISOMETRIC VIEW

The three dimensional view of an object obtained on a plane of projection in which all the projectors are parallel but inclined at an angle of 30° to the plane of projection is known as Isometric View. In this type of projection, the object is placed in such a way that all the three axis make equal angle with the plane of projection.

5.2.1 How to Draw Isometric View

It is important to know how an isometric view is obtained on the plane of projection. Consider a cube. Let the three corners of the cube meeting at the forward corner is equally inclined to the plane of projection so that the projection to be obtained should give equal length. Therefore, any line so inclined is known as isometric line and the projection thus obtained, represent isometric projection or view. As three edges meeting at a point are equally inclined to each other, hence, the angle between two adjacent edges is equal to 360/3 = 120°. The figure 5.1 shows the isometric view of a cube in which the edges are parallel to the three main axis.

It is seen from the following figure that when a line is inclined to the plane of projection, it does not give the exact size. It is very essential to know the proportion by which various edges of the cube are shortened. This can be found by means of a scale known as isometric scale.

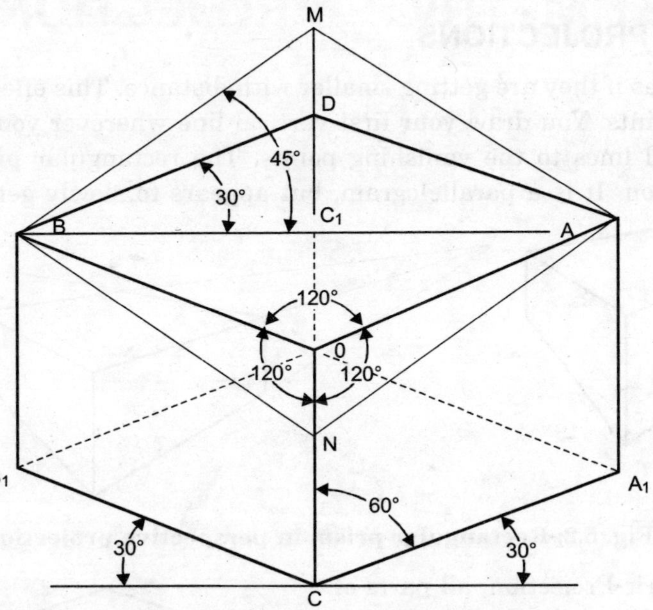

Fig. 5.1: Isometric projection of a cube

5.3 TYPES OF AXONOMETRIC PROJECTION

The three types of axonometric projection are isometric projection,

Diametric projection, and trimetric projection, depending on the exact angle at which the view deviates from the orthogonal. Typically in axonometric drawing, as in other types of pictorials, one axis of space is shown as the vertical.

The types of axonometric projections are:

- **Isometric Projection:** It is the most commonly used form of axonometric projection in engineering drawing, the direction of viewing is such that the three axis of space appear equally foreshortened, and there is a common angle of 120° between them. As the distortion caused by foreshortening is uniform the proportionality of all sides and lengths are preserved, and the axis share a common scale. This enables measurements to be read or taken directly from the drawing. 120° angles are more easily constructed using only a compass and straight edge.

- **Dimetric Projection:** In this projection the direction of viewing is such that two of the three axes of space appear equally foreshortened, of which the attendant scale and angles of presentation are determined according to the angle of viewing; the scale of the third direction(vertical) is determined separately. Dimensional approximations are common in diametric drawings.
- **Trimetric Projection:** In this projection the direction of viewing is such that all of the three axes of space appear unequally foreshortened. The scale along each of the three axes and the angles among them are determined separately as dictated by the angle of viewing. Dimensional approximations in trimetric drawings are common.

5.4 ISOMETRIC PROJECTIONS

Objects appear as if they are getting smaller with distance. This effect is usually achieved by using Vanish Points. You draw your first vertical line wherever you want, and then just extend the diagonal lines to the vanishing points. The rectangular prism below is in this perspective projection. It is a parallelogram, but appears to slowly get smaller.

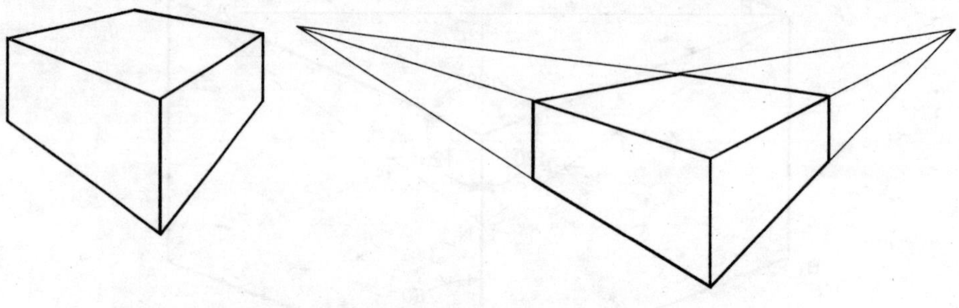

Fig. 5.2: Rectangular prism in perspective projection

With an Isometric Projection, all parts of the object are proportionally displayed; regardless of depth. It uses an angle of 30° for its diagonals. You just make sure that your diagonals are 30°, and the opposite ends are guaranteed to be equal lengths. Here is a different rectangular prism, but this one is in Isometric Projection. This time, you could pull out your ruler, and prove that it is a parallelogram if you wanted to.

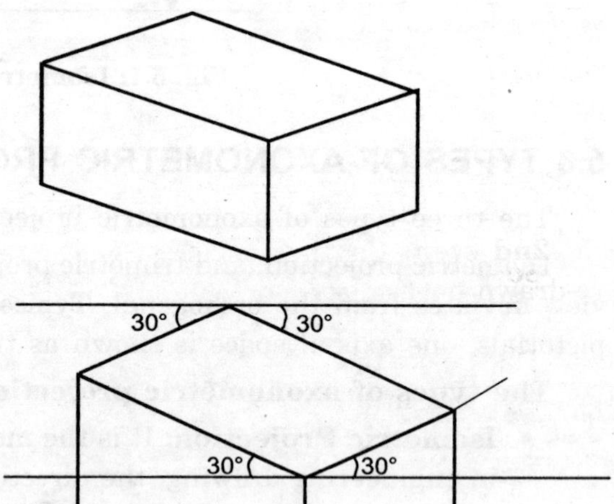

Fig. 5.3: Isometric projection

5.4.1 How To Draw Isometric Projection Using Orthographic Views

Following steps are used to convert orthographic projection into isometric projection :

Orthographic Projection

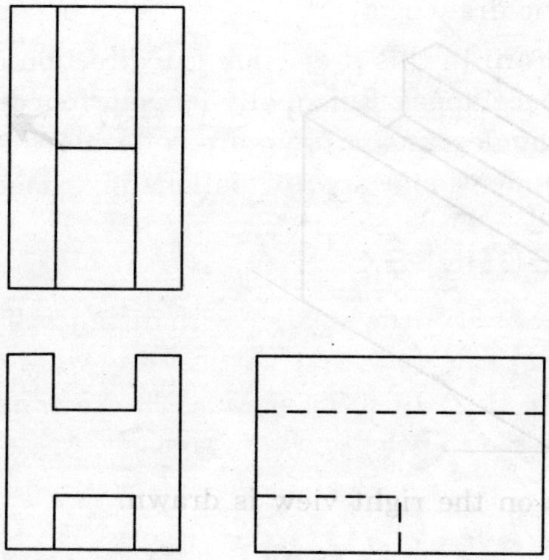

1st step: To draw a rectangular prims shape with a proportional length, width and height. A front view is determined by yourself. In this example, a front view is shown by an arrow line.

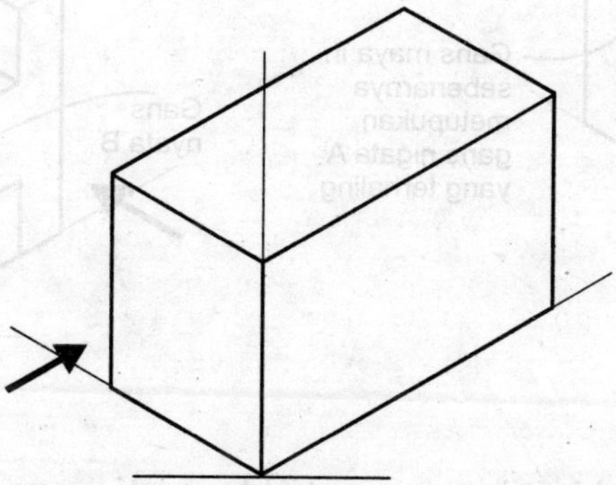

2nd step: Continuous lines (object line) on a front view of orthographic projection is re-drawn on isometric projection.

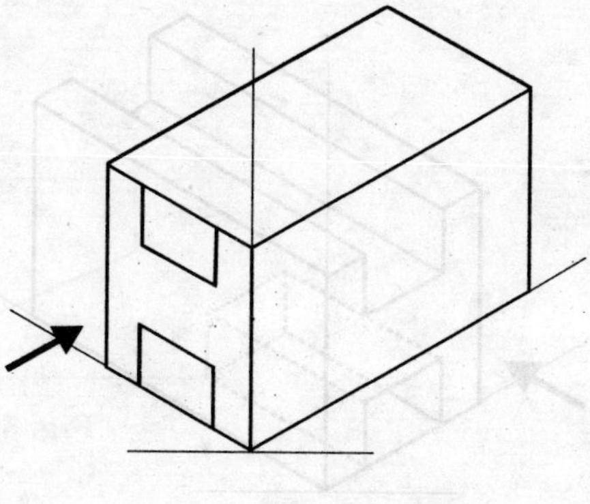

3rd step: Continuous lines (object line) on a top view of orthographic projection is re-drawn on isometric projection.

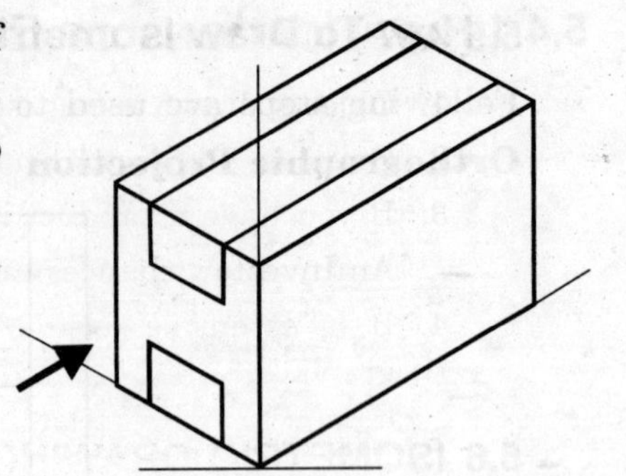

4th step: Other continuous lines on the front and top view is completed.

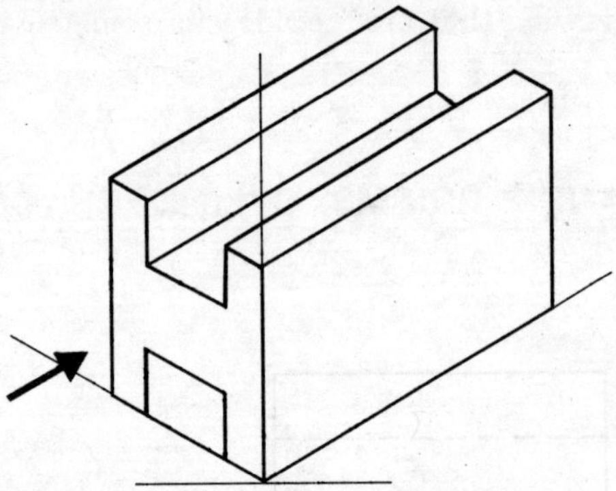

5th step: Hidden lines on the right view is drawn.

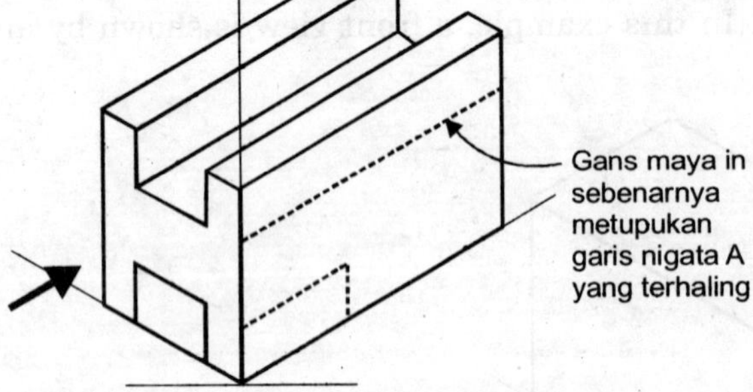

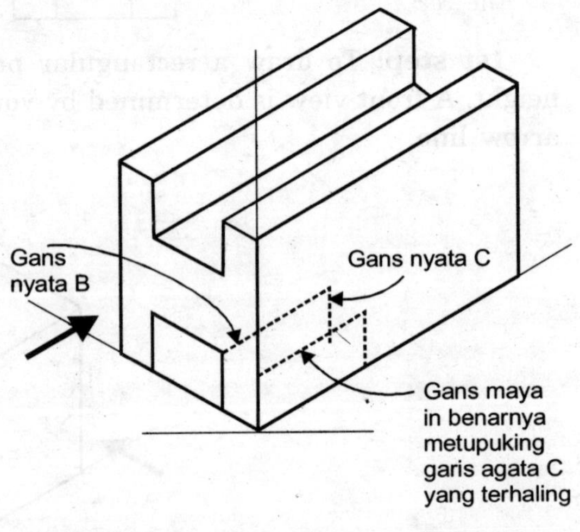

A hollow part below the object is represented by continuous line B and line C. Continuous line C has a role as the end of hollow part. After all the lines are completed, the final object is shown below:

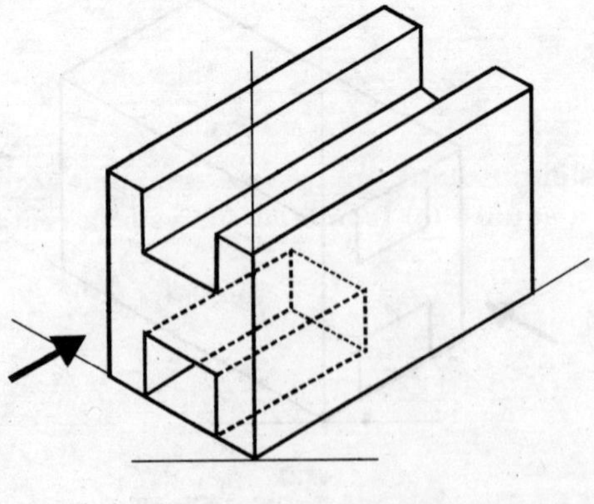

5.5 APPLICATIONS OF ISOMETRIC PROJECTIONS

1. Objects drawn using isometric projection gives clear understanding of the object.
2. One can take actual measurements of the object from an Isometric drawing.
3. If you have a product idea in mind and if you want to show it to your Boss or An Investor, chances are you'll be drawing that product idea in isometric projection.
4. It is extensive used by Architects, Interior designers, Game Designers and Info graphics.

5.6 ISOMETRIC DRAWING FOR PLANES, PRISMS, CYLINDERS, CONES AND SPHERES

5.6.1 Isometric Views of Rectangle/Square

Consider a square or a rectangle ABCD. If a square lies in the vertical plane, it will appear as a rhombus with side equal to side of square in isometric view, depending on its orientation, i.e., right-hand vertical face or left-hand vertical face. A rectangle appears as a parallelogram in isometric view. Three versions are possible depending on the orientation of the rectangle, i.e., right-hand vertical face, left-hand vertical face or horizontal face.

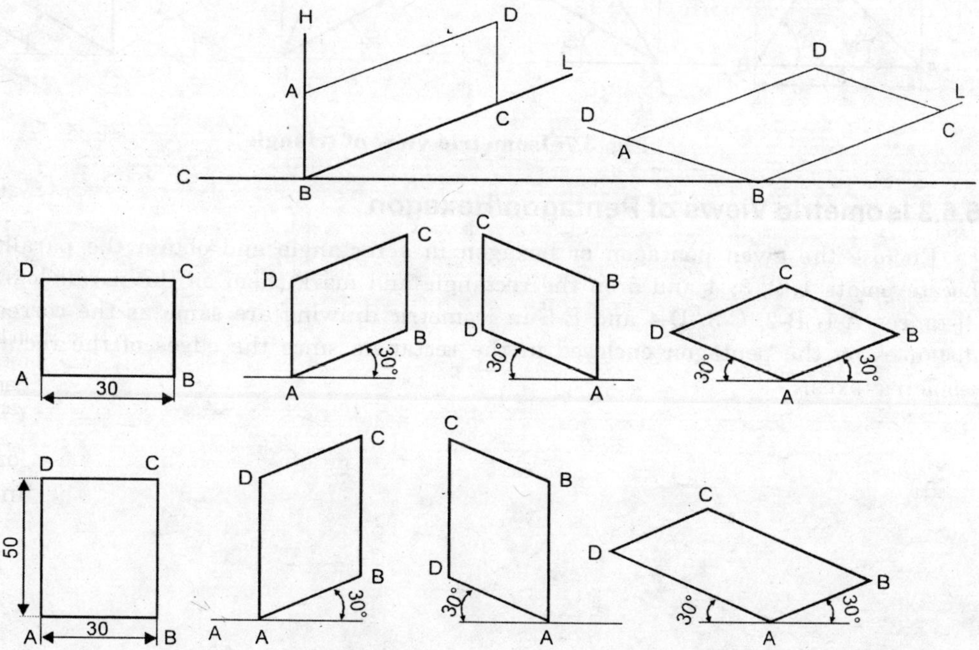

Fig. 5.6: (*a*) orientation of square and rectangle (*b*) Isometric views of a square (*c*) Isometric views of a rectangle

5.6.2 Isometric Views of a Triangle

A triangle of any type can be easily obtained in isometric view. Enclose the triangle in a rectangle ABCD and obtain parallelogram ABCD of the rectangle. Then locate point 1 in the parallelogram such that C-1 in the parallelogram is equal to C-1 in the rectangle. A-B-1 represents the isometric view of the triangle.

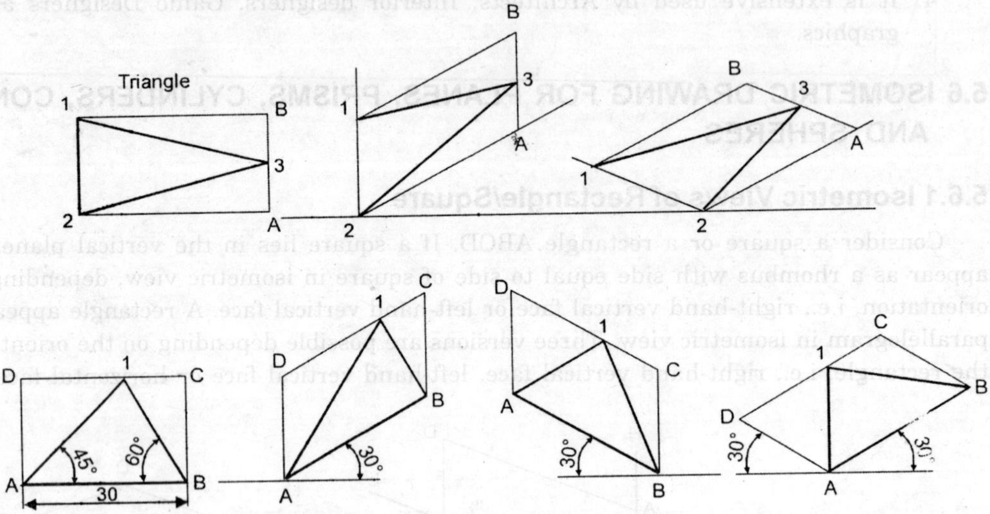

Fig. 5.7: Isometric view of triangle

5.6.3 Isometric Views of Pentagon/Hexagon

Enclose the given pentagon or hexagon in a rectangle and obtain the parallelogram. Locate points 1, 2, 3, 4 and 5 on the rectangle and mark them on the parallelogram. The distances A-1, B-2, C-3, D-4 and E-5 in isometric drawing are same as the corresponding distances on the pentagon enclosed in the rectangle since the edges of the rectangle are isometric axes.

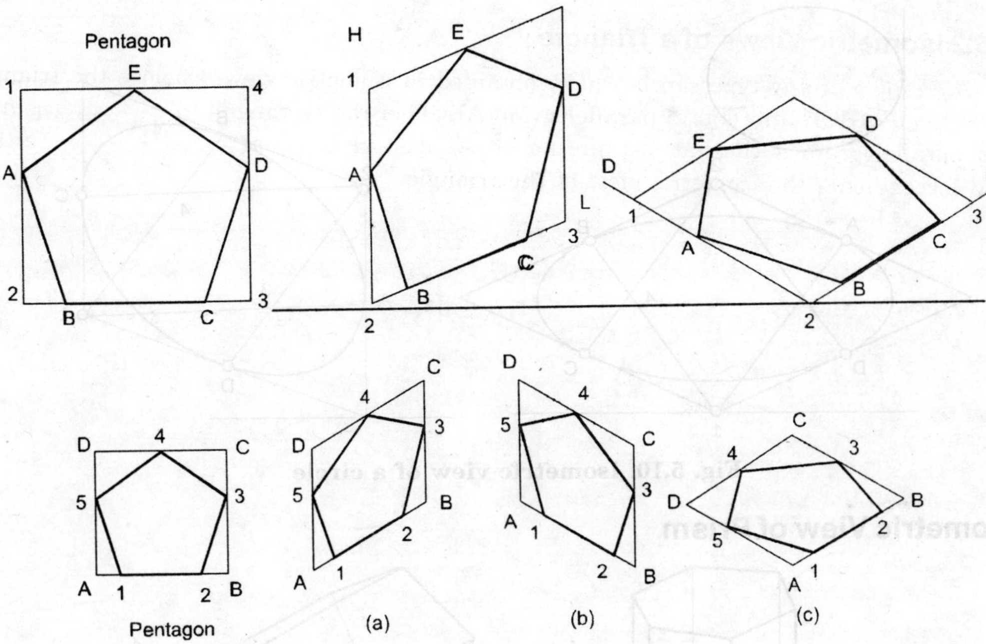

Fig. 5.8: Isometric view of pentagon

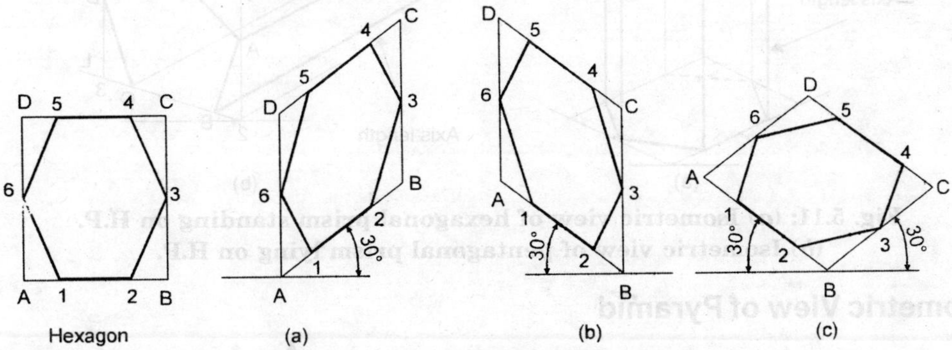

Fig. 5.9: Isometric view of pentagon

5.6.4 Isometric Views of Circle

The isometric view or isometric projection of a circle is an ellipse. First, enclose the given circle into a square ABCD. Draw rhombus ABCD as an isometric view of the square. Join the farthest corners of the rhombus. Obtain midpoints 3 and 4 of sides CD and AD respectively. Locate points 1 and 2 at the intersection of AC with B-3 and B-4 respectively. Now with 1 as a center and radius 1-3, draw a small arc 3-5. Draw another arc 4-6 with same radius but 2 as a center. With B as a center and radius B-3, draw an arc 3-4. Draw another arc 5-6 with same radius but with D as a center.

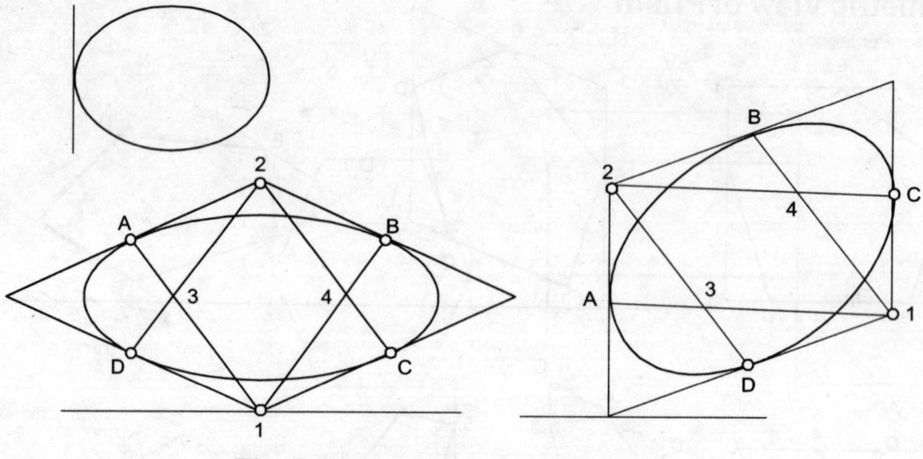

Fig. 5.10: Isometric view of a circle

5.6.5 Isometric View of Prism

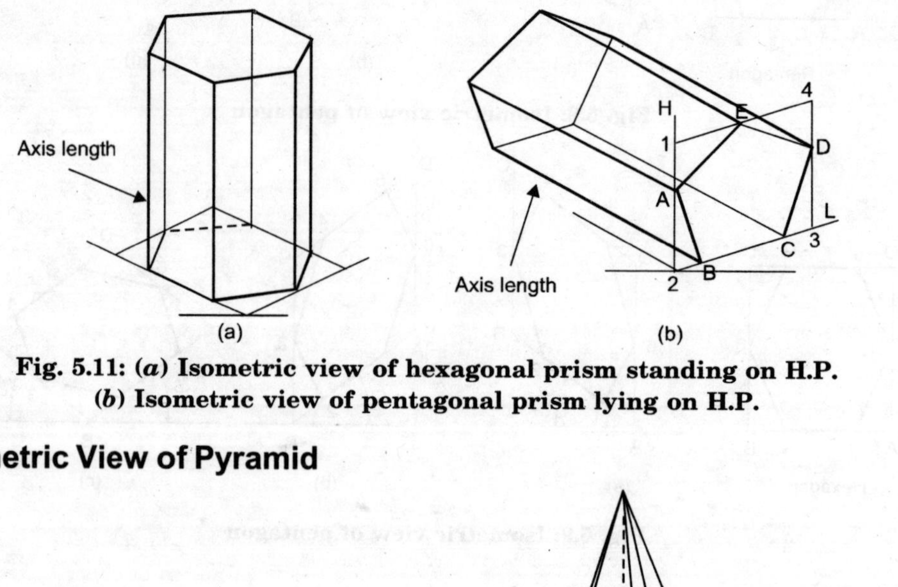

Fig. 5.11: (*a*) Isometric view of hexagonal prism standing on H.P.
(*b*) Isometric view of pentagonal prism lying on H.P.

5.6.6 Isometric View of Pyramid

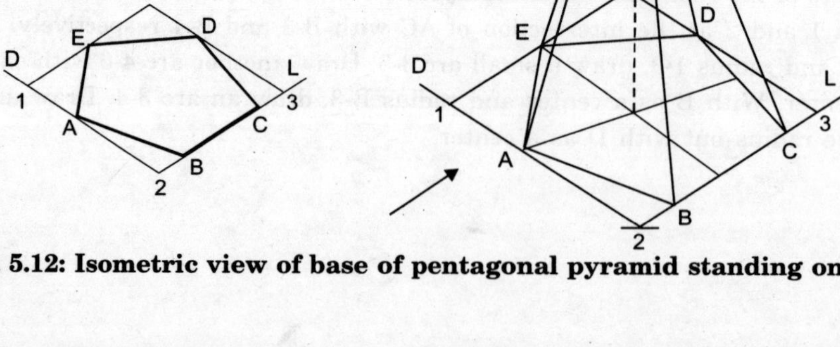

Fig. 5.12: Isometric view of base of pentagonal pyramid standing on H.P.

5.6.7 Isometric View of Prism

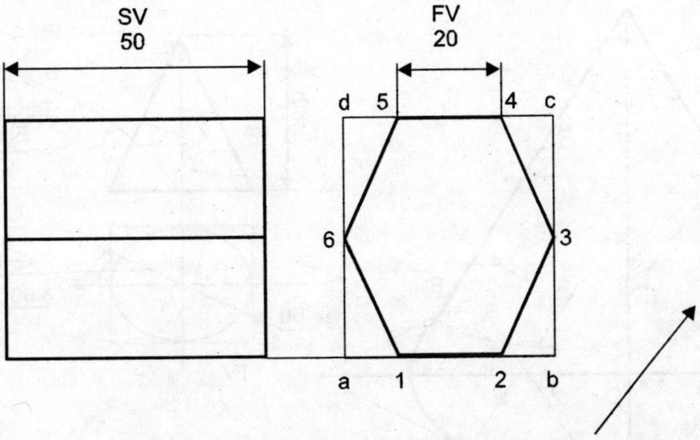

Fig. 5.13: Isometric view of prism with hexagonal base of side 20 mm

5.6.8 Isometric View of Pyramid

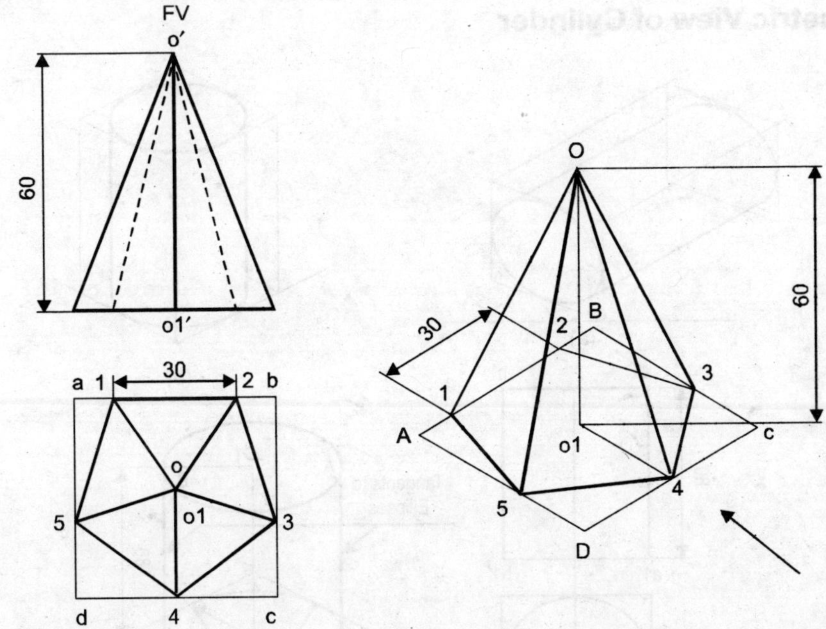

Fig. 5.14: Isometric view of pentagonal pyramid of axis length 60 mm and base 30 mm

5.6.9 Isometric View of Cone

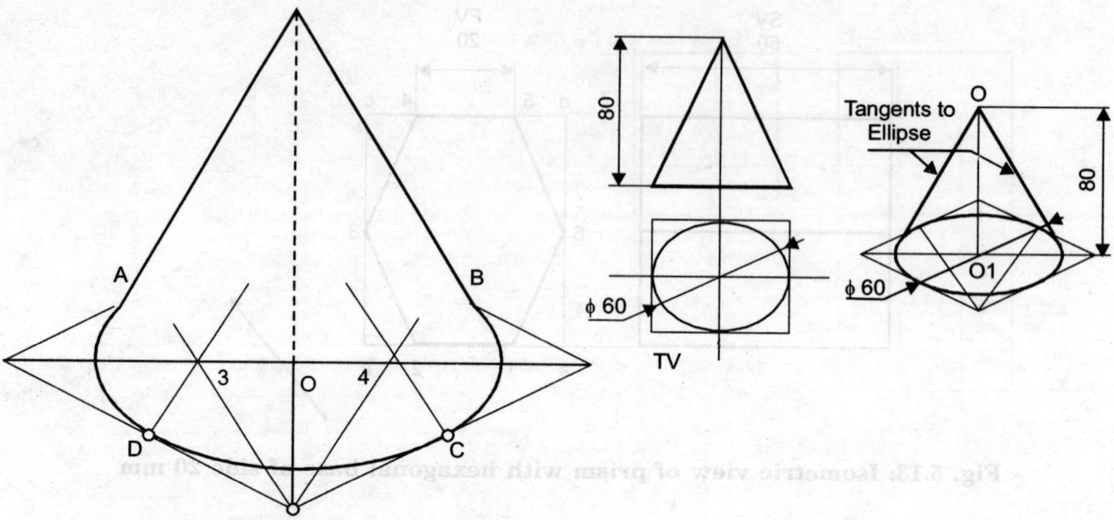

Fig. 5.15: Isometric view of a cone of height 80 mm and diameter 60 mm

5.6.10 Isometric View of Cylinder

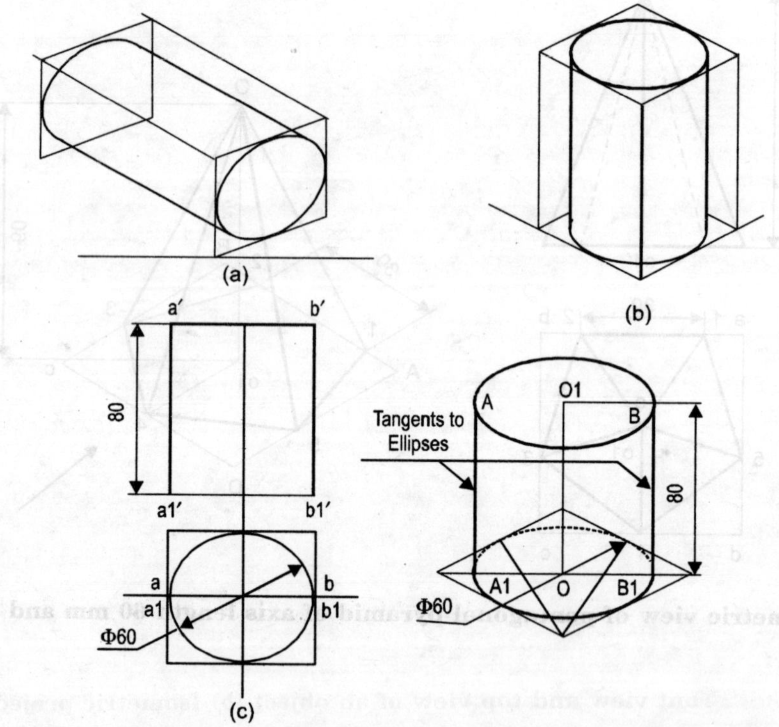

Fig. 5.16: (*a*) Cylinder lying on H.P. (*b*) Cylinder standing on H.P.
(*c*) Isometric view of cylinder of height 80 mm and base diameter 60 mm

Problem 1: F.V. , TV. And S.V. of a solid is given. Draw its isometric views.

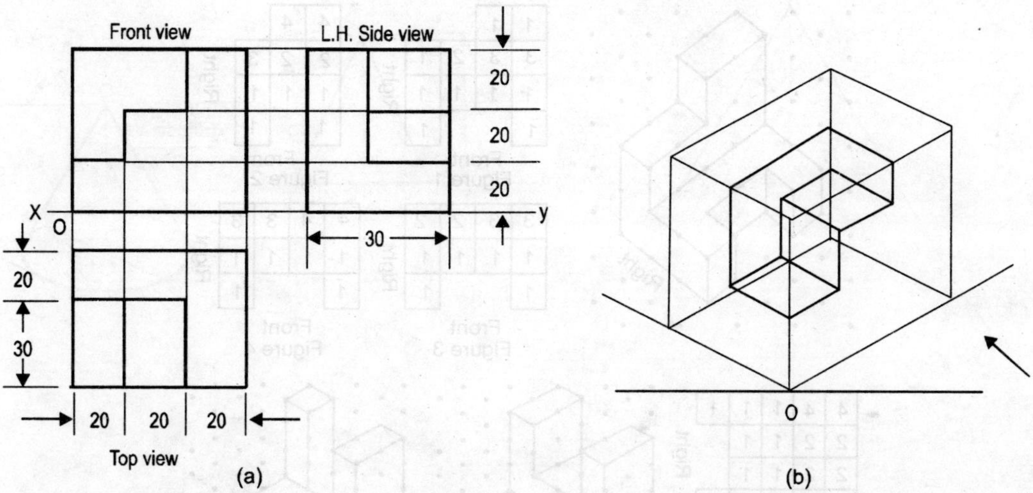

Fig. 5.17: (*a*) Front view, top view and side view of an object (*b*) Isometric projection

Problem 2: F.V. And T.V. of a solid is given. Draw its isometric views.

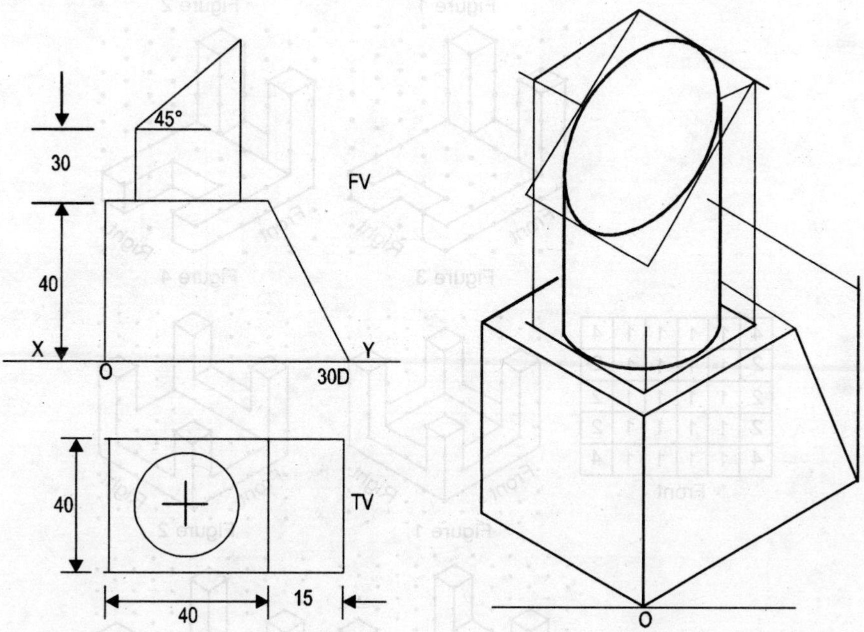

Fig. 5.18: (*a*) Front view and top view of an object (*b*) Isometric projection

5.7 CONVERSION OF ISOMETRIC VIEWS TO ORTHOGRAPHIC VIEWS

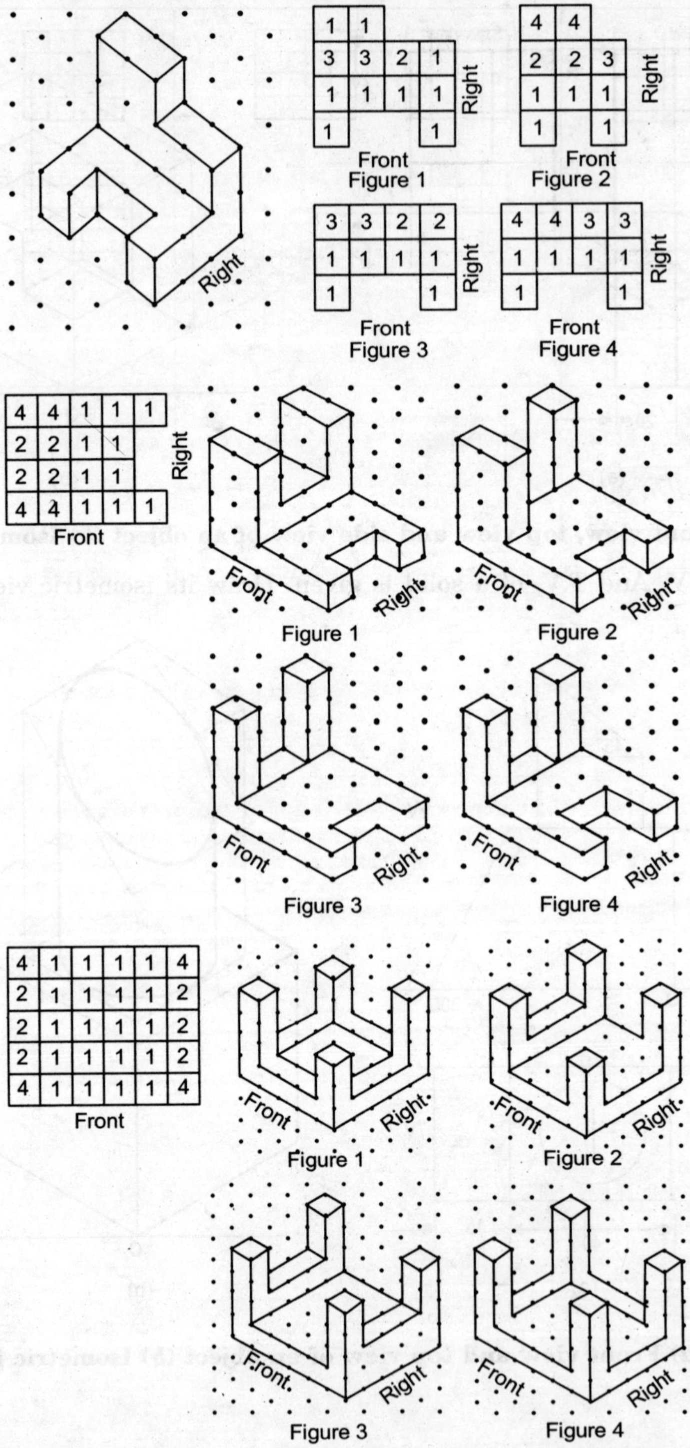

Fig. 5.19: Conversion of Isometric views to Orthographic views for various objects

5.8 CONVERSION OF ORTHOGRAPHIC VIEWS TO ISOMETRIC VIEWS

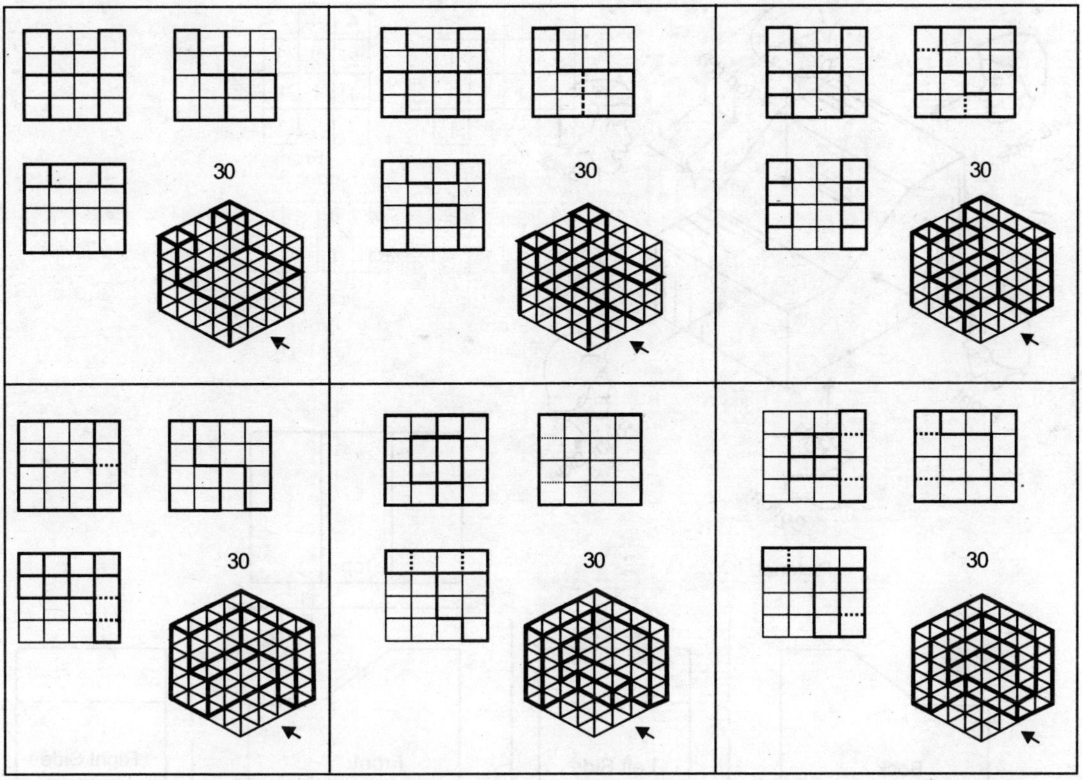

Fig. 5.20: Conversion of Orthographic views to Isometric views for 6 Objects

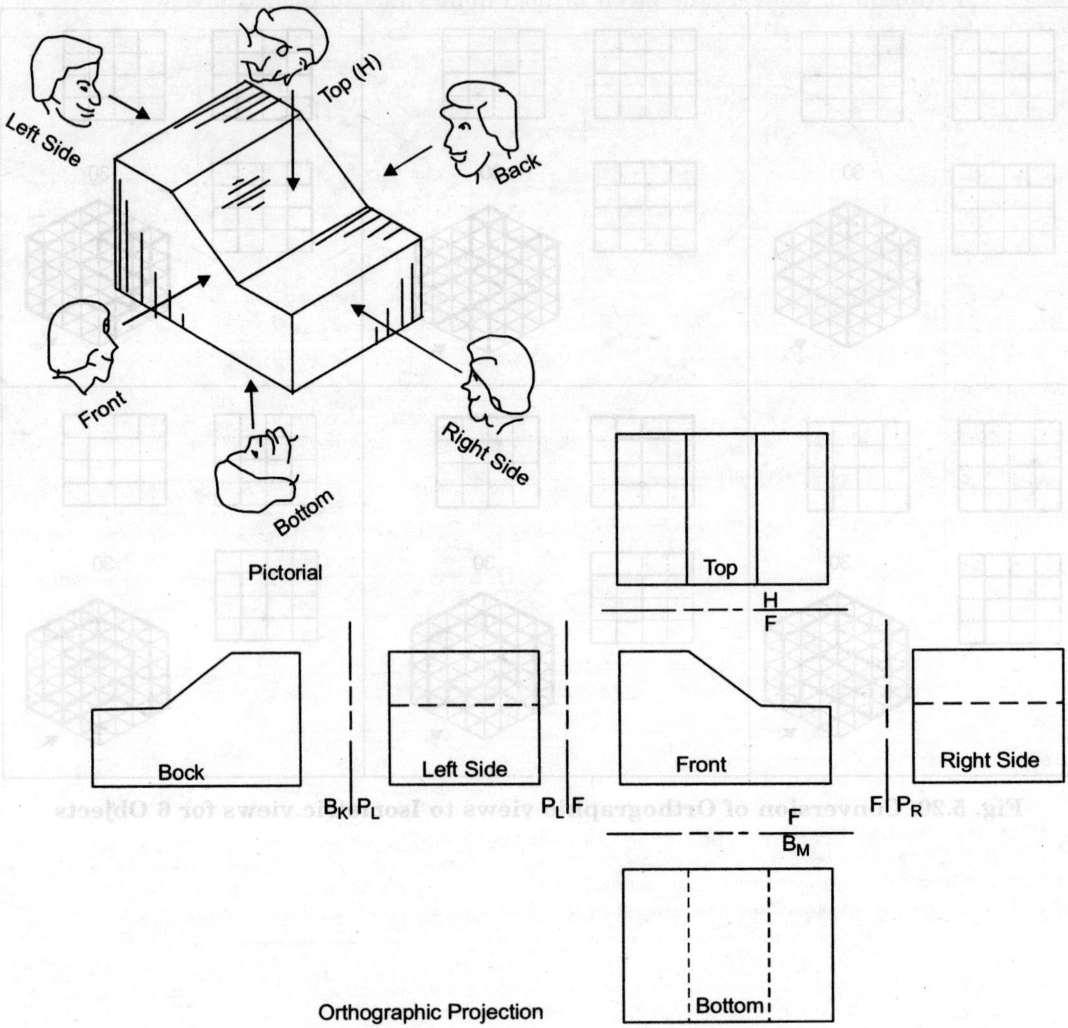

Fig. 5.21: Orthographic views of an object

Problem 3: Convert the given orthographic views to isometric views

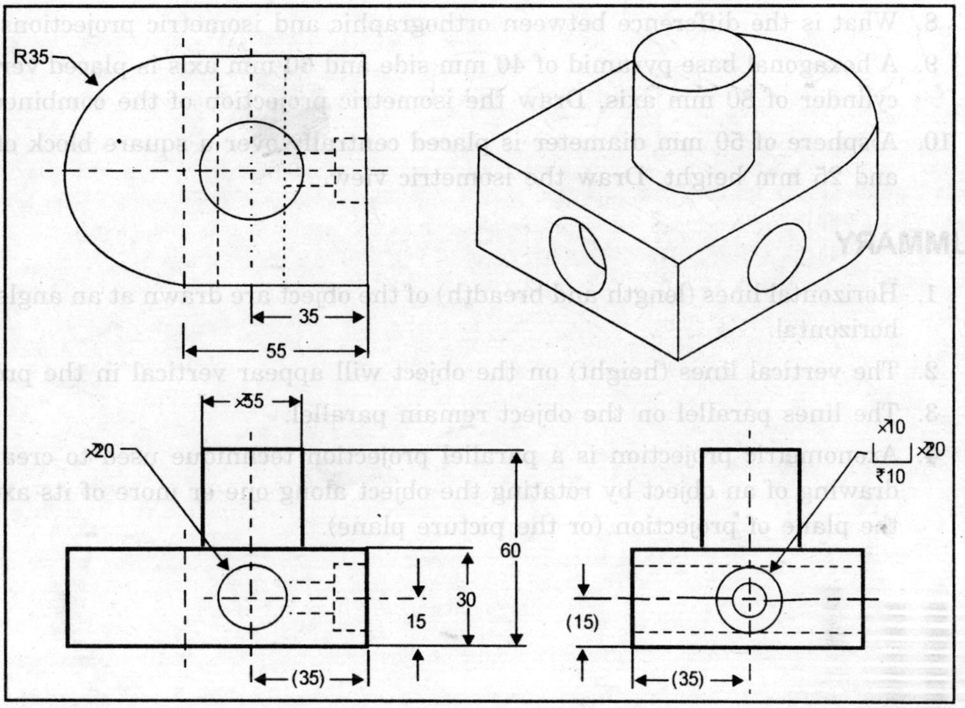

EXERCISE

1. What are the types of isometric projections?
2. A pentagonal pyramid, with edge of base 35 mm and 60 mm height stands on HP such that an edge of the base is parallel to VP. It is cut by a plane perpendicular to VP and inclined at 30° to HP cuts the pyramid passing through a point on the axis at a height to 30 mm from the base. Draw the isometric view of the truncated pyramid showing the cut surface.
3. An object consists of a hemispherical vessel of 60 mm diameter which is placed centrally over a truncated cone of 70 mm bottom base diameter, 40 mm top face diameter and height of 50 mm. The cylinder in turn is placed centrally over a square prism of 60 mm base side and 20 mm height. Draw the isometric projection of the object.
4. A cone of diameter 40 mm base and height 60 mm rests centrally on top of a square block of 80 mm side and 20 mm thick. Draw the isometric projection of the two solids.
5. Draw an isometric projection of a Hexagonal prism of base edges 25 mm. The height of the Hexagonal prism is 50 mm, having two of its rectangle faces parallel to V.P. and axis perpendicular to H.P.
6. Draw an isometric projection of a Hexagonal pyramid of base edges = 30 mm. The

height of the Hexagonal pyramid is 70 mm standing on H.P.

7. Sketch a closed cell phone in isometric view.
8. What is the difference between orthographic and isometric projections?
9. A hexagonal base pyramid of 40 mm side and 60 mm axis is placed vertically over a cylinder of 80 mm axis. Draw the isometric projection of the combined solids.
10. A sphere of 50 mm diameter is placed centrally over a square block of 80 mm side and 25 mm height. Draw the isometric view.

SUMMARY

1. Horizontal lines (length and breadth) of the object are drawn at an angle 30° with the horizontal.
2. The vertical lines (height) on the object will appear vertical in the projection.
3. The lines parallel on the object remain parallel.
4. Axonometric projection is a parallel projection technique used to create a pictorial drawing of an object by rotating the object along one or more of its axes relative to the plane of projection (or the picture plane).

CHAPTER

6

OVERVIEW OF COMPUTER AIDED DESIGN

Learning Objectives

- To Learn Basic working knowledge of AutoCAD
- Understanding the concepts and techniques with AutoCAD
- Understanding the AutoCAD workspace and user interface
- To develop a capability of both hand sketching and computer aided drawing in engineering area.
- To Use common drafting tools to construct engineering drawings

6.1 COMPUTER AIDED DESIGN

Computer-aided design (CAD) is the use of computer systems (or workstations) to aid in the creation, modification, analysis, or optimization of a design. CAD software is used to increase the productivity of the designer, improve the quality of design, improve communications through documentation, and to create a database for manufacturing. CAD output is often in the form of electronic files for print, machining, or other manufacturing operations. The term CADD (for Computer Aided Design and Drafting) is also used.

Its use in designing electronic systems is known as electronic design automation, or EDA. In mechanical design it is known as mechanical design automation (MDA) or computer-aided drafting (CAD), which includes the process of creating a technical drawing with the use of computer software.

CAD software for mechanical design uses either vector-based graphics to depict the objects of traditional drafting, or may also produce raster graphics showing the overall appearance of designed objects. However, it involves more than just shapes. As in the manual drafting of technical and engineering drawings, the output of CAD must convey information, such as materials, processes, dimensions, and tolerances, according to application-specific conventions.

6.2 OVERVIEW OF CAD SOFTWARE

Starting around the mid 1960s, with the IBM Drafting System, computer-aided design systems began to provide more capability than just an ability to reproduce manual drafting with electronic drafting, the cost-benefit for companies to switch to CAD became apparent. The benefits of CAD systems over manual drafting are the capabilities one often takes for granted from computer systems today; automated generation of Bill of Material, auto layout in integrated circuits, interference checking, and many others. Eventually, CAD provided the designer with the ability to perform engineering calculations. During this transition, calculations were still performed either by hand or by those individuals who could run computer programs. CAD was a revolutionary change in the engineering industry, where draftsmen, designers and engineering roles begin to merge. It did not eliminate departments, as much as it merged departments and empowered draftsman, designers and engineers. CAD is just another example of the pervasive effect computers were beginning to have on industry. Current computer-aided design software packages range from 2D vector-based drafting systems to 3D solid and surface modelers. Modern CAD packages can also frequently allow rotations in three dimensions, allowing viewing of a designed object from any desired angle, even from the inside looking out. Some CAD software is capable of dynamic mathematical modeling.

CAD technology is used in the design of tools and machinery and in the drafting and design of all types of buildings, from small residential types (houses) to the largest commercial and industrial structures (hospitals and factories).

CAD is mainly used for detailed engineering of 3D models or 2D drawings of physical components, but it is also used throughout the engineering process from conceptual design and layout of products, through strength and dynamic analysis of assemblies to definition of manufacturing methods of components. It can also be used to design objects such as jewelry, furniture, appliances, etc. Furthermore, many CAD applications now offer advanced rendering and animation capabilities so engineers can better visualize their product designs. 4D BIM is a type of virtual construction engineering simulation incorporating time or schedule related information for project management.

CAD has become an especially important technology within the scope of computer-aided technologies, with benefits such as lower product development costs and a greatly shortened design cycle. CAD enables designers to layout and develop work on screen, print it out and save it for future editing, saving time on their drawings.

6.3 INTRODUCTION AUTOCAD

AutoCAD is a software package used for drafting and designing almost anything you wish. The software has evolved massively over the years and now offers various versions catering for the specific requirements of different engineering disciplines, such as mechanical engineering, civil engineering and architecture.

Basically AutoCAD provides a 3D space (model space) in which you construct or draw your design. It provides drawing layout tools which enable you to prepare and plot conventional 2D drawings of your model. More recent versions also include advanced rendering tools, enabling you to create photo-realistic images of your design.

6.3.1 The interface

Standard Windows type command buttons, pull down menus etc. are provided to use AutoCAD. One strong feature of AutoCAD is the continued support of its **command line**. Any function or command you can give AutoCAD can be given by typing in the appropriate command on the screen. AutoCAD supports legacy commands, from previous versions, so learning to use AutoCAD through the command line means you should be able to use any subsequent or previous releases of the software without getting lost in the menu systems.

This manual will encourage you to use the command line but may also involve use of tool bars and pull down menus.

6.3.2 AutoCAD Display screen

(*a*) AutoCAD classic mode

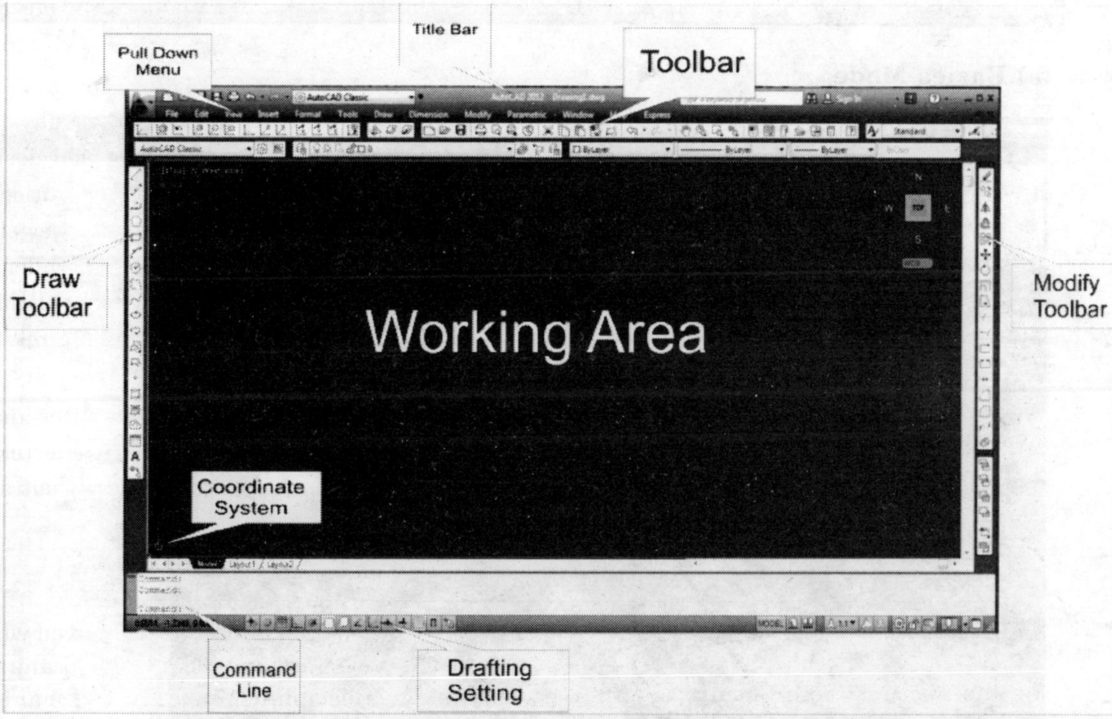

(b) Modeling Mode

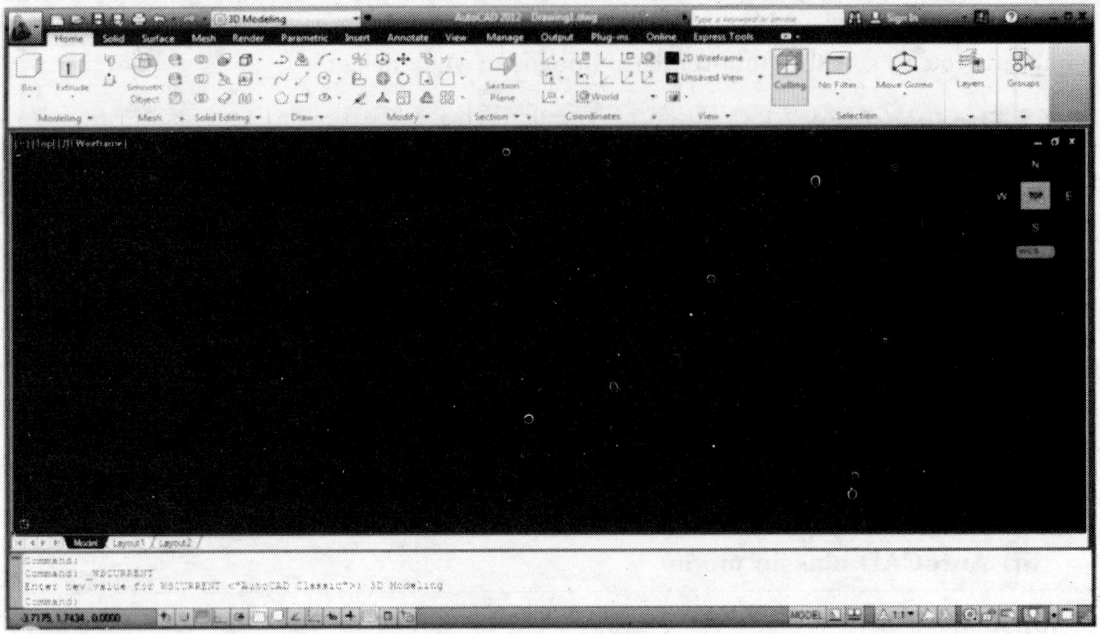

(c) Basics Mode

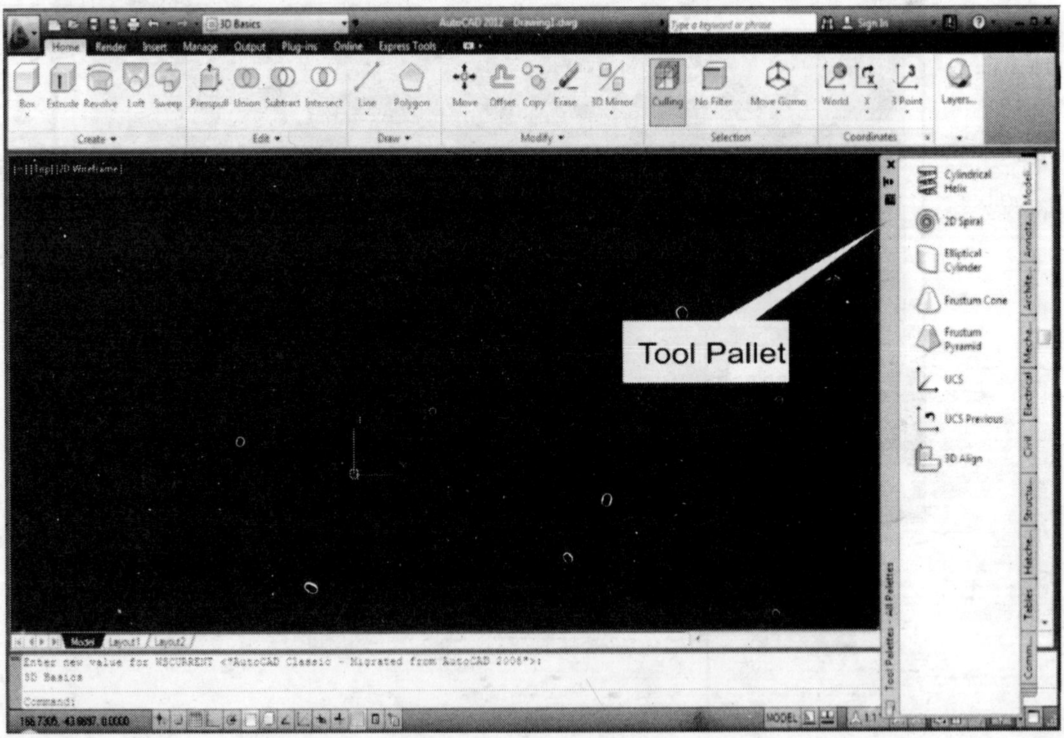

(*d*) Drafting Annotation Mode

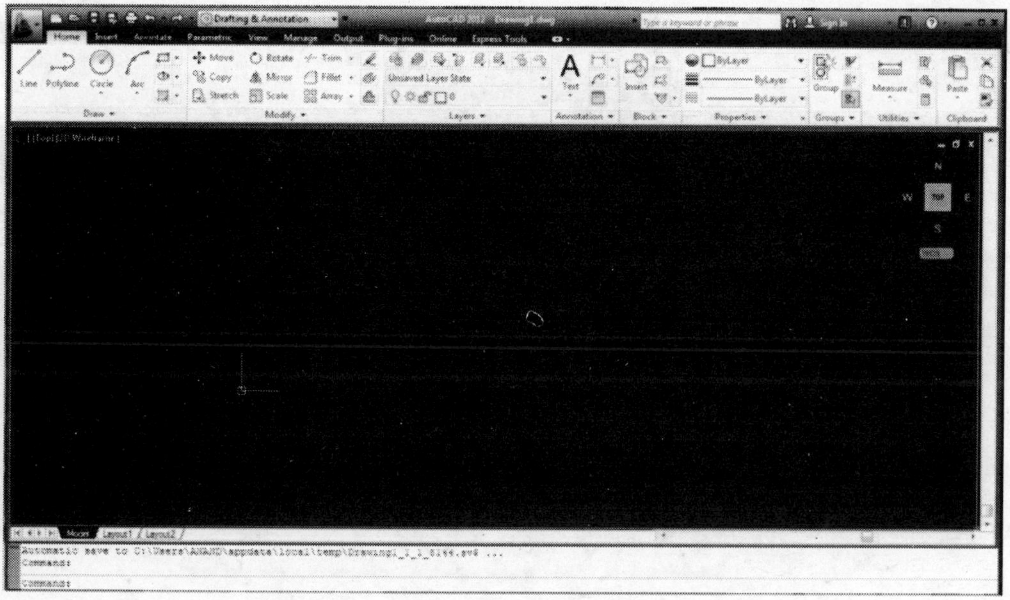

(*e*) Search for Commands: Perform a real-time search for commands on the Quick Access toolbar, in the application menu, and on the ribbon.

The Search field displays at the top of the application menu. Search results can include menu commands, basic tooltips, and command prompt text strings. You can enter a search term in any language.

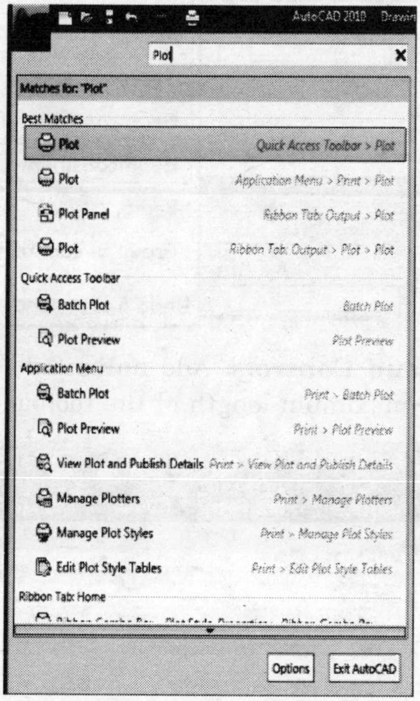

***(f)* Quick Access Toolbar:** Display frequently used tools with the Quick Access toolbar.

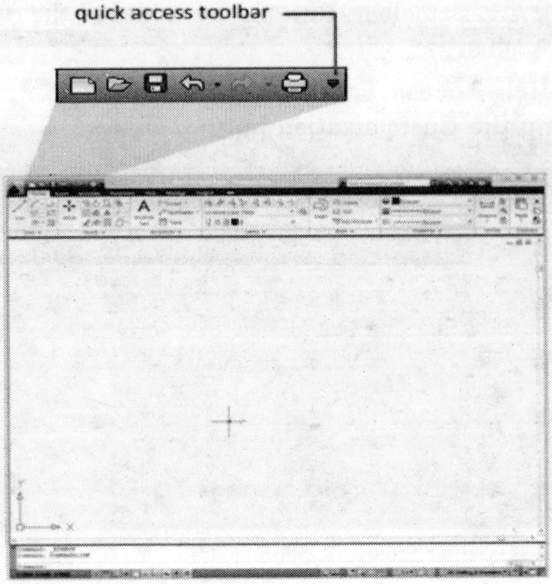

***(g)* View Undo and Redo History:** The Quick Access toolbar displays options to undo and redo changes to your file. To undo or redo a less recent change, click the drop-down button to the right of the Undo and Redo buttons.

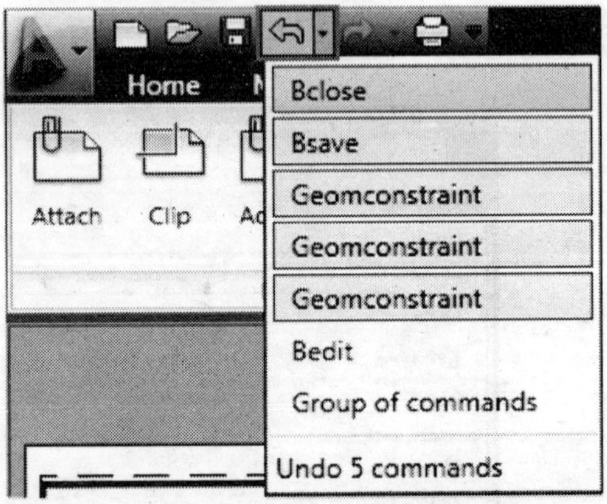

***(h)* Add Commands and Controls:** Add unlimited tools to the Quick Access toolbar. Tools that extend past the maximum length of the toolbar are displayed in a flyout button.

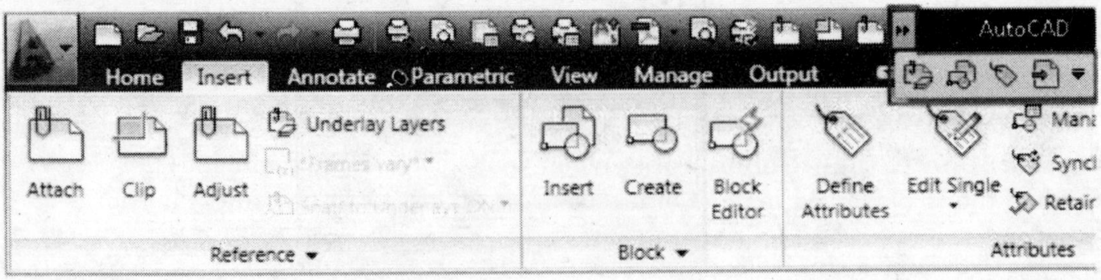

To add a ribbon button to the Quick Access toolbar, right-click the button on the ribbon and click Add to Quick Access toolbar. Buttons are added to the right of the default commands on the Quick Access toolbar.

***(i)* Move the Quick Access Toolbar:** Place the Quick Access toolbar either above or below the ribbon using the Customization button.

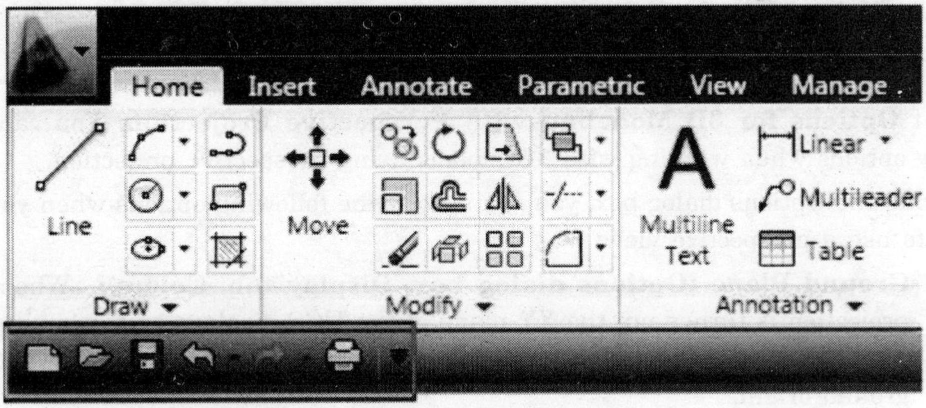

***(j)* Expanded Panels:** An arrow to the right of a panel title indicates that you can expand the panel to display additional tools and controls. By default, an expanded panel closes automatically when you click another panel. To keep a panel expanded, click the push pin icon in the bottom-left corner of the expanded panel.

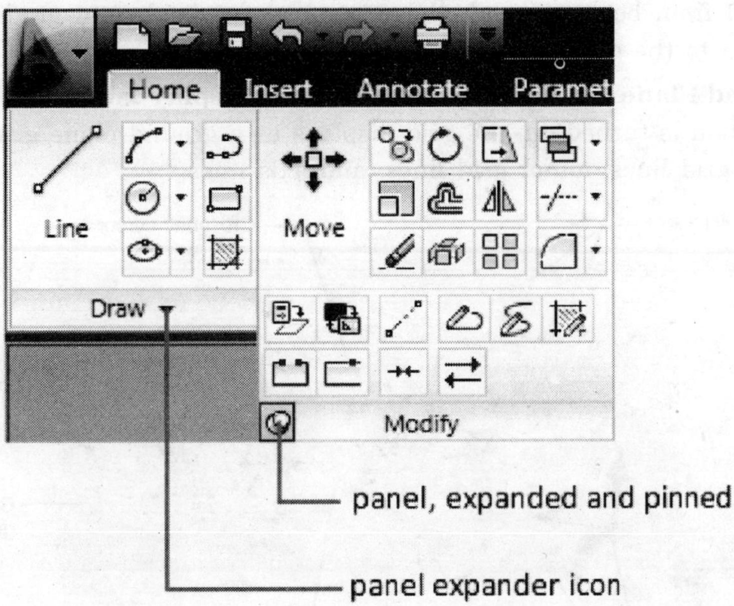

***(k)* Contextual Ribbon Tabs**: When you execute some commands, a special contextual ribbon tab is displayed instead of a toolbar or dialog box. The contextual tab is closed when you end the command.

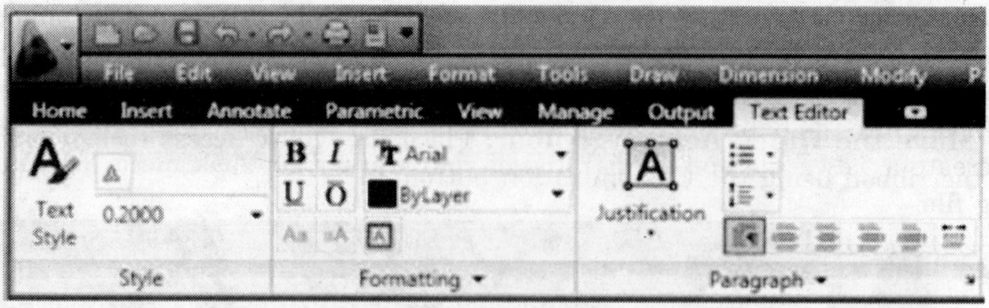

Set Options for 3D Modeling with Perspective Projection: You can set specific display options when working with 3D models using perspective projection.

Using the Options dialog box, you can specify the following options when your 3D model is set to use a perspective view:

- **Ground Plane (Options dialog box, Display tab, Colors):** When perspective projection is turned on, the XY plane of the UCS displays as a ground plane with a gradient color. The ground plane displays a gradient from the ground horizon to the ground origin.
- **Sky (Options dialog box, Display tab, Colors):** The area not covered by the ground plane is the sky, which displays a gradient color from the sky horizon to the sky zenith.
- **Underground (Options dialog box, Display tab, Colors):** If the ground plane is viewed from below ground, the ground plane displays a gradient from the earth horizon to the earth azimuth.
- **Ground Plane Grid (Options dialog box, Display tab, Colors):** When perspective projection is turned on, the grid displays as a ground plane grid. Colors are set for major grid lines, minor grid lines, and axis lines.

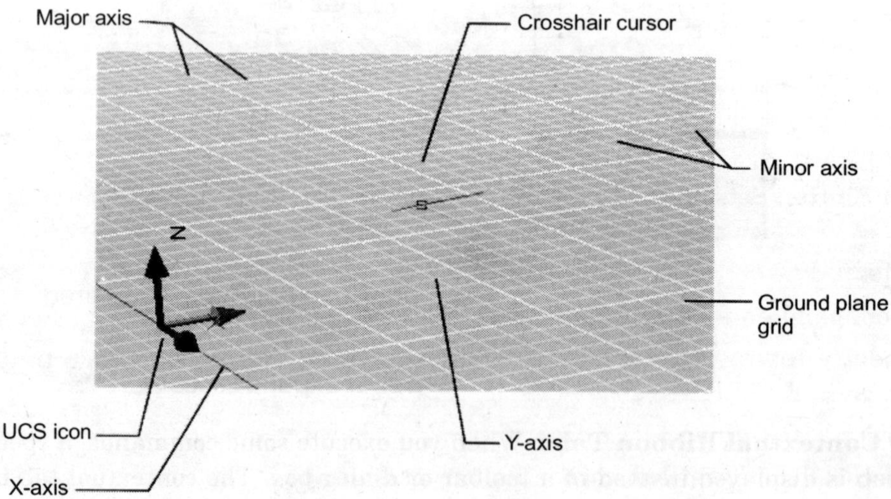

***(l)* Saving your AutoCAD work:** You are responsible for the safe keeping of your work with AutoCAD. Work saved on your network U:\ drive is safe, but the drive is limited in storage capacity. You may wish to use other removable media such as a USB flash stick (pen drive). Get into the habit of creating a folder for each project. When you save work, AutoCAD will create a *.dwg file and a *.bak file of your work. The *.bak file is a backup copy of the *.dwg file.

6.4 CREATING A NEW DRAWING

In this section you will use AutoCAD to create and set up a new drawing file. You will also become familiar with the toolbars, menus and command line.

***(a)* Activating AutoCAD:** Only certain rooms contain computers with access to AutoCAD. You will find a shortcut installed under the University Software list, under A for AutoDesk. Click the appropriate shortcut and wait for the program to load. Video clips will require use of headphones.

***(b)* Limits:** Sets and controls the limits of the grid display in the current Model or layout tab.

Menu: Format ➢ Drawing Limits

Command entry: ***limits*** (or ***'limits*** for transparent use)

Specify lower left corner or [ON/ OFF] <*current*>: Specify a point, enter ***on*** *or* ***off****, or press* ENTER

Lower-Left Corner

Specifies the lower-left corner for the grid limits

Specify upper right corner <*current*>: *Specify a point or press* ENTER

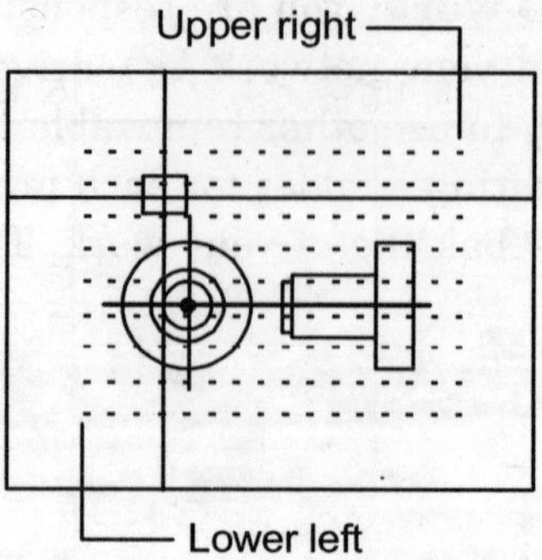

On: Turns on limits checking when limits checking are on, you cannot enter points outside the grid limits. Because limits checking tests only points that you enter, portions of objects such as circles can extend outside the grid limits.

Off: Turns off limits checking but maintains the current values for the next time you turn on limits checking.

(*c*) MVSETUP: When the TILEMODE system variable is on (the default), the following prompt is displayed:

Enable paper space? [No/Yes] <Y>: *Enter n or press Enter*

Pressing Enter turns off TILEMODE and proceeds as described in the following section, MVSETUP on a Layout Tab

Entering n displays the following prompt:

Enter units type [Scientific/Decimal/Engineering/Architectural/Metric]: *Enter an option*

A list of available units and prompts for the scale factor and paper size are displayed.

Enter the scale factor: ***Enter a value***

Enter the paper width: ***Enter a value***

Enter the paper height: ***Enter a value***

A bounding box is drawn and the command ends.

6.5 CO-ORDINATE SYSTEM

There are three type of co-ordinate system

(*a*) Absolute co-ordinate system

(*b*) Relative rectangular co-ordinate system

(*c*) Relative polar co-ordinate system

Absolute co-ordinate system: In this type of coordinate system, the coordinate of all point taken from the origin.

Syntax: *x*, *y*

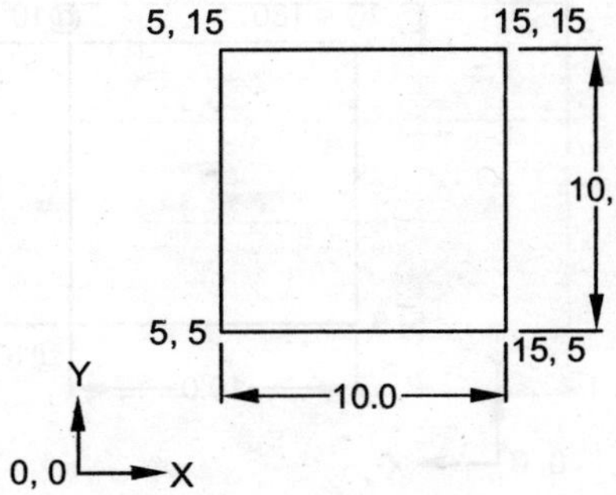

For the figure:

Command: line or L (Enter)

Line specifies first point: 5, 5 (Enter)

Line specifies Next point: 1 5, 5 (Enter)

Line specifies Next point: 15, 15 (Enter)

Line specifies Next point: 5, 15 (Enter)

Line specifies Next point: 5, 5 (Enter) (Enter)

Relative Rectangular coordinate system: In this type of coordinate system, the coordinate of next point is taken from previous point.

Syntax: @x, y

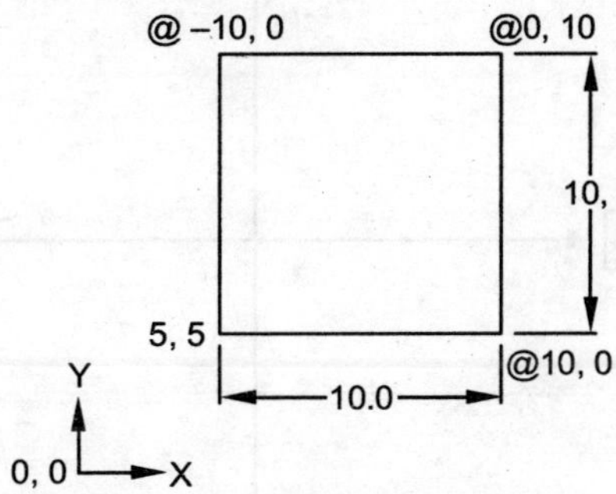

For the figure:

Command: line or L (Enter)

Line specifies first point: 5, 5 (Enter)

Line specifies Next point: @10,0 (Enter)

Line specifies Next point: @0,10(Enter)

Line specifies Next point: @-10,0 (Enter)

Line specifies Next point: 5, 5 (Enter) (Enter)

Relative polar co-ordinate system: In this system, we can draw the angular line from the specific point.

Syntax: @distance<?

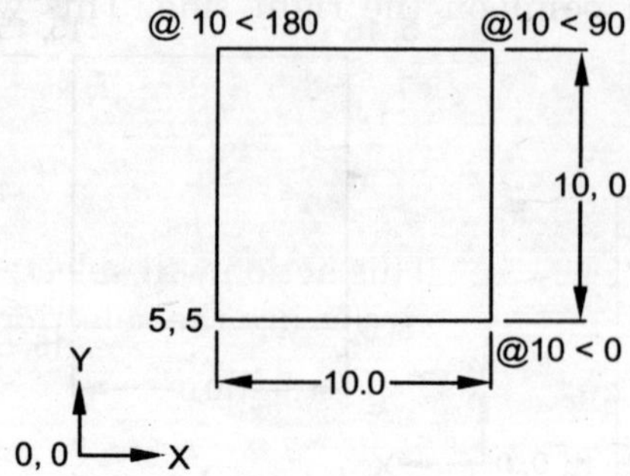

For the figure:

Command: line or L (Enter)

Line specifies first point: 5, 5 (Enter)

Line specifies Next point: @10 < 0 (Enter)

Line specifies Next point: @10 < 90(Enter)

Line specifies Next point: @10 < 180 (Enter)

Line specifies Next point:@10 < 270 or 5, 5 (Enter) (Enter)

Angle direction:

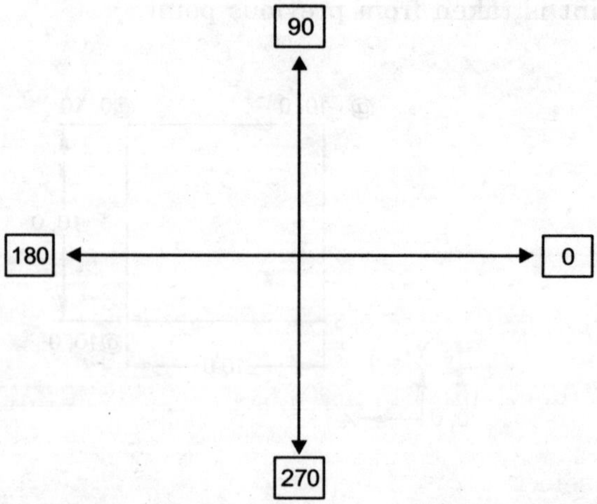

6.6 WINDOW SELECTION

1. Pick Single: You can pick single object when AutoCAD ask you to select objects.

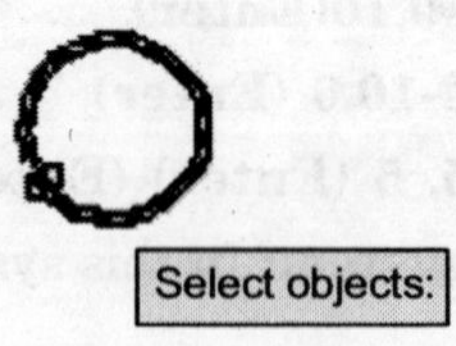

2. Window: You can define a window selection by clicking at two points. Click first point on the left side, and 2nd point on the right side. This will select all elements **inside** the window.

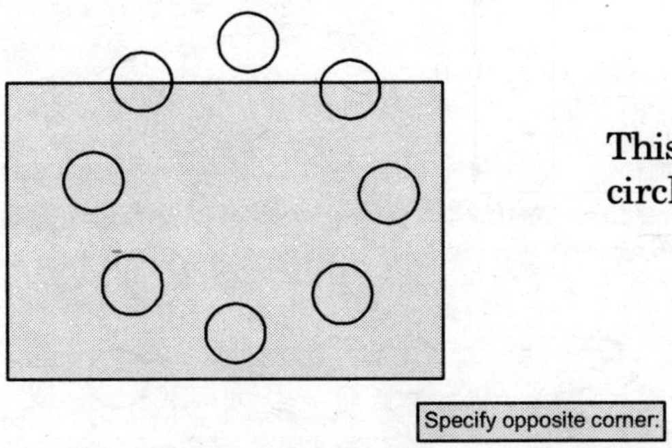

This action will select: (dotted circles are the selection

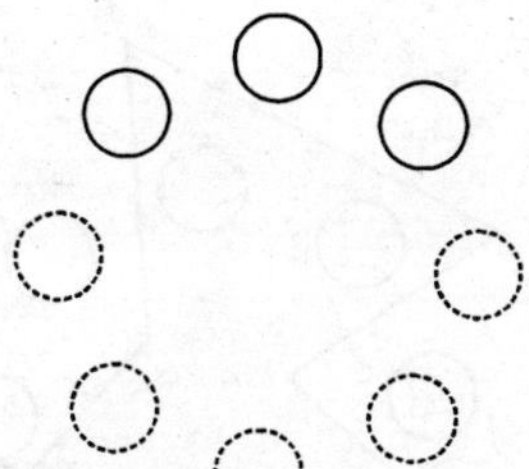

If you pick your points right then left, then it will be cross window selection (will be explained next). This two are the default window selection.

3. Cross Window: Basically if your window selection defined from right to left, then it will be cross window selection. It will select all elements **inside and touching** the window.

This action will select:

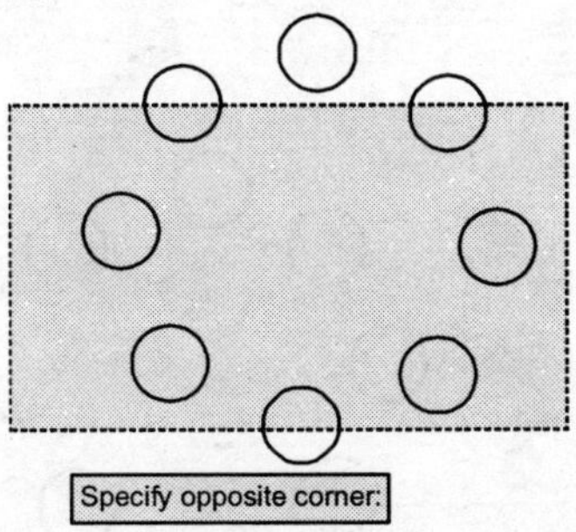

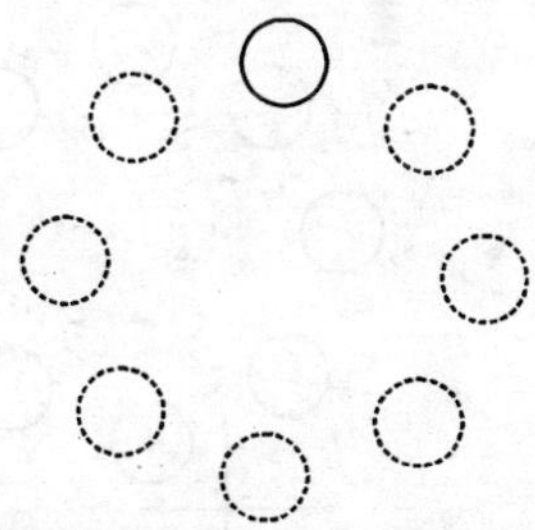

You can override to activate cross window selection by typing C then [enter].

4. Fence: You can select drawing elements by defining 'a fence' that touched objects you wanted. Just type F then [enter] to activate it. After you finish defining your fence, press [enter] to select.

This action will select:

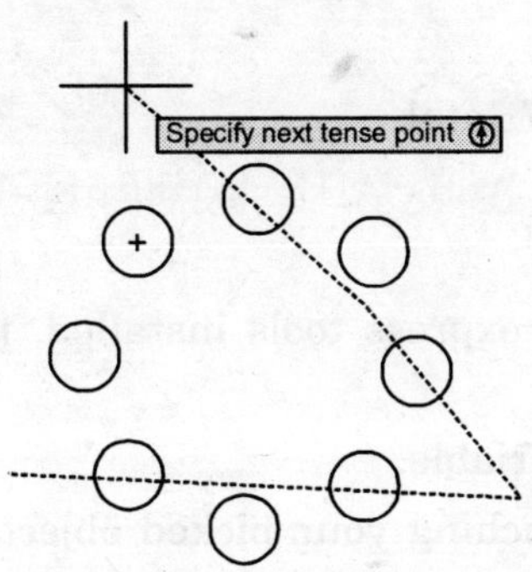

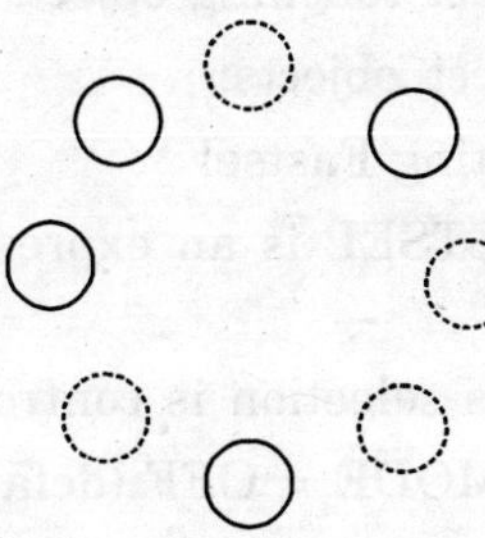

5. Window Polygon: This option behave as window selection, but you can define unlimited points for that window. Just type WP then [enter] to activate it. All elements **inside** the window will be selected.

This action will select:

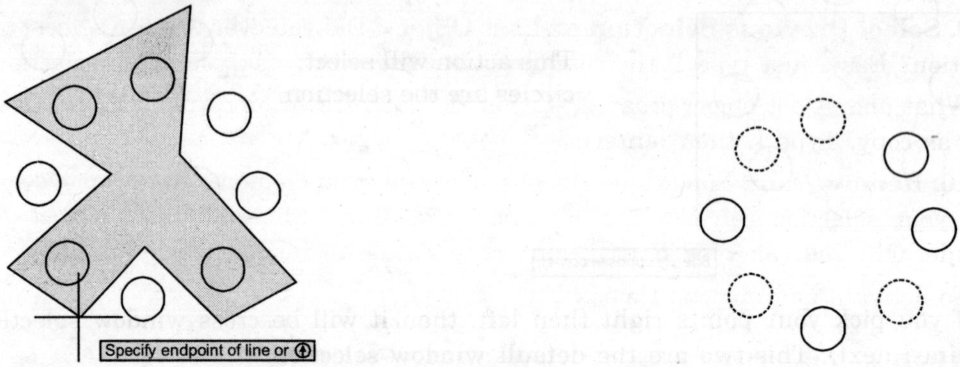

6. Crossing Polygon: Similar to cross window selection. CP

This action will select:

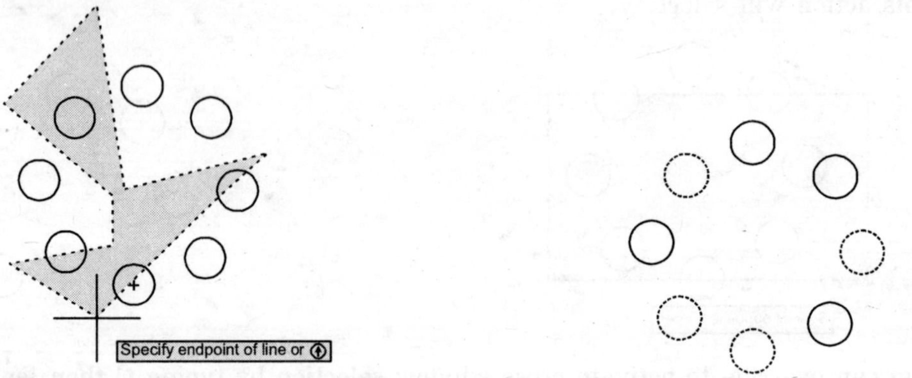

7. Touching Objects: When AutoCAD ask you to select object, type 'FS then [enter]. We use' for transparent use. Pick your object, then all objects touching it will be selected.

Select objects: 'FS

Use 'FSMODE to control chain selection.

FSMODE = OFF

Select touching object: 3 object(s) found. 3 found, 7 total

Select objects:

Exiting Fastsel

FASTSEL is an express tool. So if you don't have express tools installed, it might not work.

This selection is controlled by FSMODE system variable.

FSMODE = OFF (default), it will select objects touching your picked object only.

FSMODE = ON, it will select objects touching it, and all objects touching them too.

8. Cycle Overlapping Objects: If you have many overlapped drawing elements or close to each other, and you are too lazy to zoom in, you can cycle between them by pressing [shift] + [space]. Hold your [shift] key, and press [space] several times until you found your object. Release your keyboard key, then click your mouse.

9. Select Previous Selection or Last Object: Did you ever want to select your previous selection? Easy, just type P then [enter] when AutoCAD ask you to do selection!

What about last object created? Either the last drawn or result from modify operations such as copy. Type L then [enter]!

10. Remove/Add: Sometimes it's easier to select all elements then remove some of them from your selection set. You can do this by holding [shift] and select object you want to exclude. OR, you can type R then [enter] to switch to remove mode.

To activate add objects, type A then [enter].

6.7 UNITS (OR 'UNITS FOR TRANSPARENT USE)

Controls the displayed precision and format for coordinates and angles.

Length: Specifies the current unit of measurement and the precision for the current units.

Type: Sets the current format for units of measure. The values include Architectural, Decimal, Engineering, Fractional, and Scientific. The Engineering and Architectural formats produce feet-and-inches displays and assume that each drawing unit represents one inch. The other formats can represent any real-world unit.

Precision: Sets the number of decimal places or fractional size displayed for linear measurements.

Angle: Specifies the current angle format and the precision for the current angle display.

Type: Sets the current angle format.

Precision: Sets the precision for the current angle display.

The following conventions are used for the various angle measures: decimal degrees appear as decimal numbers, grads appear with a lowercase *g* suffix, and radians appear with a lowercase *r* suffix. The degrees/minutes/seconds format uses *d* for degrees, 'for minutes, and" for seconds; for example:

123d45'56.7"

Surveyor's units show angles as bearings, using N or S for north or south, degrees/ minutes/seconds for how far east or west the angle is from direct north or south, and E or W for east or west; for example:

N 45d0'0" E

The angle is always less than 90 degrees and is displayed in the degrees/minutes/seconds format. If the angle is precisely north, south, east, or west, only the single letter representing the compass point is displayed.

Clockwise: Calculates positive angles in the clockwise direction. The default direction for positive angles is counterclockwise.

When prompted for an angle, you can point in the desired direction or enter an angle regardless of the setting specified for Clockwise.

Insertion Scale: Controls the unit of measurement for blocks and drawings that are inserted into the current drawing. A block or a drawing that is created with units that are different from the units specified with this option is scaled when inserted. The insertion scale is the ratio of the units used in the source block or drawing and the units used in the target drawing. Select Unitless to insert the block without scaling it to match the specified units.

Note: In the Options dialog box, User Preferences tab, the Source Content Units and Target Drawing Units settings are used when, either in the source block or the target drawing, Insertion Scale is set to Unitless.

Sample Output: Displays an example of the current settings for units and angles.

Lighting: Controls the unit of measurement for the intensity of photometric lights in the current drawing.

Note: In order to create and use photometric lights, a unit other than Generic must be specified from the list of options. If Insertion Scale is set to Unitless, a warning message is displayed informing you that the rendered output might not be as expected.

EXERCISE

1. How you can create a user interface in Autocad?
2. What are coordinates systems used in AutoCAD?
3. Explain the method of object selection method.
4. How to set the unit of drawing.

SUMMARY

- Computer-aided design (CAD) is the use of computer systems (or workstations) to aid in the creation, modification, analysis, or optimization of a design.
- The benefits of CAD systems over manual drafting are the capabilities one often takes for granted from computer systems today; automated generation of Bill of Material, auto layout in integrated circuits, interference checking, and many others.
- AutoCAD provides a 3D space (model space) in which you construct or draw your design. It provides drawing layout tools which enable you to prepare and plot conventional 2D drawings of your model.
- Standard Windows type command buttons, pull down menus etc. are provided to use AutoCAD. One strong feature of AutoCAD is the continued support of its command line. Any function or command you can give AutoCAD can be given by typing in the appropriate command on the screen.
- There are three type of co-ordinate system Absolute co-ordinate system Relative rectangular co-ordinate system and Relative polar co-ordinate system.

CHAPTER

7

DRAWING USING AUTOCAD

Learning Objectives

- Using basic drawing, editing, and viewing tools in AutoCAD
- Create a basic 2D drawing in AutoCAD
- Create Lines , circles using AutoCAD
- Create arcs and rectangles using AutoCAD

7.1 DRAWING A LINE USING AUTOCAD

Specify first point: *Specify a point or press* ENTER *to* continue *from the last drawn line or arc*

Specify next point or [Close/ Undo]:

Continue: Continues a line from the endpoint of the most recently drawn line.

before pressing ENTER

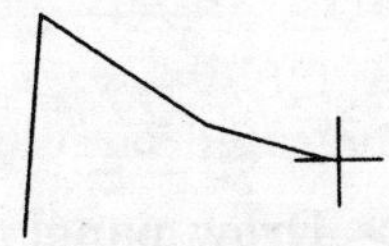

after pressing ENTER

If the most recently drawn object is an arc, its endpoint defines the starting point of the line, and the line is drawn tangent to the arc.

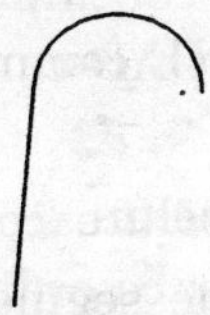

before pressing ENTER

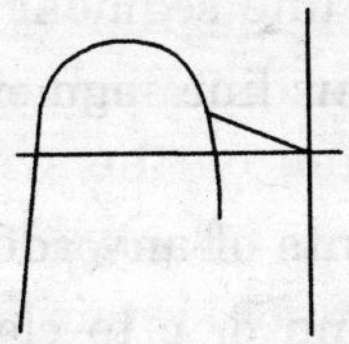

after pressing ENTER

Close: Ends the last line segment at the beginning of the first line segment, which forms a closed loop of line segments. You can use Close after you have drawn a series of two or more segments.

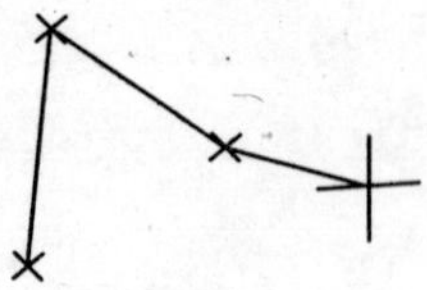

before entering

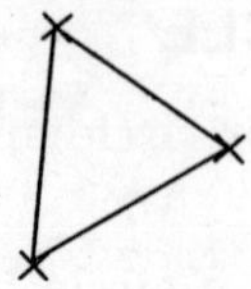

after entering

Undo: Erases the most recent segment of a line sequence.

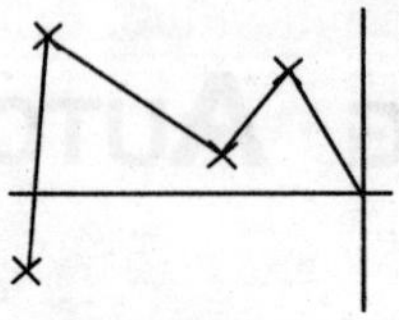

before entering

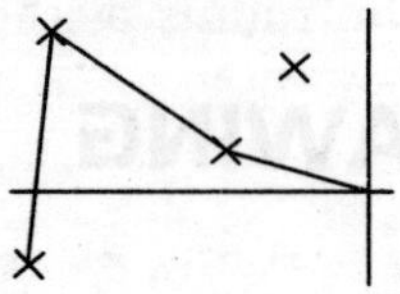

after entering

Entering *u* more than once backtracks through line segments in the order you created them.

With LINE, you can create a series of continuous line segments. Each segment is a line object that can be edited separately.

Specify the locations that define the endpoints of each line with precision.

- Enter the coordinate values for an endpoint, using either absolute or relative coordinates
- Specify an object snap relative to an existing object. For example, you can specify the center of a circle as one endpoint of the line
- Turn grid snap on and snap to a location

There are other methods for creating precise lines. A highly efficient technique is to offset a line from an existing line, and then trim or extend it to the desired length.

Use polyline objects instead of line objects if you want the segments to be connected as a single object.

To draw lines

1. Click Home tab ➤ Draw panel ➤ Line .
2. Specify the start point.
   You can use the pointing device or enter coordinate values at the command prompt.
3. Complete the first line segment by specifying the endpoint.
   To undo the previous line segment during the LINE command, enter u or click Undo on the toolbar.
4. Specify the endpoints of any additional line segments.
5. Press ENTER to end or c to close a series of line segments.

To start a new line at the endpoint of the last line drawn, start the LINE command again and press ENTER at the Specify Start Point prompt.

7.2 DRAWING A CIRCLE

Specify <u>center point</u> for circle or [<u>3P</u>/ <u>2P</u>/ <u>Ttr</u> (<u>tan tan radius</u>)]: *Specify a point or enter an option*

Centre Point: Draws a circle based on a center point and a diameter or a radius.

Specify radius of circle or [Diameter]: *Specify a point, enter a value, enter d, or press* Enter

Radius: Defines the radius of the circle. Enter a value, or specify a point.

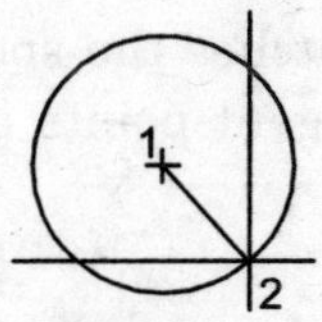

radius

Diameter: Defines the diameter of the circle. Enter a value, or specify a second point.

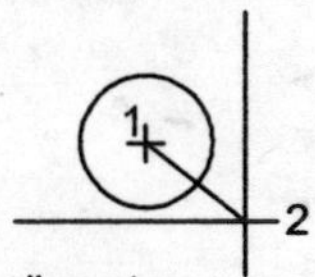

diameter

Specify diameter of circle *<current>: Specify a point (2), enter a value, or press* Enter

3P (three Point)

Draws a circle based on three points on the circumference.

Specify first point on circle: *Specify a point (1)*

Specify second point on circle: *Specify a point (2)*

Specify third point on circle: *Specify a point (3)*

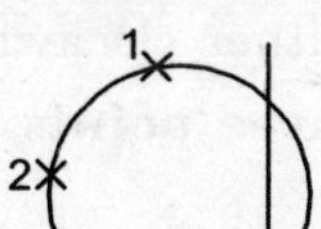

3P

2P (two Points)

Draws a circle based on two endpoints of the diameter.

Specify first endpoint of circle's diameter: *Specify a point (1)*

Specify second endpoint of circle's diameter: *Specify a point (2)*

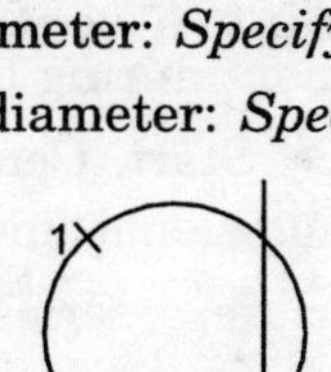

2P

TTR (Tangent, Tangent, Radius)

Draws a circle with a specified radius tangent to two objects.

Specify point on object for first tangent of circle: *Select a circle, arc, or line*

Specify point on object for second tangent of circle: *Select a circle, arc, or line*

Specify radius of circle *<current>*:

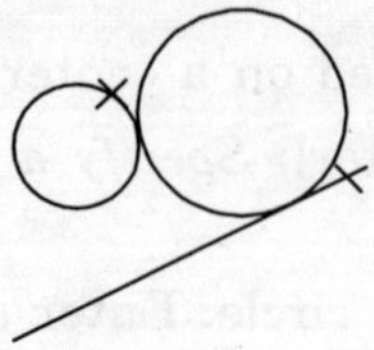

Sometimes more than one circle matches the specified criteria. The program draws the circle of the specified radius whose tangent points are closest to the selected points.

7.3 DRAWING A ARC

Specifies the starting point of the arc.

Second Point

Draws an arc using three specified points on the arc's circumference. The first point is the start point (1). The third point is the endpoint (3). The second point (2) is a point on the circumference of the arc.

Specify end point of arc: *Specify a point (3)*

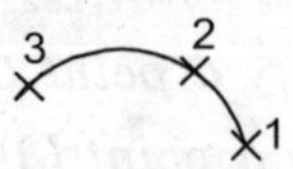

You can specify a three-point arc either clockwise or counterclockwise.

To draw an arc by specifying three points

1. Click Home tab ➤ Draw panel ➤ 3-Point. 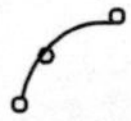
2. Specify the start point.
3. Specify a point on the arc.
4. Specify the endpoint.

To draw an arc using a start point, a center point, and an endpoint

1. Click Home tab ➤ Draw panel ➤ Start, Center, End. 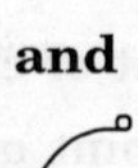
2. Specify a start point.
3. Specify the center point.
4. Specify the endpoint.

To continue an arc with a tangential line

1. Complete the arc.
2. Click Home tab ➤ Draw panel ➤ Line. 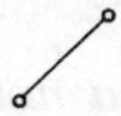
3. Press ENTER at the first prompt.
4. Enter the length of the line and press ENTER.

To continue an arc with a tangential arc

1. Complete the arc.
2. Click Home tab ➤ Draw panel ➤ Continue. 

Specify the second endpoint of the tangent arc.

7.4 DRAWING A RECTANGLE

Current settings: Rotation = 0

Specify first corner point or [Chamfer/ Elevation/ Fillet/ Thickness/ Width]: *Specify a point or enter an option*

With this command, you can specify the rectangle parameters (length, width, rotation) and control the type of corners (fillet, chamfer, or square).

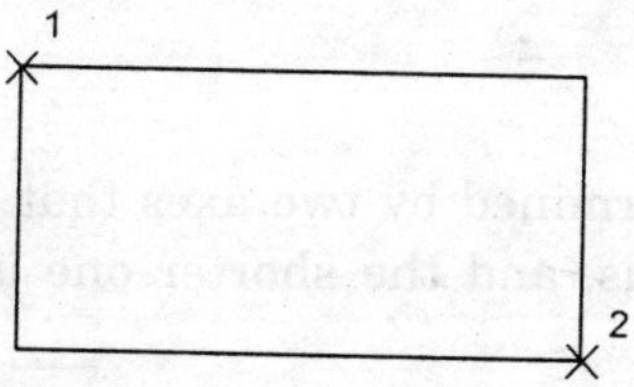

First Corner Point

Specifies a corner point of the rectangle.

Specify other corner point or [Area/Dimensions/Rotation]: *Specify a point or enter an option.*

Other Corner Point

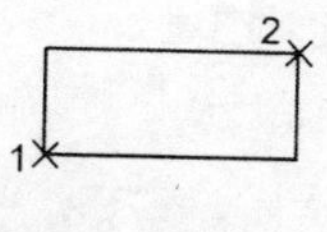

Creates a rectangle using the specified points as diagonally opposite corners.

Area

Creates a rectangle using the area and either a length or a width. If the Chamfer or Fillet option is active, the area includes the effect of the chamfers or fillets on the corners of the rectangle.

Enter area of rectangle in current units <100>: *Enter a positive value.*

Calculate rectangle dimensions based on [Length/Width] <Length>: *Enter L or w*

Enter rectangle length <10>: *Enter a non-zero value*

or

Enter rectangle width <10>: *Enter a non-zero value*

Specify other corner point or [Area/Dimensions/Rotation]: *Move the cursor to display one of four possible locations for the rectangle and click the one that you want*

Dimensions

Creates a rectangle using length and width values.

Specify length for rectangles <0.0000> *Enter a non-zero value*

Specify width for rectangles <0.0000> *Enter a non-zero value*

Specify other corner point or [Area/Dimensions/Rotation]: *Move the cursor to display one of four possible locations for the rectangle and click the one that you want.*

Rotation

Creates a rectangle at a specified rotation angle.

Specify rotation angle or [Points] <0> Specify an angle by entering a value, specifying a point, or entering p and specifying two points

Specify other corner point or [Area/Dimensions/Rotation]: *Move the cursor to display one of four possible locations for the* rectangle *and click the one that you want*

7.5 DRAWING A ELLIPSE

The shape of an ellipse is determined by two axes that define its length and width. The longer axis is called the major axis, and the shorter one is the minor axis.

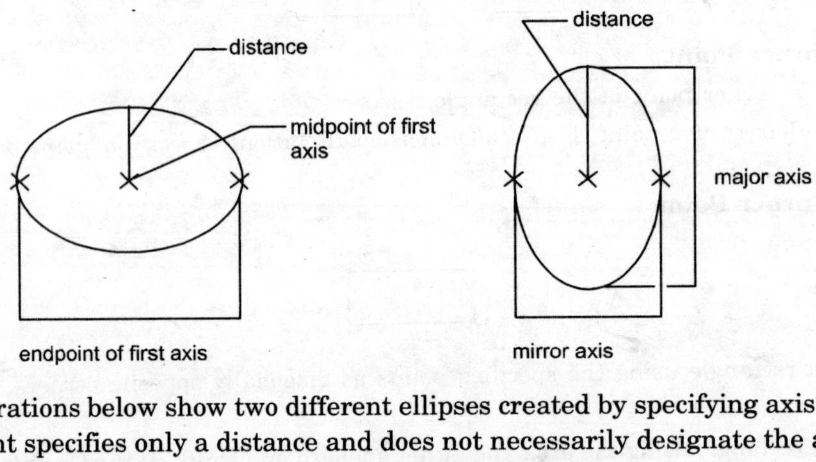

The illustrations below show two different ellipses created by specifying axis and distance. The third point specifies only a distance and does not necessarily designate the axis endpoint.

To draw a true ellipse using endpoints and distance

1. Click Home tab ➤ Draw panel ➤ Axis, End. 
2. Specify the first endpoint of the first axis (1).
3. Specify the second endpoint of the first axis (2).
4. Drag the pointing device away from the midpoint, and click to specify a distance (3) for half the length of the second axis.

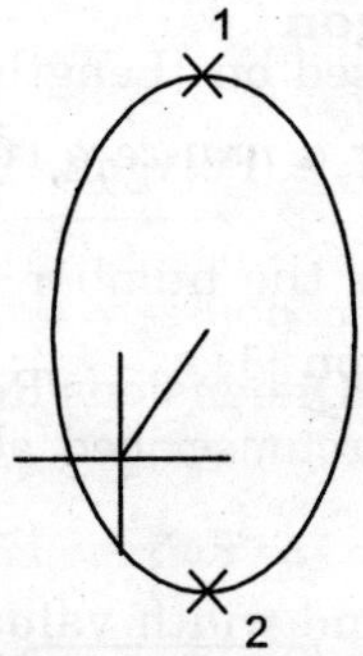

To draw an elliptical arc using start and end angles

1. Click Home tab ➤ Draw panel ➤ Arc.
2. Specify endpoints for the first axis (1 and 2).
3. Specify a distance to define half the length of the second axis (3).
4. Specify the start angle (4).
5. Specify the end angle (5).

The elliptical arc is drawn counterclockwise between the start point and endpoint.

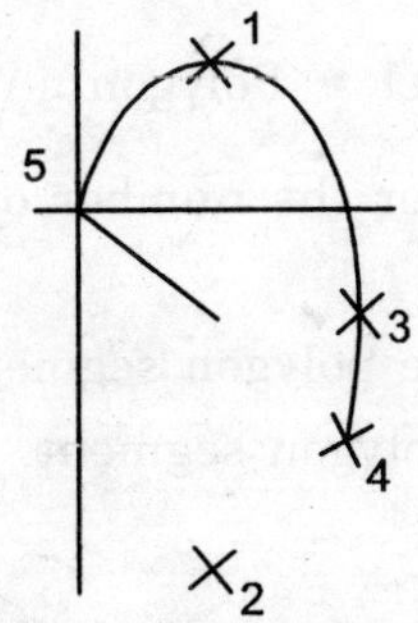

7.6 DRAW REGULAR POLYGONS

Use POLYGON to create closed polylines with between 3 and 1,024 equal-length sides. The following illustrations show polygons created using three methods. In each case, two points are specified.

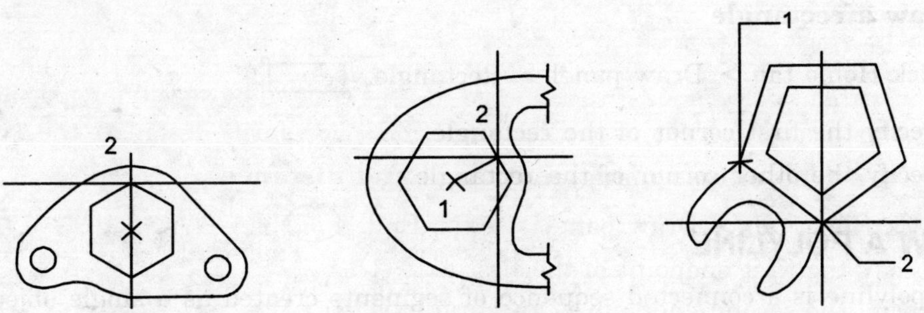

To draw a circumscribed polygon

1. Click Home tab ➤ Draw panel ➤ Polygon. 
2. At the command prompt, enter the number of sides.
3. Specify the center of the polygon (1).
4. Enter c to specify a polygon circumscribed about a circle.
5. Enter the radius length (2).

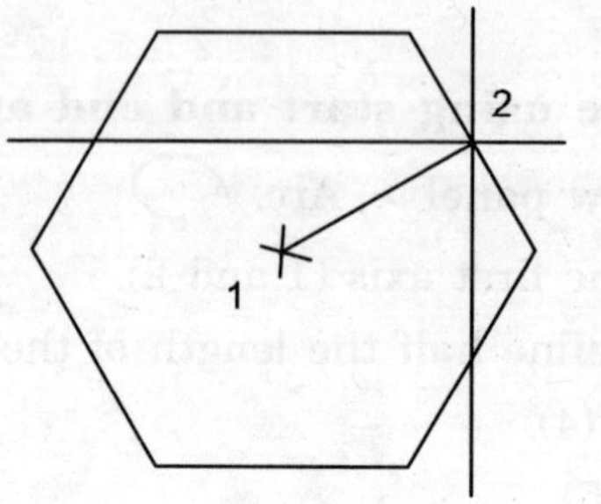

To draw a polygon by specifying one edge

1. Click Home tab ➤ Draw panel ➤ Polygon. 
2. At the command prompt, enter the number of sides.
3. Enter e (Edge).
4. Specify the start point for one polygon segment.
5. Specify the endpoint of the polygon segment.

To draw an inscribed polygon

1. Click Home tab ➤ Draw panel ➤ Polygon. 
2. At the command prompt, enter the number of sides.
3. Specify the center of the polygon.
4. Enter i to specify a polygon inscribed within a circle of specified points.
5. Enter the radius length.

To draw a rectangle

1. Click Home tab ➤ Draw panel ➤ Rectangle. 
2. Specify the first corner of the rectangle.
3. Specify the other corner of the rectangle.

7.7 DRAW A POLYLINE

A 2D polyline is a connected sequence of segments created as a single object. You can create straight line segments, arc segments, or a combination of the two.

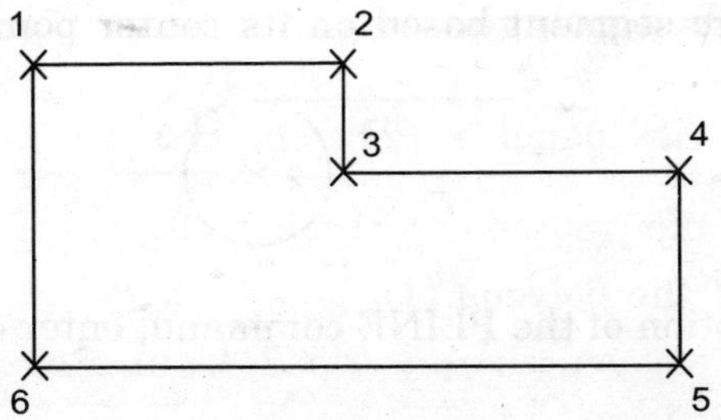

Note: A temporary plus-shaped marker displays at the first point. This marker can be useful when you create long and complicated polylines. It is removed when you complete the polyline.

The PLINEGEN system variable controls the linetype pattern display around and the smoothness of the vertices of a 2D polyline. Setting PLINEGEN to 1 generates new polylines in a continuous pattern around the vertices of the completed polyline. Setting PLINEGEN to 0 starts and ends the polyline with a dash at each vertex. PLINEGEN does not apply to polylines with tapered segments.

Plinegen set to 0

The following prompts are displayed.

Start point

Sets the starting point for the polyline.

- If you specify a second point, you draw line segments.
- If you enter a (for Arc), you draw arc segments.

Arc: The following prompts are displayed when you draw arc segments.

Endpoint of arc: Completes an arc segment. The arc segment is tangent to the previous segment of the polyline.

Angle: Specifies the included angle of the arc segment from the start point.

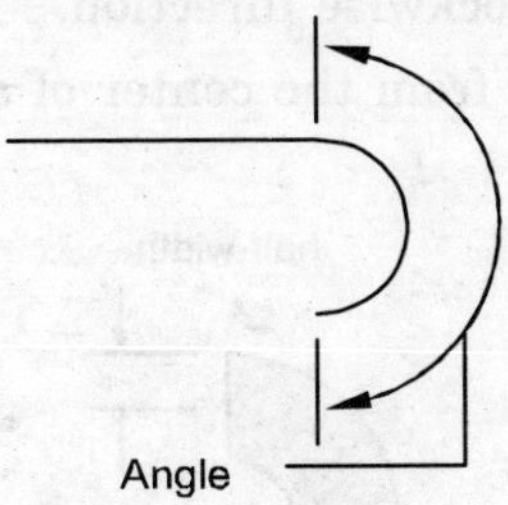

Entering a positive number creates counterclockwise arc segments. Entering a negative number creates clockwise arc segments.

Center: Specifies an arc segment based on its center point.

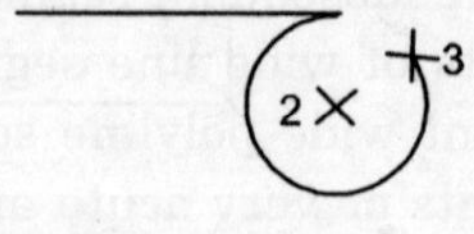

Note: For the Center option of the PLINE command, enter ce; for the Center object snap, enter **cen** or **center**.

- Center point of arc. Specifies the center point of the polyline arc.
- Endpoint of arc. Specifies the endpoint and draws the arc segment.
- Angle. Specifies the included angle of the arc segment from the start point.
- Length. Specifies the chord length (the distance from endpoint to endpoint of the arc segment. If the previous segment is an arc, the new arc segment is drawn tangent to the previous arc segment.

Close: Draws an arc segment from the last point specified to the starting point, creating a closed polyline. At least two points must be specified to use this option.

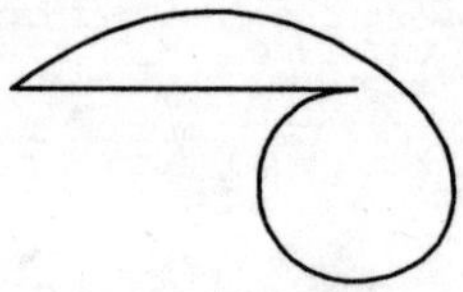

Direction: Specifies the tangent for the arc segment.

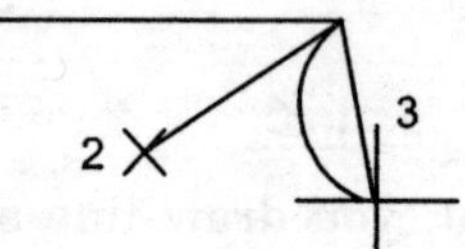

- **Tangent direction from the start point of the arc:** Specifies a point that establishes a tangency of the curve to the start point. The arc curves away from the vector between the start point and the tangent point.
- **Endpoint of the arc:** Specifies the endpoint of the arc segment.

Tip: Press Ctrl to draw in a clockwise direction.

Halfwidth: Specifies the width from the center of a wide polyline segment to one of its edges.

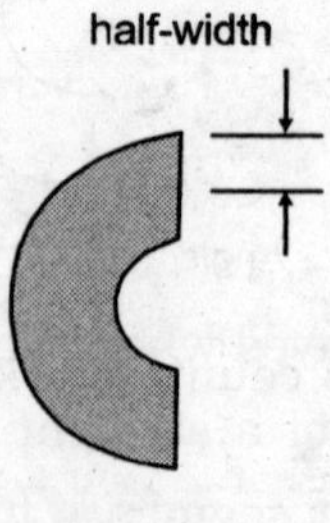

The starting half-width becomes the default ending half-width. The ending half-width becomes the uniform half-width for all subsequent segments until you change the half-width again. The starting and ending points of wide line segments are at the center of the line. Typically, the intersections of adjacent wide polyline segments are beveled. No beveling is performed for nontangent arc segments or very acute angles or when a dot-dash linetype is used.

- **Starting half-width:** Specifies the width and the start point of the segment.
- **Ending half-width:** Specifies the width at the endpoint.

Line: Switches from drawing arc segments to drawing straight segments.

Radius: Specifies the radius of the arc segment.

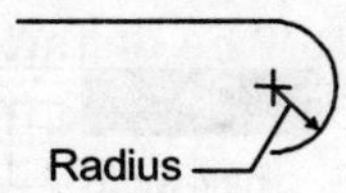

- **Radius of arc:** Specify a distance.
- **Endpoint of arc:** Specifies the endpoint and draws the arc segment.

Second pt: Specifies the second point and endpoint of a three-point arc.

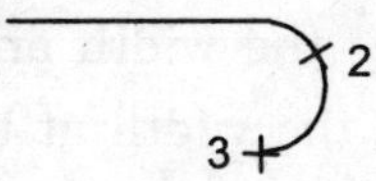

Undo: Removes the most recent arc segment added to the polyline.

Width: Specifies the width of the next arc segment.

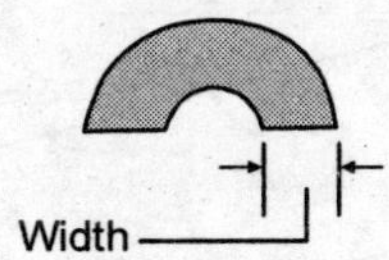

The starting width becomes the default ending width. The ending width becomes the uniform width for all subsequent segments until you change the width again. The starting and ending points of wide line segments are at the center of the line.

Typically, the intersections of adjacent wide polyline segments are beveled. No beveling is performed for nontangent arc segments, very acute angles, or when a dot-dash linetype is used.

- Starting width.
- Ending width.

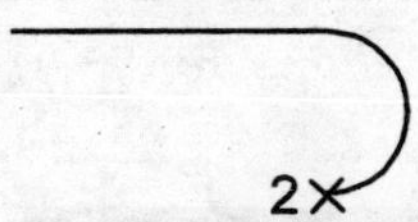

Line: The following prompts are displayed when you are drawing straight polyline segments.

Next point: Specifies the endpoint of a straight polyline segment.

Arc: Switches from drawing straight segments to drawing arc segments.

Close: Draws a line segment from the last point specified to the starting point, creating a closed polyline. At least two points must be specified to use this option.

Halfwidth: Specifies the width from the center of a wide polyline line segment to one of its edges.

The starting half-width becomes the default ending half-width. The ending half-width becomes the uniform half-width for all subsequent segments until you change the half-width again. The starting and ending points of wide line segments are at the center of the line.

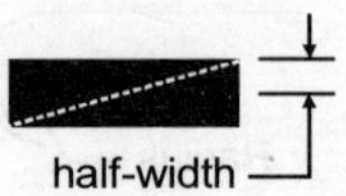

Typically, the intersections of adjacent wide polyline segments are beveled. No beveling is performed for nontangent arc segments or very acute angles or when a dot-dash linetype is used.

- **Starting half-width:** Specifies the width and the start point of the segment.
- **Ending half-width:** Specifies the width at the endpoint.

Length: Draws a line segment of a specified length at the same angle as the previous segment. If the previous segment is an arc, the new line segment is drawn tangent to that arc segment.

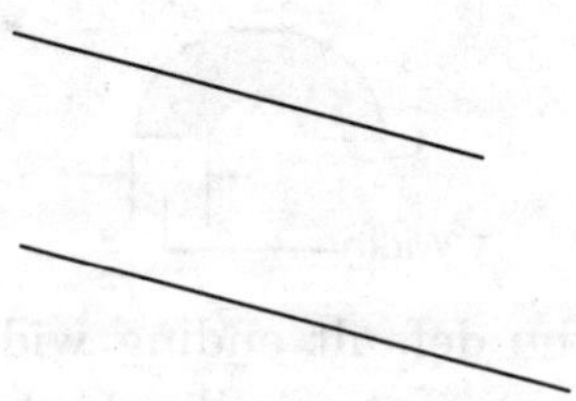

Undo: Removes the most recent line segment added to the polyline.

Width: Specifies the starting and ending width of the next line segment.

The starting width becomes the default ending width. The ending width becomes the uniform width for all subsequent segments until you change the width again. The starting and ending points of wide line segments are at the center of the line.

Typically, the intersections of adjacent wide polyline segments are beveled. No beveling is performed for nontangent arc segments or very acute angles or when a dot-dash linetype is used.

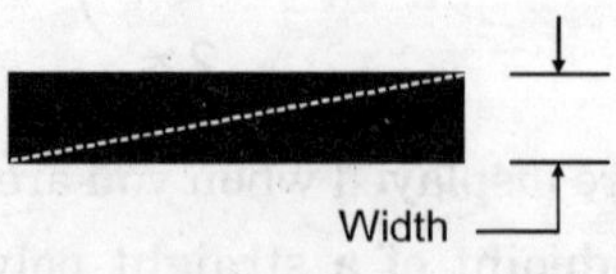

EXERCISE

1. When in absolute Cartesian coordinates have points A (10.8) and B (6.5), then to make a line from A -> B with relative polar coordinates will write
   (a) @ -5 < 36.88 (b) @ 4 < 30
   (c) @ 5 < 216, 88 (d) @ 3 < 60
2. To call the Line tool from the command line, the following abbreviation can be entered:
   (a) lin (b) li
   (c) l (d) ln
3. The command line abbreviation for the Circle tool is:
   (a) cir (b) c
   (c) ci (d) cl
4. What are the differences between outlines drawn using the Line tool and those drawn using the Polyline tool ?
5. The Arc tool can be called by entering the following abbreviation at the command line:
   (a) ar (b) a
   (c) ac;
   (d) There is no abbreviation. The name arc must be entered in full.
6. How many points do you need to define for the Rectangle command?
   (a) One (b) Two
   (c) Three (d) Four
7. How long will line from 1, 5 to @5<10 be?
   (a) One unit (b) Four units
   (c) Five units (d) Ten units
8. Draw the following using Line command

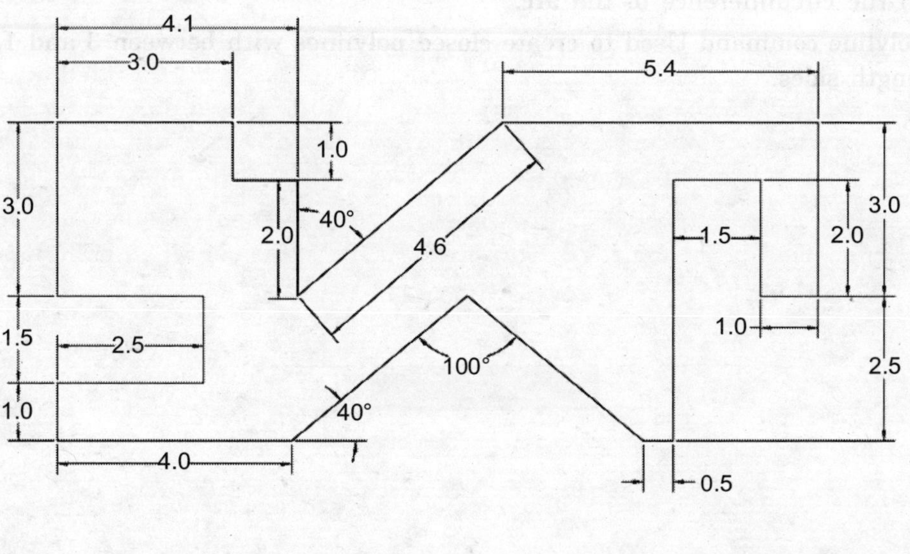

9. Draw the following using AutoCAD :

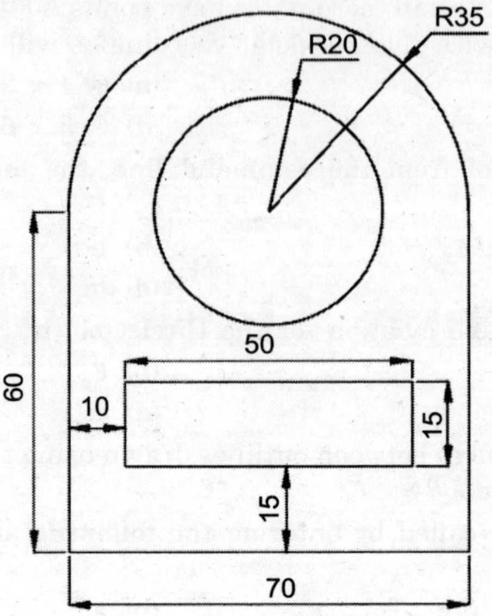

SUMMARY

- To Draw a line Specify first point: *Specify a point or press* ENTER to continue *from the last drawn line or arc* Specify next point.
- Continue command is used for Continuous a line from the endpoint of the most recently drawn line.
- Draws a circle based on a center point and a diameter or a radius.
- Draws an arc using three specified points on the arc's circumference. The first point is the start point (1). The third point is the endpoint (3). The second point (2) is a point on the circumference of the arc.
- Polyline command Used to create closed polylines with between 3 and 1,024 equal-length sides.

CHAPTER

8

MODIFYING USING AUTOCAD

Learning Objectives

- Modify the existing drawings.
- Manipulate the drawings using move and copy.
- Learn to offsetting the objects.
- Transform the object using scale and rotation.
- Learn to modify the object using chamfer and fillet.

8.1 MOVE COMMAND

Moves objects a specified distance in a specified direction.

Use coordinates, grid snaps, object snaps, and other tools to move objects with precision.

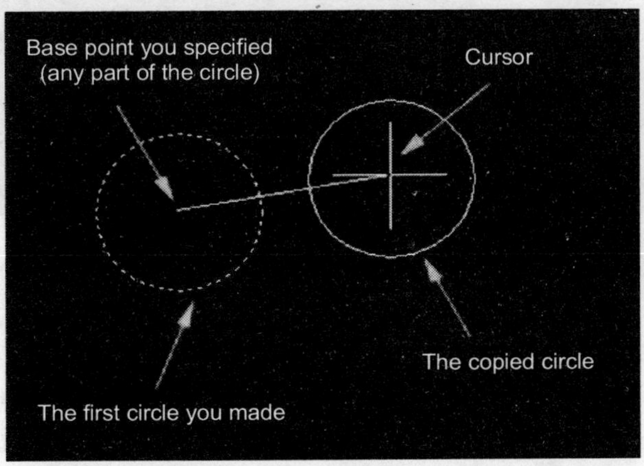

The following prompts are displayed.

Select objects: Specifies which objects to move.

Base point: Specifies the start point for the move.

Second point: In combination with the first point, specifies a vector that indicates how far, and in what direction, the selected objects are moved.

If you press Enter to accept the Use first point as displacement value, the first point is interpreted as a relative X, Y, Z displacement. For example, if you specify 2, 3 for the base point and press Enter at the next prompt, the objects move 2 units in the X direction and 3 units in the Y direction from their current position.

Displacement: Specifies a relative distance and direction.

The two points you specify define a vector that indicates how far from the original the copied objects are to be placed and in what direction.

8.2 COPY COMMAND

Copies objects a specified distance in a specified direction.

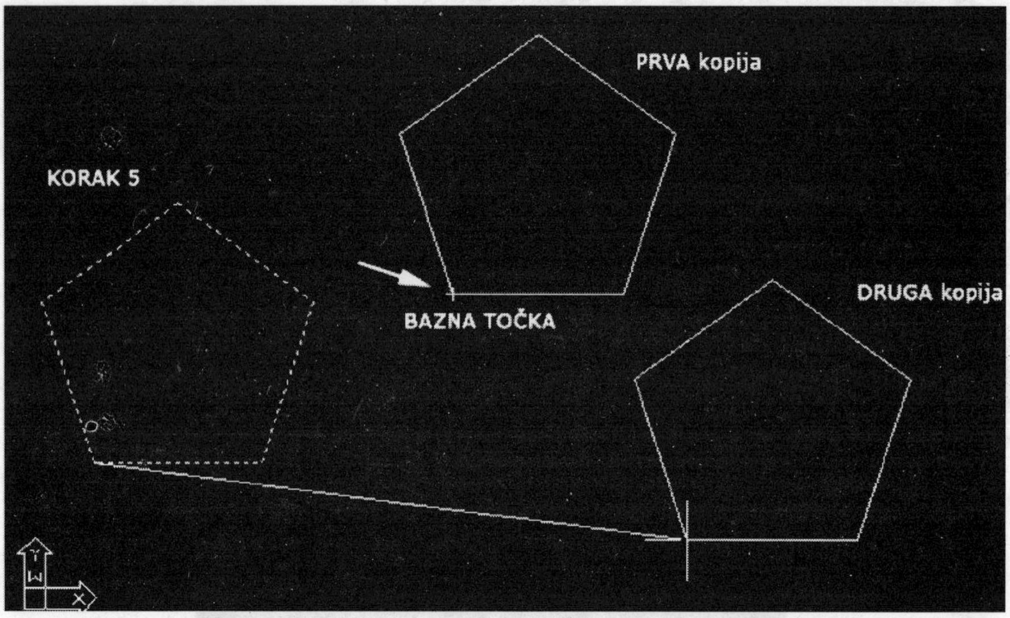

With the COPYMODE system variable, you can control whether multiple copies are created automatically.

The following prompts are displayed.

Select objects: *Use an object selection method and press Enter when you finish*

Specify base point or [Displacement/mOde/Multiple] <Displacement>: *Specify a base point or enter an option*

Specify second point or [Array] <use first point as displacement>: *Specify a second point or enter an option*

Displacement: Specifies a relative distance and direction using coordinates.

The two points you specify define a vector that indicates how far from the original the copied objects are to be placed and in what direction.

If you press Enter at the Specify Second Point prompt, the first point is interpreted as a relative X, Y, Z displacement. For example, if you specify 2, 3 for the base point and press Enter at the next prompt, the objects are copied 2 units in the X direction and 3 units in the Y direction from their current location.

Mode: Controls whether the command repeats automatically (COPYMODE system variable).

Single: Creates a single copy of selected objects and ends the command.

Multiple: Overrides the Single mode setting. The COPY command is set to repeat automatically for the duration of the command.

8.3 OFFSETTING

Creates concentric circles, parallel lines, and parallel curves.

You can offset an object at a specified distance or through a point. After you offset objects, you can trim and extend them as an efficient method to create drawings containing many parallel lines and curves.

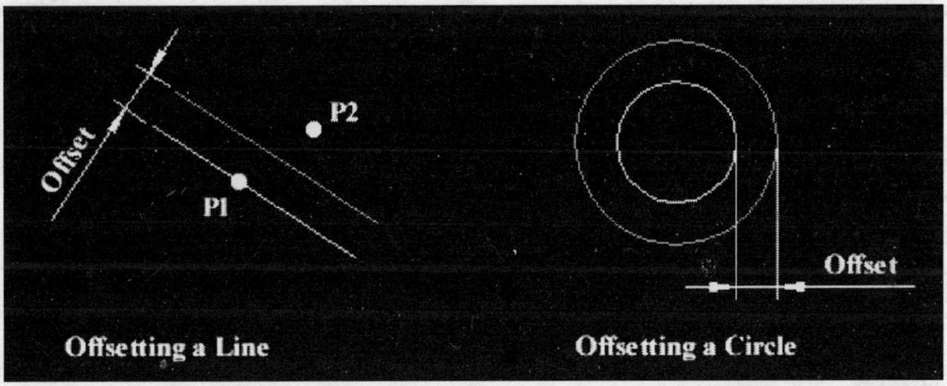

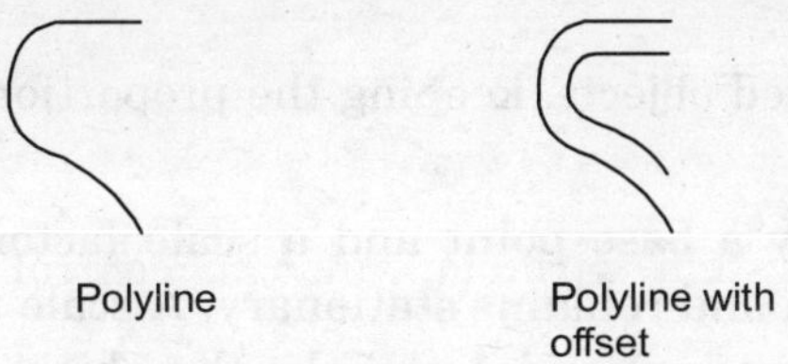

The OFFSET command repeats for convenience. To exit the command, press Enter.

The following prompts are displayed.

Offset Distance: Creates an object at a specified distance from an existing object.

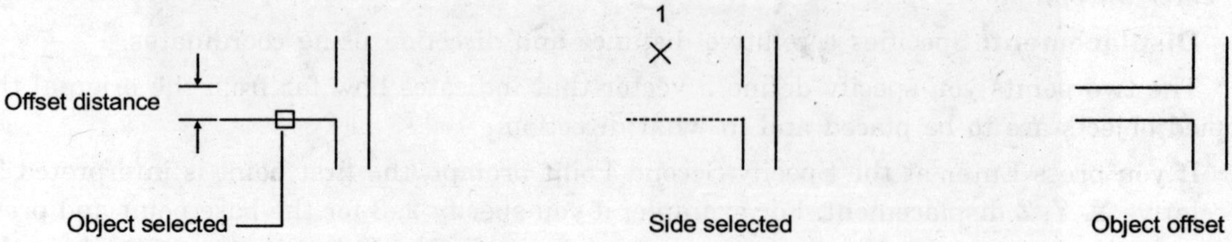

Exit: Exits the OFFSET command.

Multiple: Enters the Multiple offset mode, which repeats the offset operation using the current offset distance.

Undo: Reverses the previous offset.

Through: Creates an object passing through a specified point.

Note: For best results when you offset a polyline with corners, specify the through point near the midpoint of a line segment, not near a corner.

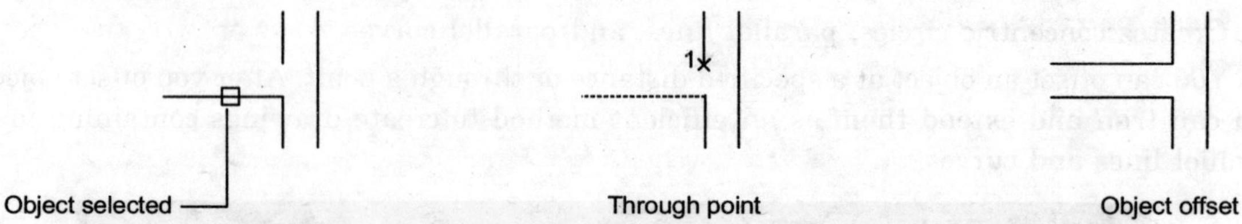

- Exit
- Multiple
- Undo

Erase: Erases the source object after it is offset.

Layer: Determines whether offset objects are created on the current layer or on the layer of the source object.

8.4 SCALING

Enlarges or reduces selected objects, keeping the proportions of the object the same after scaling.

To scale an object, specify a base point and a scale factor. The base point acts as the center of the scaling operation and remains stationary. A scale factor greater than 1 enlarges the object. A scale factor between 0 and 1 shrinks the object.

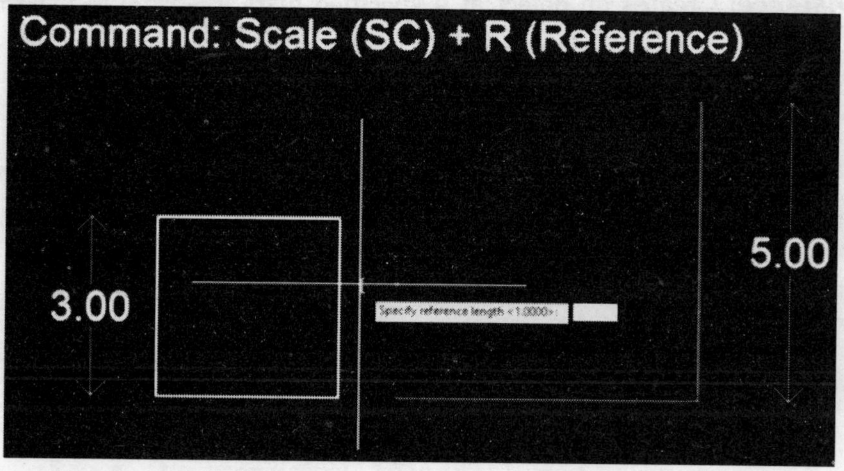

The following prompts are displayed.

Select objects: Specifies which objects you want to resize.

Base point: Specify a base point for the scale operation.

The base point you specify identifies the point that remains in the same location as the selected objects change size (and thus move away from the stationary base point).

Note: When you use the SCALE command with *annotative* objects, the position or location of the object is scaled relative to the base point of the scale operation, but the size of the object is not changed.

Scale Factor: Multiplies the dimensions of the selected objects by the specified scale. A scale factor greater than 1 enlarges the objects. A scale factor between 0 and 1 shrinks the objects. You can also drag the cursor to make the object larger or smaller.

Copy: Creates a copy of the selected objects for scaling.

Reference: Scales the selected objects based on a reference length and a specified new length.

8.5 ROTATE COMMAND

Rotates objects around a base point.

You can rotate selected objects around a base point to an absolute angle.

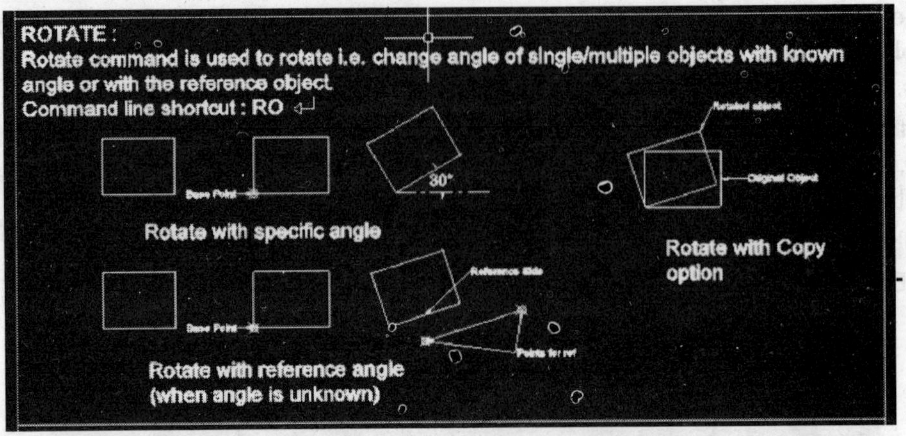

The following prompts are displayed.

Select objects: Use an object selection method and press Enter when you finish.

Specify base point: Specify a point.

Specify rotation angle: Enter an angle, specify a point, enter c , or enter r.

- **Rotation Angle:** Determines how far an object rotates around the base point. The axis of rotation passes through the specified base point and is parallel to the Z axis of the current UCS.
- **Copy:** Creates a copy of the selected objects for rotation.
- **Reference:** Rotates objects from a specified angle to a new, absolute angle. When you rotate a viewport object, the borders of the viewport remain parallel to the edges of the drawing area.

8.6 FILLET COMMAND

Rounds or fillets the edges of two 2D objects or the adjacent faces of a 3D solid.

A round or fillet is

- an arc that is created tangent between two 2D objects.
- a curved transition between two surfaces or adjacent faces on a 3D solid.

In this example, an arc is created tangent to the selected lines, which are trimmed to meet the endpoints of the arc.

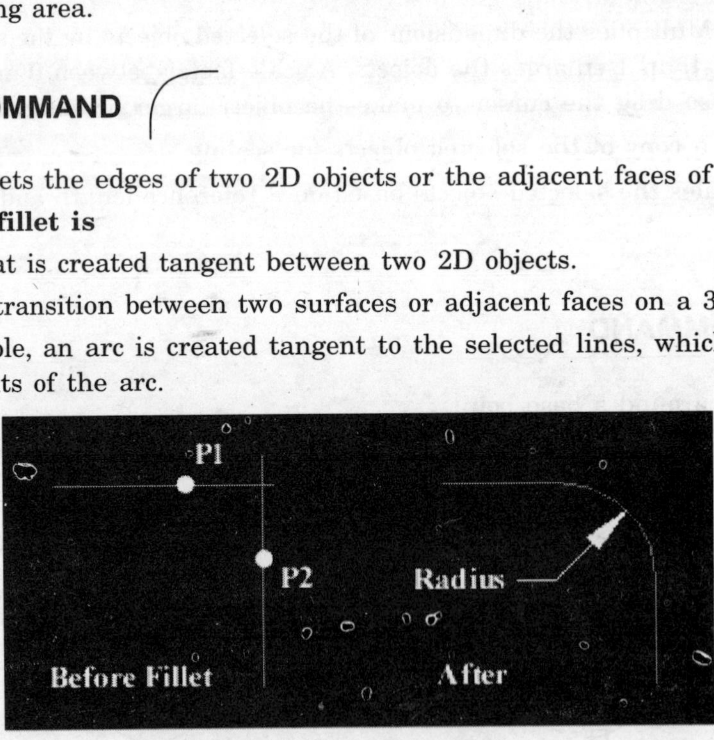

Create 2D Fillets: A round or fillet can be created between two objects of the same or different object types: 2D polylines, arcs, circles, ellipses, elliptical arcs, lines, rays, splines, and xlines.

If the two selected objects are on the same layer, the arc defined is created on that layer. Otherwise, the arc is created on the current layer. The layer affects object properties including color and linetype.

The following prompts are displayed when creating a 2D fillet.

First Object: Select the first of two objects or the first line segment of a 2D polyline to define the fillet.

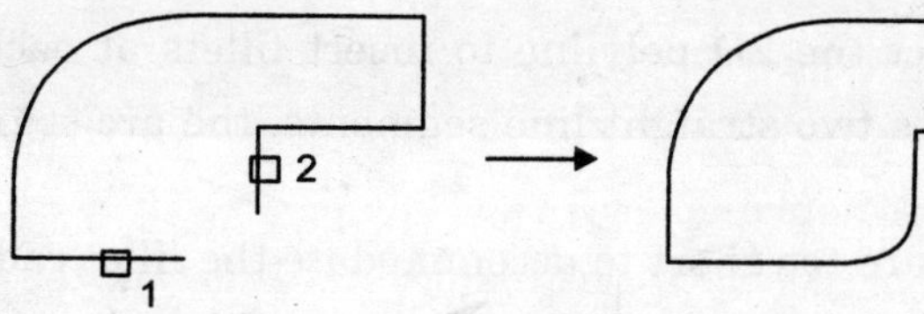

Second object or shift-select to apply corner: Select the second object or line segment of a 2D polyline to define the fillet.

You can also hold down the Shift key before selecting the second object or line segment of a 2D polyline to extend or trim the selected objects to form a sharp corner. While Shift is held down, a temporary value of zero is assigned to the current fillet radius value.

If the selected objects are straight line segments of a 2D polyline, the line segments can be adjacent to each other or separated by one other segment. When the selected segments are separated by a segment, the segment that separates them is removed and replaced with the fillet.

The direction and length of the arc created is determined by the points picked to select the objects. Always select an object closest to where you want the endpoints of the fillet to be drawn.

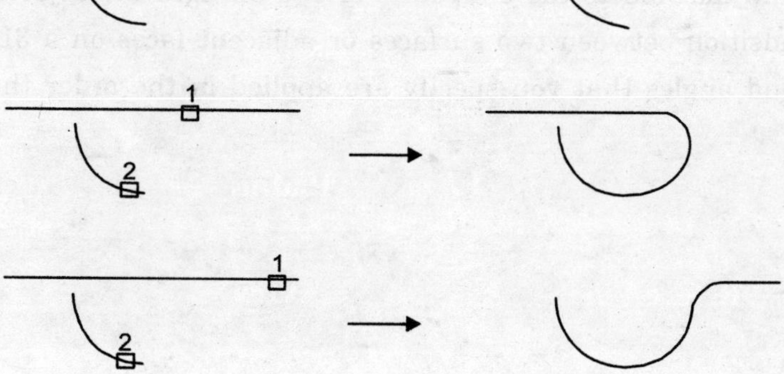

Undo: Reverses the previous action in the command.

Polyline: Inserts a fillet at each vertex of a 2D polyline where two straight line segments meet. The fillets become new segments of the polyline, unless the Trim option is set to No Trim.

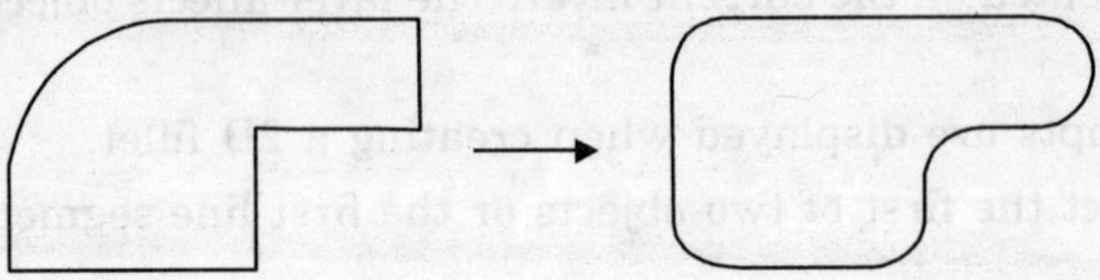

Select 2D polyline: Select the 2D polyline to insert fillets at each vertex.

If an arc segment separates two straight line segments, the arc segment is removed and replaced with the fillet.

Note: Line segments that are too short to accommodate the fillet radius are not modified.

Radius: Sets the radius for subsequent fillets; changing this value does not affect existing fillets.

Note: A radius value of zero can be used to create a sharp corner. Filleting two lines, rays, xlines, or line segments of a 2D polyline with a radius of zero extends or trims the objects so they intersect.

Trim: Controls whether the selected objects are trimmed to meet the endpoints of the fillet.

- **Trim:** Selected objects or line segments are trimmed to meet the endpoints of the fillet.
- **No Trim:** Selected objects or line segments are not trimmed before the fillet is added.

The current value is stored in the TRIMMODE system variable.

Multiple: Allows for the rounding of more than one set of objects.

8.7 CHAMFER COMMAND 

Bevels or chamfers the edges of two 2D objects or the adjacent faces of a 3D solid.

A bevel or chamfer is

- an angled line that meets the endpoints of two straight 2D objects.
- a sloped transition between two surfaces or adjacent faces on a 3D solid.

The distances and angles that you specify are applied in the order that you select the objects.

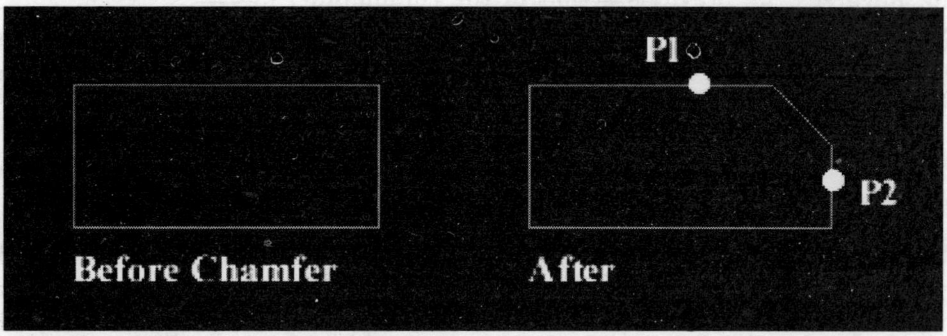

Create 2D Chamfer: A bevel or chamfer can be defined by selecting two objects of the same or different object types: lines, polylines, rays, and xlines.

If the two selected objects are on the same layer, the line defined is created on that layer. Otherwise, the line is created on the current layer. The layer affects object properties including color and linetype.

The following prompts are displayed when creating a 2D chamfer.

First Line: Select the first of two objects or the first line segment of a 2D polyline to define the chamfer.

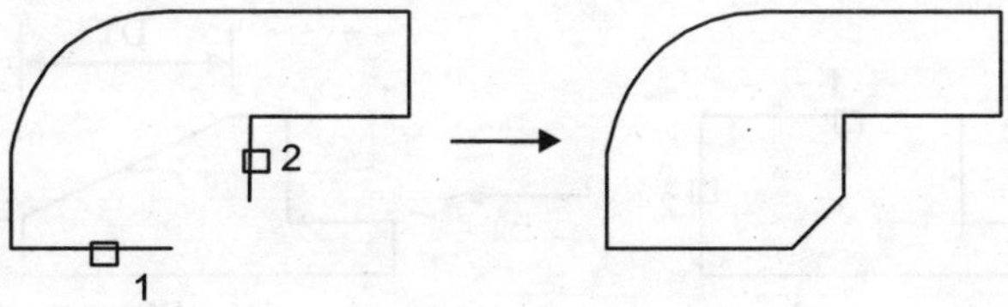

Second line or shift-select to apply corner

Select the second object or line segment of a 2D polyline to define the chamfer.

You can also hold down the Shift key before selecting the second object or line segment of a 2D polyline to extend or trim the selected objects to form a sharp corner. While Shift is held down, a temporary value of zero is assigned to the current chamfer distance and angle values.

If the selected objects are straight line segments of a 2D polyline, the line segments can be adjacent to each other or separated by one other segment. When the selected segments are separated by a segment, the segment that separates them is removed and replaced with the chamfer.

Note: Adding a chamfer or bevel to a hatch boundary that was defined with individual objects results in the removal of hatch associativity. If the hatch boundary was defined from a polyline, associativity is maintained.

Undo: Reverses the previous action in the command.

Polyline: Inserts a chamfer line at each vertex of a 2D polyline where two straight line segments meet. The chamfer lines become new segments of the polyline, unless the Trim option is set to No Trim.

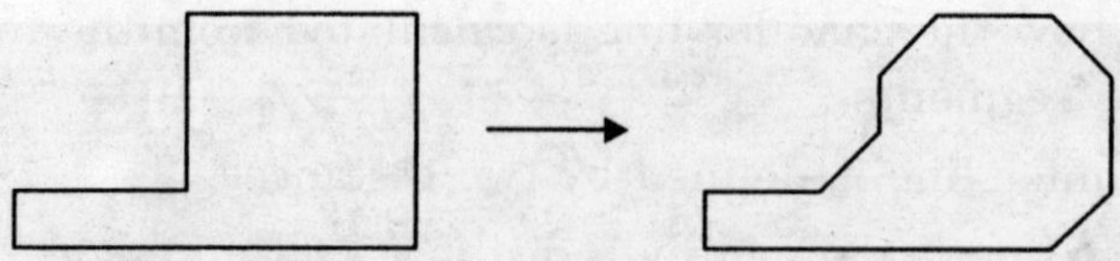

Note: Line segments that are too short to accommodate the chamfer distance are not modified.

Distance: Sets the chamfer distances from the intersecting points of the first and second objects.

If both distances are set to zero, the selected objects or line segments are extended or trimmed so they intersect.

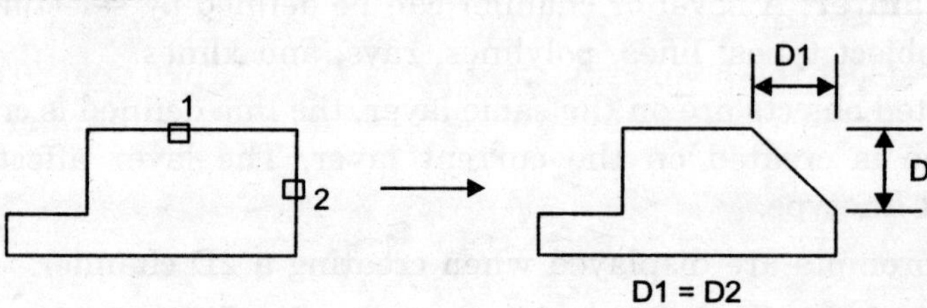

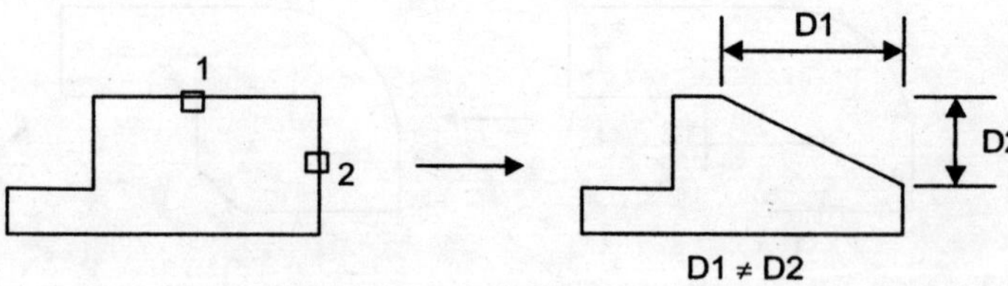

Angle: Sets the chamfer distance from the intersecting point of the selected objects and the XY angle from the first object or line segment.

If both values are set to zero, the selected objects or line segments are extended or trimmed so they intersect.

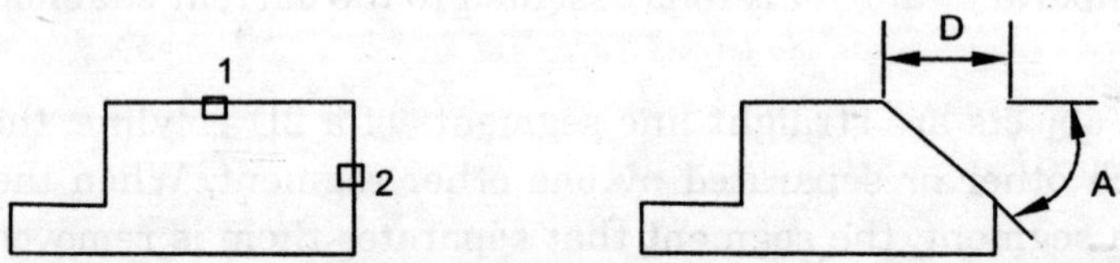

Trim: Controls whether the selected objects are trimmed to meet the endpoints of the chamfer line.

- **Trim:** Selected objects or line segments are trimmed to meet the endpoints of the chamfer line. If the selected objects or line segments do not intersect with the chamfer line, they are extended or trimmed before the chamfer line is added.
- **No Trim:** Selected objects or line segments are not trimmed before the chamfer line is added.

Note: The current value is stored in the TRIMMODE system variable.

Method: Controls how the chamfer line is calculated from the intersecting point of the selected objects or line segments.

- **Distance:** Chamfer line is defined by two distances.
- **Angle:** Chamfer line is defined by a distance and an angle.

The current value is stored in the CHAMMODE system variable.

Multiple: Allows for the beveling of more than one set of objects.

8.8 TRIMMING

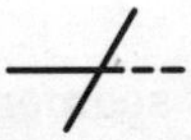

Trims objects to meet the edges of other objects.

To trim objects, select the boundaries and press Enter. Then select the objects that you want to trim. To use all objects as boundaries, press Enter at the first Select Objects prompt.

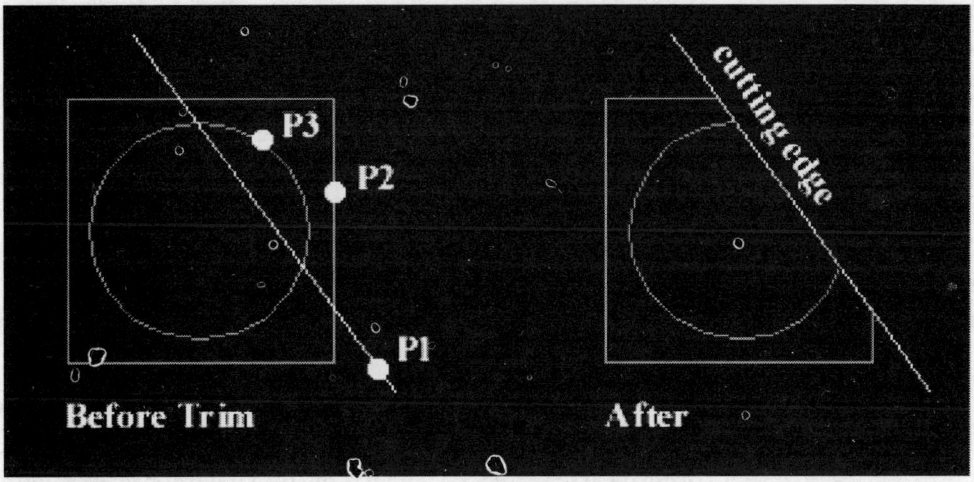

The following prompts are displayed.

Select cutting edges: Specifies one or more objects to be used as a boundary for the trim. TRIM projects the cutting edges and the objects to be trimmed onto the XY plane of the current user coordinate system (UCS).

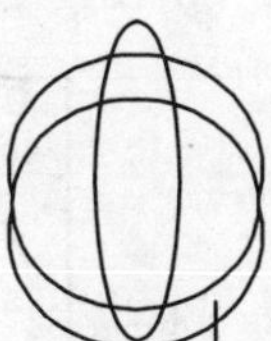

Note: To select cutting edges that include blocks, you can use only the single selection, Crossing, Fence, and Select All options.

Select objects: Specifies objects individually.

Select all: Specifies that all objects in the drawing can be used as a trim boundary.

Object to Trim: Specifies the object to trim. If more than one trim result is possible, the location of the first selection point determines the result.

Shift-Select to Extend: Extends the selected objects rather than trimming them. This option provides an easy method to switch between trimming and extending.

Fence: Selects all objects that cross the selection fence. The selection fence is a series of temporary line segments that you specify with two or more fence points. The selection fence does not form a closed loop.

Crossing: Selects objects within and crossing a rectangular area defined by two points.

Note: Some crossing selections of objects to be trimmed are ambiguous. TRIM resolves the selection by following along the rectangular crossing window in a clockwise direction from the first point to the first object encountered.

Project: Specifies the projection method used when trimming objects.

None: Specifies no projection. The command trims only objects that intersect with the cutting edge in 3D space.

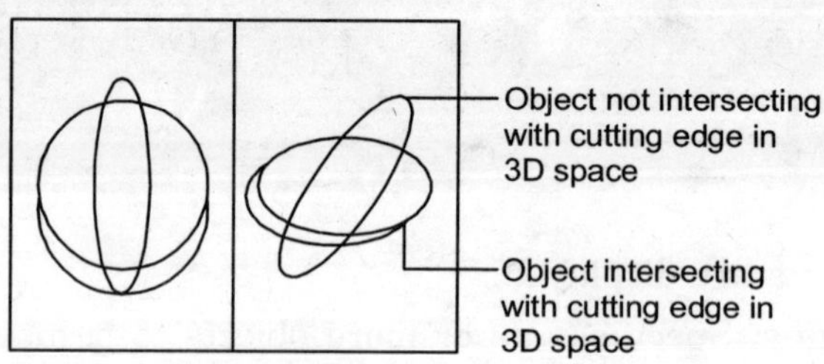

UCS: Specifies projection onto the XY plane of the current UCS. The command trims objects that do not intersect with the cutting edge in 3D space.

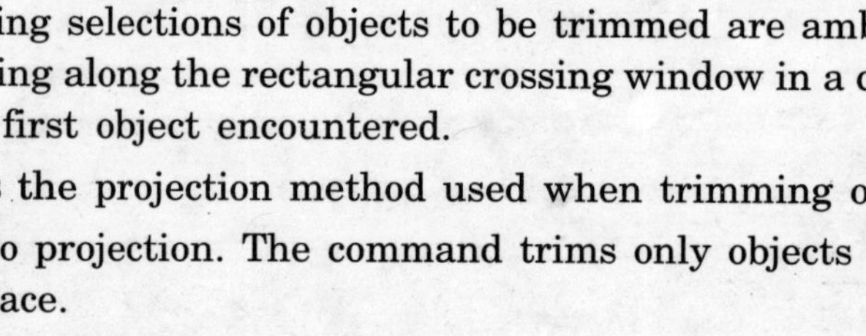

View: Specifies projection along the current view direction. The command trims objects that intersect the boundary in the current view.

Left viewpoint Right viewpoint

Edge: Determines whether an object is trimmed at another object's extrapolated edge or only to an object that intersects it in 3D space.

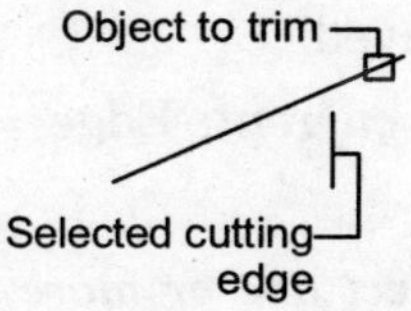

Extend: Extends the cutting edge along its natural path to intersect an object in 3D space.

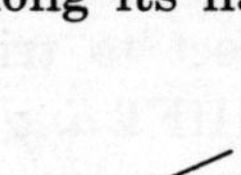

Extend

No Extend: Specifies that the object is trimmed only at a cutting edge that intersects it in 3D space.

Note: When trimming hatches, do not set Edge to Extend. If you do, gaps in the trim boundaries will not be bridged when trimming hatches, even when the gap tolerance is set to a correct value.

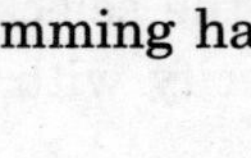

No extend

Erase: Deletes selected objects. This option provides a convenient method to erase unneeded objects without leaving the TRIM command.

Undo: Reverses the most recent change made by TRIM.

8.9 EXTEND COMMAND 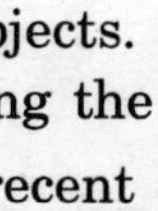

Extends objects to meet the edges of other objects.

To extend objects, first select the boundaries. Then press Enter and select the objects that you want to extend. To use all objects as boundaries, press Enter at the first Select Objects prompt.

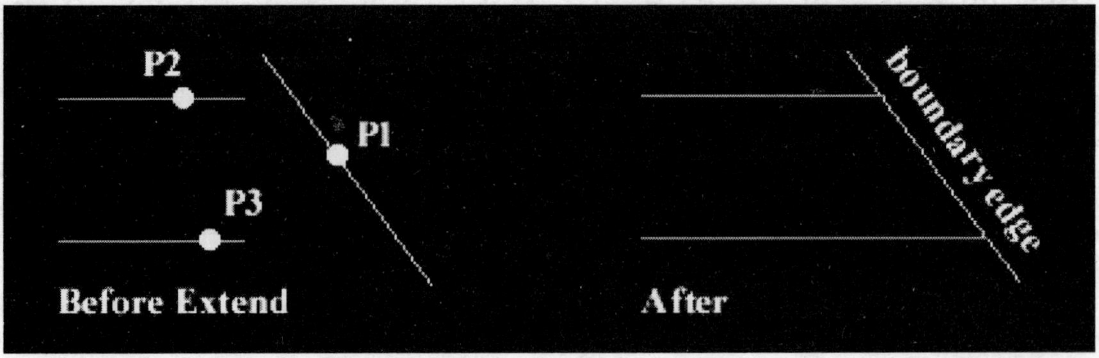

The following prompts are displayed.

Current settings: Projection = *current*, Edge = *current*

Select boundary edges...

Select objects or <select all>: *Select one or more objects and press* Enter, *or press* Enter *to select all displayed objects.*

Select object to extend or shift-select to trim or [Fence/Crossing/Project/Edge/Undo]: *Select objects to extend, or hold down* SHIFT *and select an object to trim, or enter an option*

Boundary Object Selection: Uses the selected objects to define the boundary edges to which you want to extend an object.

Object to Extend: Specifies the objects to extend. Press Enter to end the command.

Shift-Select to Trim: Trims the selected objects to the nearest boundary rather than extending them. This is an easy method to switch between trimming and extending.

Fence: Selects all objects that cross the selection fence. The selection fence is a series of temporary line segments that you specify with two or more fence points. The selection fence does not form a closed loop.

Crossing: Selects objects within and crossing a rectangular area defined by two points.

8.10 LENGTHEN 

Changes the length of objects and the included angle of arcs.

You can specify changes as a percentage, an increment, or as a final length or angle. LENGTHEN is an alternative to using TRIM or EXTEND.

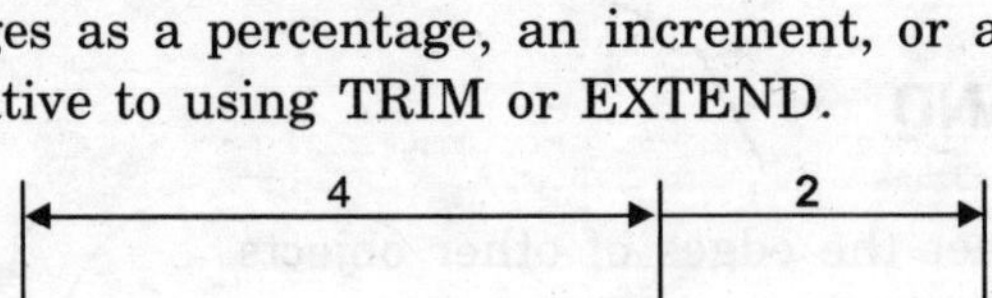

The following prompts are displayed.

Object Selection: Displays the length and, where applicable, the included angle of the object.

LENGTHEN does not affect closed objects. The extrusion direction of the selected object need not be parallel to the Z axis of the current user coordinate system (UCS).

Delta: Changes the length of an object by a specified increment, measured from the endpoint that is closest to the selection point. Delta also changes the angle of an arc by a specified increment, measured from the endpoint that is closest to the selection point. A positive value extends the object; a negative value trims it.

Delta Length: Changes the length of the object by the specified increment.

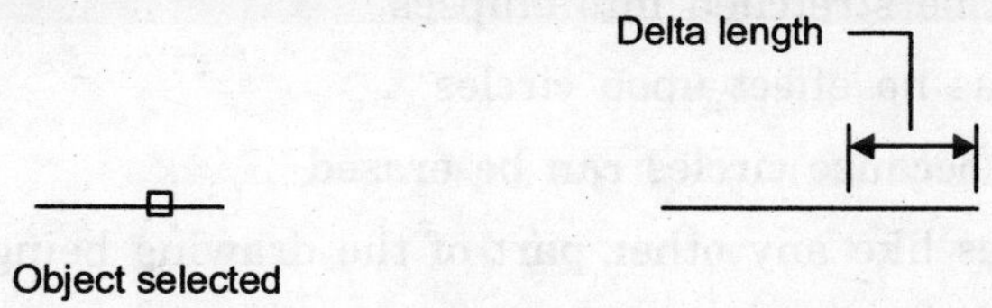

Angle: Changes the included angle of the selected arc by the specified angle.

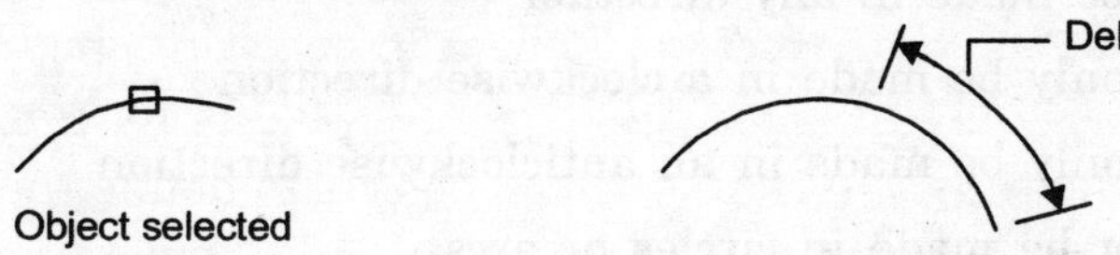

Percent: Sets the length of an object by a specified percentage of its total length.

Total: Sets the length of a selected object by specifying the total absolute length from the fixed endpoint. Total also sets the included angle of a selected arc by a specified total angle.

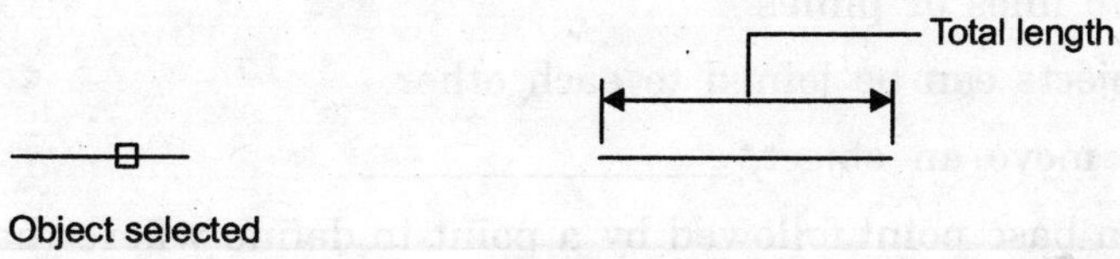

Total Length: Lengthens the object to the specified value from the endpoint that is closest to the selection point.

Angle: Sets the included angle of the selected arc.

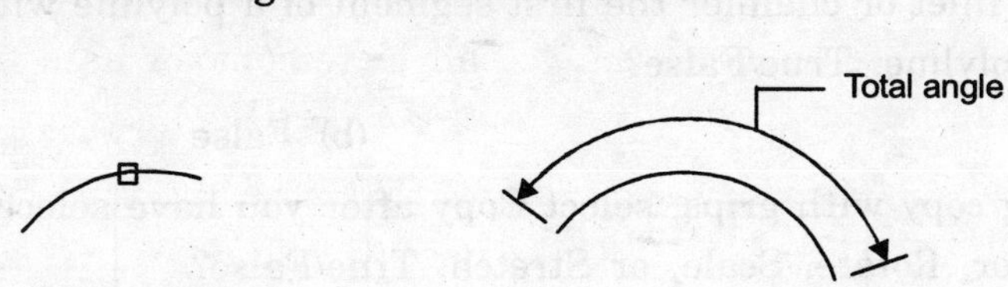

Dynamic: Turns on Dynamic Dragging mode. You change the length of a selected object by dragging one of its endpoints. The other end remains fixed.

EXERCISE

1. When using the Rotate tool the angle of rotation is in the following direction:
   (a) Clockwise
   (b) Anticlockwise
   (c) The direction in which the cursor is moved
   (d) There is no fixed rotation direction.
2. When using the Stretch tool:
   (a) Circles can be stretched into ellipses
   (b) The tool has no effect upon circles
   (c) Be careful because circles can be erased
   (d) Treat circles like any other part of the drawing being stretched.
3. When using the Break tool on circles and/or arcs:
   (a) Breaks can be made in any direction
   (b) Breaks can only be made in a clockwise direction
   (c) Breaks can only be made in an anticlockwise direction
   (d) Break cannot be made in circles or arcs.
4. The Join tool can be used for joining:
   (a) Any two arcs
   (b) Only arcs or lines in line with each other
   (c) Any two lines or plines
   (d) Any objects can be joined to each other.
5. In order to move an object, _______________
   (a) Select a base point,followed by a point to define where the object is moving to.
   (b) Use the Displacement option
   (c) Select object for grips, then move
   (d) Above all
6. You cannot fillet or chamfer the first segment of a polyline with the last segment of the same polyline. True/False?
   (a) True (b) False
7. When using copy with grips, select Copy after you have selected an option, such as Move, Mirror, Rotate, Scale, or Stretch. True/False?
   (a) True (b) False

8. By using Copy, rotate & scale command draw the following

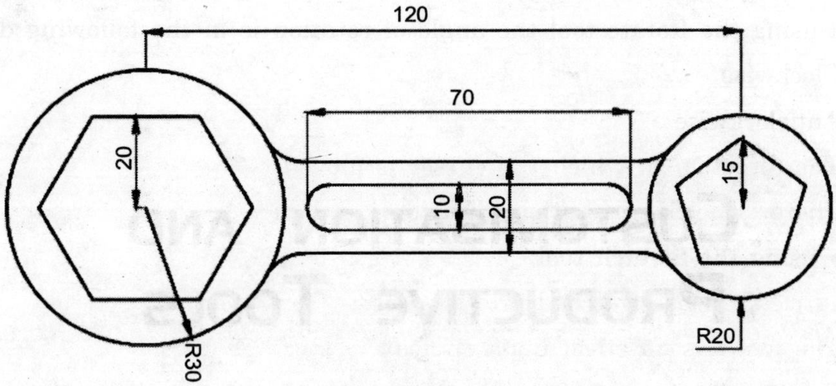

SUMMARY

- Move command used for translate objects a specified distance in a specified direction.
- Copy command used to Copies objects a specified distance in a specified direction.
- offsetting Creates concentric circles, parallel lines, and parallel curves.
- Scaling is used to Enlarges or reduces selected objects, keeping the proportions of the object the same after scaling.
- Rotate command is to used Rotates objects around a base point.
- Lengthen Command is used to changes the length of objects and the included angle of arcs.

CHAPTER

9

CUSTOMISATION AND PRODUCTIVE TOOLS

Learning Objectives

- How to customize the drawing
- How to Plot the drawing
- Learn to Apply Tolerances
- Learn different geometric Constructions
- Learn to Apply SNAP and GRID

9.1 PLOT COMMAND

Specifies device and media settings, and plots your drawing.

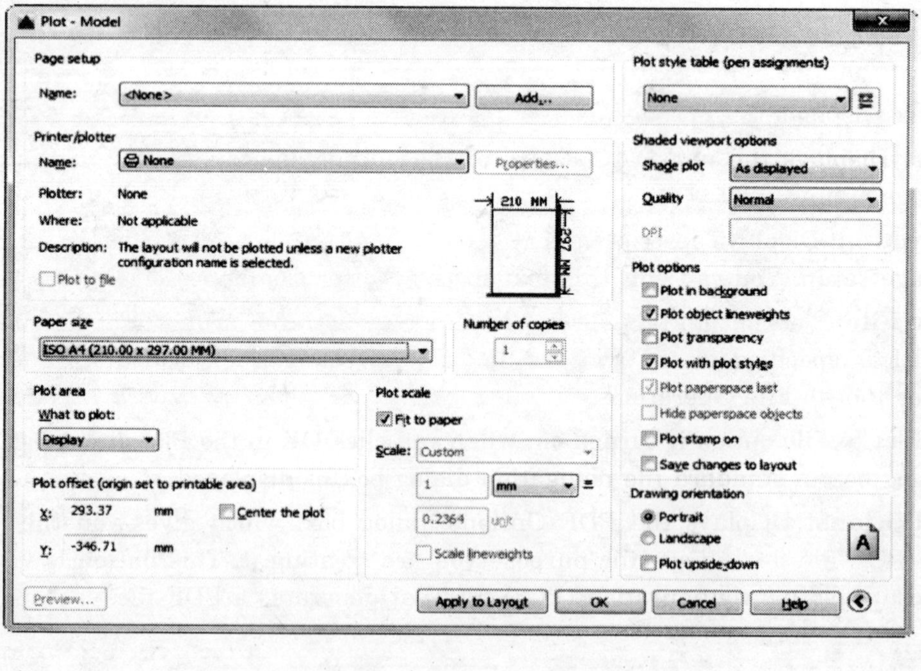

You can display more options by clicking the More Options button. 

List of Options

The following options are displayed.

Page Setup: Displays a list of any named and saved page setups in the drawing. You can base the current page setup on a named page setup saved in the drawing, or you can create a new named page setup based on the current settings in the Plot dialog box by clicking Add.

Name: Displays the name of the current page setup.

Add: Displays the Add Page Setup dialog box, in which you can save the current settings in the Plot dialog box to a named page setup. You can modify this page setup through the Page Setup Manager.

Printer/Plotter: Specifies a configured plotting device to use when plotting layouts.

If the selected plotter doesn't support the layout's selected paper size, a warning is displayed and you can select the plotter's default paper size or a custom paper size.

Name: Lists the available PC3 files or system printers from which you can select to plot the current layout. An icon in front of the device name identifies it as a PC3 file or a system printer.

- PC3 file icon. Indicates a PC3 file.

- System printer icon. Indicates a system printer.

Properties: Displays the Plotter Configuration Editor (PC3 editor), in which you can view or modify the current plotter configuration, ports, device, and media settings.

If you make changes to the PC3 file using the Plotter Configuration Editor, the Changes to a Printer Configuration File dialog box is displayed.

Plotter: Displays the plot device specified in the currently selected page setup.

Where: Displays the physical location of the output device specified in the currently selected page setup.

Description: Displays descriptive text about the output device specified in the currently selected page setup. You can edit this text in the Plotter Configuration Editor.

Plot to File: Plots output to a file rather than to a plotter or printer. The default location for plot files is specified in the Options dialog box, Plot and Publish tab, under Default Location for Plot-to-File Operations.

If the Plot to File option is turned on, when you click OK in the Plot dialog box, the Plot to File dialog box (a standard file navigation dialog box) is displayed.

PDF Options: Displays the PDF Options dialog box, which gives you the ability to optimize a PDF file for the specific purpose you are creating it. This button is visible only if you choose a plotter configuration (PC3) file that generates a PDF file.

Partial Preview: Shows an accurate representation of the effective plot area relative to the paper size and printable area. The tooltip displays the paper size and printable area.

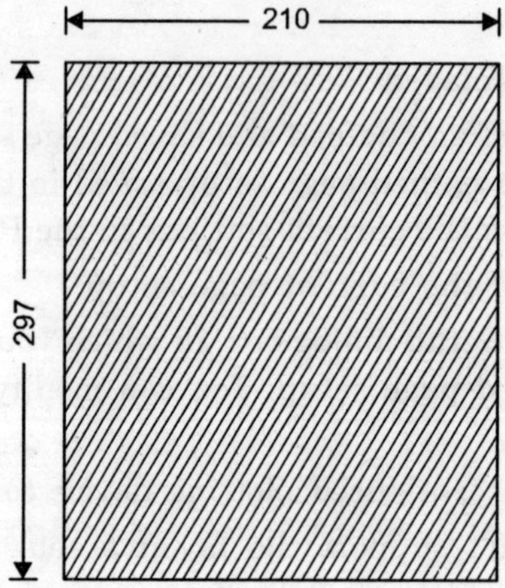

Paper Size: Displays standard paper sizes that are available for the selected plotting device. If no plotter is selected, the full standard paper size list is displayed and available for selection.

If the selected plotter doesn't support the layout's selected paper size, a warning is displayed, and you can select the plotter's default paper size or a custom paper size.

A default paper size is set for the plotting device when you create a PC3 file with the Add-a-Plotter wizard. The paper size that you select in the Page Setup dialog box is saved with the layout and overrides the PC3 file settings.

The actual printable area of the page, which is determined by the selected plotting device and paper size, is indicated in the layout by a dashed line.

If you are plotting a raster image, such as a BMP or TIFF file, the size of the plot is specified in pixels, not in inches or millimeters.

Number of Copies: Specifies the number of copies to plot. This option is not available when you plot to file.

Plot Area: Specifies the portion of the drawing to be plotted. Under What to Plot, you can select an area of the drawing to be plotted.

Layout/Limits: When plotting a layout, plots everything within the printable area of the specified paper size, with the origin calculated from 0,0 in the layout.

When plotting from the Model tab, plots the entire drawing area that is defined by the grid limits. If the current viewport does not display a plan view, this option has the same effect as the Extents option.

Extents: Plots the portion of the current space of the drawing that contains objects. All geometry in the current space is plotted. The drawing may be regenerated to recalculate the extents before plotting.

Display: Plots the view in the current viewport in the selected Model tab or the current paper space view in the layout.

View: Plots a view that was previously saved with the VIEW command. You can select a named view from the list. If there are no saved views in the drawing, this option is unavailable.

When the View option is selected, a View list is displayed that lists the named views that are saved in the current drawing. You can select a view from this list to plot.

Window: Plots any portion of the drawing that you specify. When you select Window, the Window button becomes available. Click the Window button to use the pointing device to specify the two corners of the area to be plotted, or enter coordinate values.

Plot Offset: Specifies an offset of the plot area relative to the lower-left corner of the printable area or to the edge of the paper, depending on the setting made in the Specify Plot Offset Relative To option (Options dialog box, Plot and Publish tab). The Plot Offset area of the Plot dialog box displays the specified plot offset option in parentheses.

The printable area of a drawing sheet is defined by the selected output device and is represented by a dashed line in a layout. When you change to another output device, the printable area may change.

You can offset the geometry on the paper by entering a positive or negative value in the X and Y offset boxes. The plotter unit values are in inches or millimeters on the paper.

Center the Plot: Automatically calculates the X and Y offset values to center the plot on the paper. This option is not available when Plot Area is set to Layout.

X: Specifies the plot origin in the X direction relative to the setting of the Plot Offset Definition option.

Y: Specifies the plot origin in the Y direction relative to the setting of the Plot Offset Definition option.

Plot Scale: Controls the relative size of drawing units to plotted units. The default scale setting is 1:1 when plotting a layout. The default setting is Fit to Paper when plotting from the Model tab.

Fit to Paper: Scales the plot to fit within the selected paper size and displays the custom scale factor in the Scale, Inch =, and Units boxes.

Scale: Defines the exact scale for the plot. Custom defines a user-defined scale. You can create a custom scale by entering the number of inches (or millimeters) equal to the number of drawing units.

Inch(es) =/mm =/Pixel(s) =: Specifies the number of inches, millimeters, or pixels equal to the specified number of units.

Inch/mm/pixel: Specifies inches or mm for display of units in the Plot dialog box. The default is based on the paper size and changes each time a new paper size is selected. Pixel is available only when a raster output is selected.

Units: Specifies the number of units equal to the specified number of inches, millimeters, or pixels.

Scale Lineweights: Scales lineweights in proportion to the plot scale. Lineweights normally specify the linewidth of plotted objects and are plotted with the linewidth size regardless of the plot scale.

Note: You can modify the list of scales with SCALELISTEDIT.

Preview: Displays the drawing as it will appear when plotted by starting the PREVIEW command. To exit the preview and return to the Plot dialog box, press ESC, press ENTER, or right-click and then click Exit on the shortcut menu.

Apply to Layout: Saves the current Plot dialog box settings to the current layout.

More Options: Controls display of additional options in the Plot dialog box.

- Plot Style Table (Pen Assignments)
- Shaded Viewport Options
- Plot Options
- Drawing Orientation

Plot Style Table (Pen Assignments): Sets the plot style table, edits the plot style table, or creates a new plot style table.

Name (Unlabeled): Displays the plot style table that is assigned to the current Model tab or layout tab and provides a list of the currently available plot style tables.

If you select New, the Add Plot Style Table wizard is displayed, which you can use to create a new plot style table. The wizard that is displayed is determined by whether the current drawing is in color-dependent or named mode.

Edit: Displays the Plot Style Table Editor, in which you can view or modify plot styles for the currently assigned plot style table.

Shaded Viewport Options: Specifies how shaded and rendered viewports are plotted and determines their resolution level and dots per inch (dpi). Rendered viewport options are not available in AutoCAD LT.

Note: When hardware acceleration is disabled or is enabled, but does not support Shadows, it is possible to plot a drawing that contains shaded viewports with unsupported hardware effects by the graphics card through software emulation. To enable the software emulation of hardware effects that are not supported by your graphics card, enter 3DCONFIG and click Manual Tune. In the Manual Performance Tuning dialog box, click Emulate unsupported hardware effects in software when plotting. The unsupported effects will not appear in the viewport in real-time, but will appear in the hardcopy or electronic file that is created during the plot process. (Not applicable to AutoCAD LT)

Shade Plot: Specifies how views are plotted. To specify this setting for a viewport on a layout tab, select the viewport and then, on the Tools menu, click Properties.

From the Model tab, you can select from the following options:

- **As Displayed:** Plots objects the way they are displayed on the screen. (Available in AutoCAD LT).
- **Legacy Wireframe:** Objects in wireframe regardless of the way they are displayed on the screen, using the legacy SHADEMODE command. (Available in AutoCAD LT).

- **Legacy Hidden:** Objects with hidden lines removed regardless of the way the objects are displayed on the screen, using the legacy SHADEMODE command. (Available in AutoCAD LT)
- **Conceptual:** Plots objects with the Conceptual visual style applied regardless of the way the objects are displayed on the screen.
- **Hidden:** Plots objects with hidden lines removed regardless of the way the objects are displayed on the screen.
- **Realistic:** Plots objects with the Realistic visual style applied regardless of the way the objects are displayed on the screen.
- **Shaded:** Plots objects with Shaded visual style applied regardless of the way the objects are displayed on the screen
- **Shaded with Edges:** Plots objects with Shaded with Edges visual style applied regardless of the way the objects are displayed on the screen
- **Shades of Gray:** Plots objects with Shades of Gray visual style applied regardless of the way the objects are displayed on the screen.
- **Sketchy:** Plots objects with Sketchy visual style applied regardless of the way the objects are displayed on the screen.
- **Wireframe:** Plots objects in wireframe regardless of the way they are displayed on the screen.
- **X-ray:** Plots objects with x-ray visual style applied regardless of the way the objects are displayed on the screen
- **Rendered:** Plots objects as rendered regardless of the way they are displayed on the screen.

Quality: Specifies the resolution at which shaded and rendered viewports are plotted. You can select from the following resolution options:

- **Draft:** Sets rendered and shaded model space views to be plotted as wireframe.
- **Preview:** Sets rendered and shaded model space views to be plotted at one quarter of the current device resolution, to a maximum of 150 dpi.
- **Normal:** Sets rendered and shaded model space views to be plotted at one half of the current device resolution, to a maximum of 300 dpi.
- **Presentation:** Sets rendered and shaded model space views to be plotted at the current device resolution, to a maximum of 600 dpi.
- **Maximum:** Sets rendered and shaded model space views to be plotted at the current device resolution with no maximum.
- **Custom:** Sets rendered and shaded model space views to be plotted at the resolution setting that you specify in the DPI box, up to the current device resolution.

DPI: Specifies the dots per inch for shaded and rendered views, up to the maximum resolution of the current plotting device. This option is available if you select Custom in the Quality box.

Plot Options: Specifies options for lineweights, transparency, plot styles, shaded plots, and the order in which objects are plotted.

Plot in Background: Specifies that the plot is processed in the background. (BACKGROUNDPLOT system variable)

Plot Object Lineweights: Specifies whether lineweights assigned to objects and layers are plotted.

Plot Transparency: Specifies whether object transparency is plotted. This option should only be used when plotting drawings with transparent objects.

Attention: For perfomance reasons, plotting transparency is disabled by default. To plot transparent objects, select the Plot Transparency option. This setting can be overridden by the PLOTTRANSPARENCYOVERRIDE system variable. By default, the system variable honors the setting in the Page Setup and the Plot dialog boxes.

Plot with Plot Styles: Specifies whether plot styles applied to objects and layers are plotted.

Plot Paperspace Last: Plots model space geometry first. Paper space geometry is usually plotted before model space geometry.

Hide Paperspace Objects: Specifies whether the HIDE operation applies to objects in the paper space viewport. This option is available only from a layout tab. The effect of this setting is reflected in the plot preview, but not in the layout.

Plot Stamp On: Turns on plot stamping. Places a plot stamp on a specified corner of each drawing and/or logs it to a file.

Plot stamp settings are specified in the Plot Stamp dialog box, in which you can specify the information that you want applied to the plot stamp, such as drawing name, date and time, plot scale, and so on. To open the Plot Stamp dialog box, select the Plot Stamp On option, and then click the Plot Stamp Settings button that is displayed to the right of the option.

You can also open the Plot Stamp dialog box by clicking the Plot Stamp Settings button on the Plot and Publish tab of the Options dialog box.

 **Plot Stamp Settings button:** Displays the Plot Stamp dialog box when the Plot Stamp On option is selected in the Plot dialog box.

Save Changes to Layout: Saves changes that you make in the Plot dialog box to the layout.

Drawing Orientation: Specifies the orientation of the drawing on the paper for plotters that support landscape or portrait orientation. The paper icon represents the media orientation of the selected paper. The letter icon represents the orientation of the drawing on the page.

Portrait: Orients and plots the drawing so that the short edge of the paper represents the top of the page.

Landscape: Orients and plots the drawing so that the long edge of the paper represents the top of the page.

Plot Upside-Down: Orients and plots the drawing upside down.

Icon: Indicates the media orientation of the selected paper and represents the orientation of the drawing on the page as a letter on the paper.

Note: Drawing orientation is also affected by the PLOTROTMODE system variable.

Less Options: Hides the following options in the Plot dialog box:

- Plot Style Table (Pen Assignments)
- Shaded Viewport Options
- Plot Options
- Drawing Orientation

9.2 GEOMETRIC TOLERANCES

Geometric tolerances show acceptable deviations of form, profile, orientation, location, and runout of a feature.

You add geometric tolerances in feature control frames. These frames contain all the tolerance information for a single dimension. Geometric tolerances can be created with or without leader lines, depending on whether you create them with the TOLERANCE command, or with options in the LEADER and QLEADER commands.

A feature control frame consists of two or more components. The first feature control frame contains a symbol that represents the geometric characteristic to which a tolerance is being applied, for example, location, profile, form, orientation, or runout. Form tolerances control straightness, flatness, circularity and cylindricity; profiles control line and surface. In the illustration, the characteristic is position.

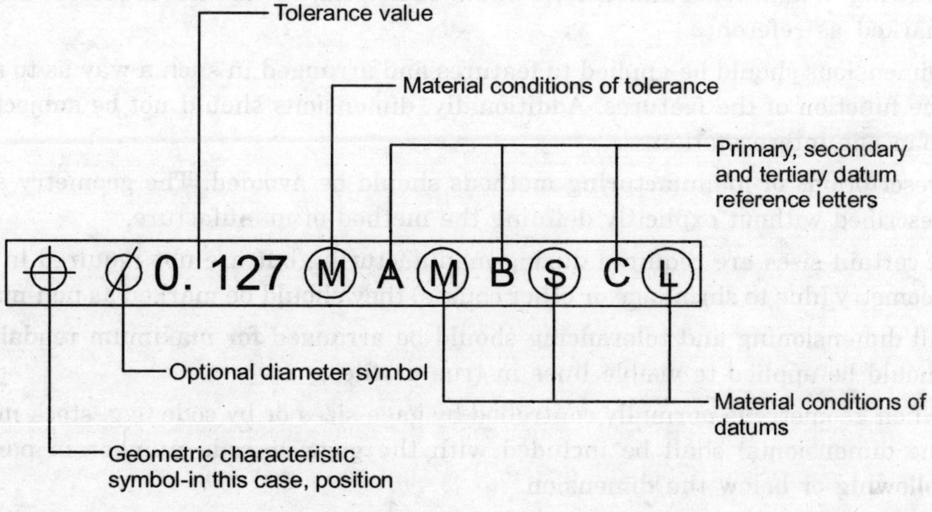

You can use most editing commands and grips to change feature control frames, and you can snap to them using object snaps.

You can also create tolerances that use *annotative scaling*. For more *information* about creating and working with an annotative tolerances, see *About Creating Annotative Dimensions and Tolerances, and About Scaling Annotations*.

Dimensioning and Tolerancing (ISO Standard)

According to the ASME Y14.5-2009 standard, the purpose of geometric dimensioning and tolerancing (GD&T) is to describe the engineering intent of parts and assemblies. The datum reference frame can describe how the part fits or functions. GD&T can more accurately define the dimensional requirements for a part, allowing over 50% more tolerance zone than coordinate (or linear) dimensioning in some cases. Proper application of GD&T will ensure that the part defined on the drawing has the desired form, fit (within limits) and function with the largest possible tolerances. GD&T can add quality and reduce cost at the same time through producibility.

There are some fundamental rules that need to be applied

- All dimensions must have a tolerance. Every feature on every manufactured part is subject to variation, therefore, the limits of allowable variation must be specified. Plus and minus tolerances may be applied directly to dimensions or applied from a general tolerance block or general note. For basic dimensions, geometric tolerances are indirectly applied in a related Feature Control Frame. The only exceptions are for dimensions marked as minimum, maximum, stock or reference.
- Dimensions define the nominal geometry and allowable variation. Measurement and scaling of the drawing is not allowed except in certain cases.
- Engineering drawings define the requirements of finished (complete) parts. Every dimension and tolerance required to define the finished part shall be shown on the drawing. If additional dimensions would be helpful, but are not required, they may be marked as reference.
- Dimensions should be applied to features and arranged in such a way as to represent the function of the features. Additionally, dimensions should not be subject to more than one interpretation.
- Descriptions of manufacturing methods should be avoided. The geometry should be described without explicitly defining the method of manufacture.
- If certain sizes are required during manufacturing but are not required in the final geometry (due to shrinkage or other causes) they should be marked as non-mandatory.
- All dimensioning and tolerancing should be arranged for maximum readability and should be applied to visible lines in true profiles.
- When geometry is normally controlled by gage sizes or by code (e.g. stock materials), the dimension(s) shall be included with the gage or code number in parentheses following or below the dimension.
- Angles of 90° are assumed when lines (including center lines) are shown at right angles, but no angular dimension is explicitly shown. (This also applies to other orthogonal angles of 0°, 180°, 270°, etc.)

- Dimensions and tolerances are valid at 20°C /101.3 kPa unless stated otherwise.
- Unless explicitly stated, all dimensions and tolerances are only valid when the item is in a free state.
- Dimensions and tolerances apply to the length, width, and depth of a feature including form variation.
- Dimensions and tolerances only apply at the level of the drawing where they are specified. It is not mandatory that they apply at other drawing levels, unless the specifications are repeated on the higher level drawing(s).

To Create Geometric Tolerances

1. Click Annotate tab Dimensions panel ➤ Tolerance. ⌖.1
2. In the Geometric Tolerance dialog box, click the first square under Sym and select a symbol to insert.
3. Under Tolerance 1, click the first black box to insert a diameter symbol.
4. In the Text box, enter the first tolerance value.
5. To add a material condition (optional), click the second black box and click a symbol in the Material Conditions dialog box to insert it.
6. In the Geometric Tolerance dialog box, add a second tolerance value (optional) in the same way as the first tolerance value.
7. Under Datum 1, Datum 2, Datum 3, enter the datum reference letter.
8. Click the black box to insert a material condition symbol for each datum reference.
9. In the Height box, enter a height.
10. Click the Projected Tolerance Zone box to insert the symbol.
11. In the Datum Identifier box, add a datum value.
12. Click OK.
13. In the drawing, specify a location for the feature control frame.

9.3 GEOMETRICAL CONSTRAINS

Geometric constraints associate geometric objects together, or specify a fixed location or angle.

For example, you can specify that a line should always be perpendicular to another one, that an arc and a circle should always remain concentric, or that a line should always be tangent to an arc.

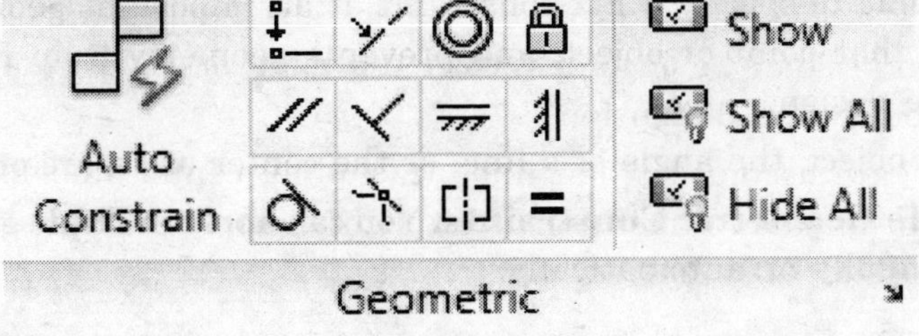

When you apply a constraint, two things occur:

- The object that you select adjusts automatically to conform to the specified constraint
- By default, a gray constraint icon displays near the constrained object as shown in the previous illustration, and a small, blue glyph displays with your cursor when you move it over a constrained object

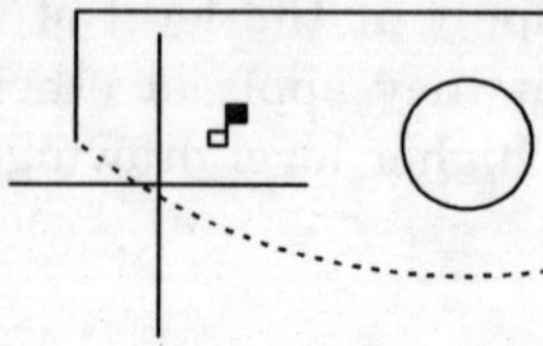

Once applied, constraints permit only those changes to the geometry that do not violate the constraints. This provides a method for exploring design options or making design changes while maintaining the requirements and specifications of the design.

Note: The order in which you select two objects when you apply a constraint is important in some cases. Normally, the second object you select adjusts to the first object. For example, when you apply a Perpendicular constraint, the second object you select will adjust to become perpendicular to the first.

You can apply geometric constraints to 2D geometric objects only. Objects cannot be constrained between model space and paper space.

1. Specify Constraint Points: With some constraints, you specify *constraint points* on objects instead of selecting the objects. This behavior is similar to that of object snaps, but the locations are limited to endpoints, midpoints, center points, and insertion points.

For example, the Coincident constraint can restrict the location of the endpoint of one line to the endpoint of another line.

The following glyph is displayed on the object as you roll over the object.

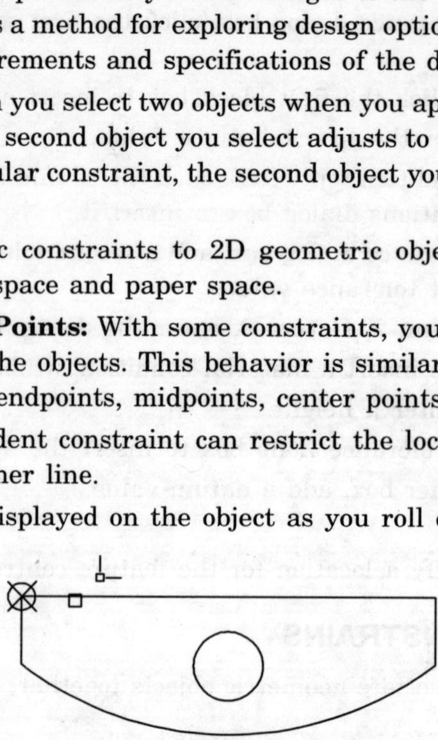

You use this glyph to confirm whether you are specifying the intended point to constrain.

2. Use Fix Constraints: A Fix constraint associates a constraint point on an object, or the object itself with a fixed location with respect to the World Coordinate System.

It is often advisable to specify a Fix constraint at an important geometric feature. This locks the location of that point or object, and prevents geometry from relocating when you make changes to the design.

When you fix an object, the angle of a line, or the center of an arc or circle is also fixed.

3. Apply Multiple Geometric Constraints: You can apply multiple geometric constraints to objects either manually or automatically.

When you want to apply all essential geometric constraints to a design automatically, you can use AUTOCONSTRAIN with the objects that you select in your drawing. This helps constrain the geometric shape of the design-depending on your design, there might be cases where you need to apply additional geometric constraints.

AUTOCONSTRAIN also provides settings in which you can specify the following options:

- What geometric constraints to apply
- What order to apply geometric constraints
- What tolerances are used to determine whether objects are horizontal, vertical, or touching

Note: Equal or Fix constraints are not applied with AUTOCONSTRAIN. You must apply these constraints individually.

To fully constrain the size and proportions of a design, you will later need to apply dimensional constraints.

(*a*) **To apply a coincident constraint**

1. Click Parametric tab ➤ Geometric panel ➤ Coincident.
2. Select the first and second points on two different objects.

   The second point is made coincident to the first.

(*b*) **To apply a collinear constraint**

1. Click Parametric tab ➤ Geometric panel ➤ Collinear.
2. Select the first and second object. You can select either a line object or a polyline sub object.

   The second object is made collinear to the first.

(*c*) **To apply a concentric constraint**

1. Click Parametric tab ➤ Geometric panel ➤ Concentric.
2. Select the first and second arc or circle object.

   The second arc or circle moves to have the same center point as the first object.

(*d*) **To apply a fix constraint**

1. Click Parametric tab ➤ Geometric panel ➤ Fix.
2. Select a point on an object.

   Applying the Fix constraint to a point on the object locks the node in place. You can still move the object.

(*e*) **To apply a parallel constraint**

1. Click Parametric tab ➤ Geometric panel ➤ Parallel.

2. Select the two objects to be made parallel. You can select either a line object or a polyline subobject.

   The second object is made parallel with the first object.

(*f*) To apply a perpendicular constraint

1. Click Parametric tab ➤ Geometric panel ➤ Perpendicular.
2. Select the two objects to be made perpendicular. You can select either a line object or a polyline subobject.

   The second object is made perpendicular to the first object.

(*b*) To apply a horizontal constraint

1. Click Parametric tab ➤ Geometric panel ➤ Horizontal.
2. Select the line object or polyline subobject to be made horizontal.

(*c*) To apply a vertical constraint

1. Click Parametric tab ➤ Geometric panel ➤ Vertical.
2. Select the line object or polyline subobject to be made vertical.

(*d*) To apply a tangent constraint

1. Click Parametric tab ➤ Geometric panel ➤ Tangent.
2. Select the two objects to be made tangential.

   The second object maintains a point of tangency with the first object.

(*e*) To apply a smooth constraint

1. Click Parametric tab ➤ Geometric panel ➤ Smooth.
2. Select the first spline curve.
3. Select the second spline, line, polyline (subobject), or arc object.

   The two objects are updated to be contiguous with one another.

(*f*) To apply a symmetric constraint

1. Click Parametric tab ➤ Geometric panel ➤ Symmetric.
2. Select the first and second object.
3. Select the symmetry line.

   The selected objects become symmetrically constrained about the selected line.

(*g*) To apply an equal constraint

1. Click Parametric tab ➤ Geometric panel ➤ Equal.
2. Select the first and second object.

   The second object is made equal to the first object.

(*h*) To apply multiple geometric constraints to an object

1. Click Parametric tab ➤ Geometric panel ➤ AutoConstrain.
2. Select the objects that you want to constrain.
3. Press Enter when you select the objects to be automatically constrained.

   The Command prompt displays the number of constraints applied.

(i) **To set the order for applying multiple geometric constraints to an object**

1. Click Parametric tab ➤ Geometric panel ➤ AutoConstrain.
2. At the Command prompt, enter s (Settings).
3. In the Constraint Settings dialog box, on the AutoConstrain tab, select a Constraint Type.
4. Click Move Up or Move Down. This changes the priority for a constraint when you use the AUTOCONSTRAIN command on an object.
5. Click OK.

4. Dimensional constraints: Dimensional constraints control the size and proportions of a design. They can constrain the following:

- Distances between objects, or between points on objects
- Angles between objects, or between points on objects
- Sizes of arcs and circles

For example, the following illustration includes linear, aligned, angular, and diameter constraints.

9.4 SPECIFIES SNAP AND GRID SETTINGS

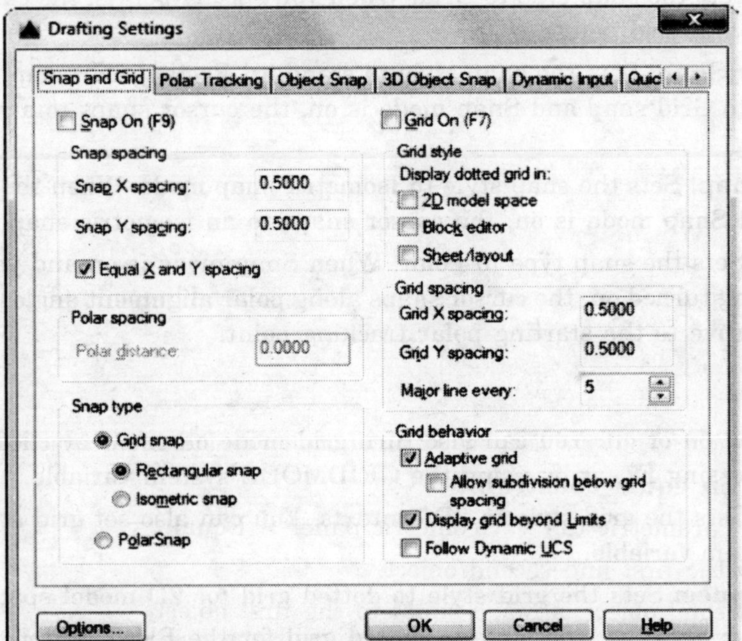

Note: The 3D Object Snap tab is not available in AutoCAD LT.

List of Options

Snap On

Turns Snap mode on or off. You can also turn Snap mode on or off by clicking Snap Mode on the status bar, by pressing F9, or by using the SNAPMODE system variable.

Snap Spacing: Controls an invisible, rectangular grid of snap locations that restricts cursor movement to specified X and Y intervals.

Snap X Spacing: Specifies the snap spacing in the X direction. The value must be a positive real number.

Snap Y Spacing: Specifies the snap spacing in the Y direction. The value must be a positive real number.

Equal X and Y Spacing: Forces the X and Y spacing to the same values for snap spacing and for grid spacing. The snap spacing intervals can be different from the grid spacing intervals.

Polar Spacing: Controls the PolarSnap increment distance.

Polar Distance: Sets the snap increment distance when PolarSnap is selected under Snap Type & Style.

If this value is 0, the PolarSnap distance assumes the value for Snap X Spacing. The Polar Distance setting is used in conjunction with polar tracking and/or object snap tracking. If neither tracking feature is enabled, the Polar Distance setting has no effect.

Snap Type: Sets the snap style and snap type.

Grid Snap: Sets the snap type to Grid. When you specify points, the cursor snaps along vertical or horizontal grid points.

Rectangular Snap: Sets the snap style to standard Rectangular snap mode. When the snap type is set to Grid snap and Snap mode is on, the cursor snaps to a rectangular snap grid.

Isometric Snap: Sets the snap style to Isometric snap mode. When the snap type is set to Grid snap and Snap mode is on, the cursor snaps to an isometric snap grid.

PolarSnap: Sets the snap type to Polar. When Snap mode is on and you specify points with polar tracking turned on, the cursor snaps along polar alignment angles set on the Polar Tracking tab relative to the starting polar tracking point.

Grid On

Turns the grid on or off. You can also turn grid mode on or off by clicking Grid on the status bar, by pressing F7, or by using the GRIDMODE system variable.

Grid Style: Sets the grid style in 2D contexts. You can also set grid style by using the GRIDSTYLE system variable.

2D Model Space: Sets the grid style to dotted grid for 2D model space.

Block Editor: Sets the grid style to dotted grid for the Block Editor.

Sheet/Layout: Sets the grid style to dotted grid for sheet and layout.

Grid Spacing: Controls the display of a grid that helps you visualize distances.

Grid X Spacing: Specifies the grid spacing in the X direction. If this value is 0, the grid assumes the value set for Snap X Spacing.

Grid Y Spacing: Specifies the grid spacing in the Y direction. If this value is 0, the grid assumes the value set for Snap Y Spacing.

Major Line Every: Specifies the frequency of major grid lines compared to minor grid lines.

Grid Behavior: Grid lines rather than grid dots are displayed under the following circumstances:

- **AutoCAD:** GRIDSTYLE is set to 0.
- **AutoCAD LT:** SHADEMODE is set to Hidden.

Controls the appearance of the grid lines.

Adaptive Grid: Limits the density of the grid when zoomed out.

- **Allow Subdivision Below Grid Spacing:** Generates additional, more closely spaced grid lines when zoomed in. The frequency of these grid lines is determined by the frequency of the major grid lines.

Display Grid Beyond Limits: Displays the grid beyond the area specified by the LIMITS command.

Changes the grid plane to follow the XY plane of the dynamic UCS.

EXERCISE

1. The following are some of the properties and settings that you can save in a layout, except ____________.
   (a) Plot scale (b) Paper size
   (c) Line weight (d) Drawing orientation
2. An AutoCAD sheet set is:
   (a) A number of drawings which have been printed from AutoCAD drawings and saved in an envelope A
   (b) A series of printed AutoCAD drawings A
   (c) A number of AutoCAD drawings saved in Paper Space format and held in a file A
   (d) Any drawings relating to a specific architectural, building or engineering process. Any
3. Drawing for a sheet set must be:
   (a) Saved in Model Space format Drawin
   (b) Saved in Paper Space format Saved
   (c) Saved in either Model Space or Paper Space format Saved
   (d) Saved in *.dwf format. Saved

4. DWF is an abbreviation for a file extension. The abbreviation stands for: DWF
   (a) Drawing with files Drawin
   (b) Design with file Design
   (c) Design Web File Design
   (d) Design Web Format. Design
5. DWF files can be included as attachments in an email. Is this statement: DWF
   a) True True
   b) Only partly true Only
   c) Incorrect Incorre
   d) Only when sent as a Zip file. Only

SUMMARY

- Displays standard paper sizes that are available for the selected plotting device. If no plotter is selected, the full standard paper size list is displayed and available for selection.
- According to the ASME Y14.5-2009 standard, the purpose of geometric dimensioning and tolerancing (GD&T) is to describe the engineering intent of parts and assemblies.
- Geometric tolerances show acceptable deviations of form, profile, orientation, location, and runout of a feature.
- Dimensional constraints control the size and proportions of a design. They can constrain the following:
- Distances between objects, or between points on objects, Angles between objects, or between points on objects, Sizes of arcs and circles.

CHAPTER 10

DIMENSIONING, ANNOTATIONS, LAYERING

Learning Objectives

- Using Dimensioning and annotative styles and objects
- Creating and Managing layers
- Learn to use filters
- Inserting and modifying blocks
- Learn Sectional Views

10.1 DIMENSIONING

The basic types of dimensioning are linear, radial, angular, ordinate, and arc length. Use the DIM command to create dimensions automatically according to the object type that you want to dimension.

You can control the appearance of dimensions by setting up dimension styles, or by editing individual dimensions in special cases. Dimension styles allow you to specify your conventions quickly and maintain industry or project dimensioning standards.

Tip: To simplify drawing organization and dimension scaling, you can create dimensions on layouts rather than in model space.

Linear Dimensions

Linear dimensions can be horizontal, vertical, or aligned. You can create an aligned, horizontal, or vertical dimension with the DIM command depending on how you move the cursor when placing the text.

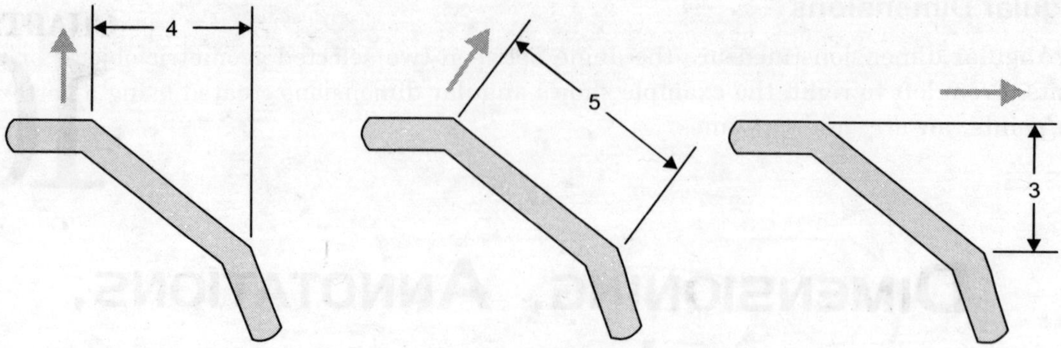

In rotated dimensions, the dimension line is placed at an angle to the extension line origin points. In this example, the angle specified for dimension rotation is equal to the angle of the slot.

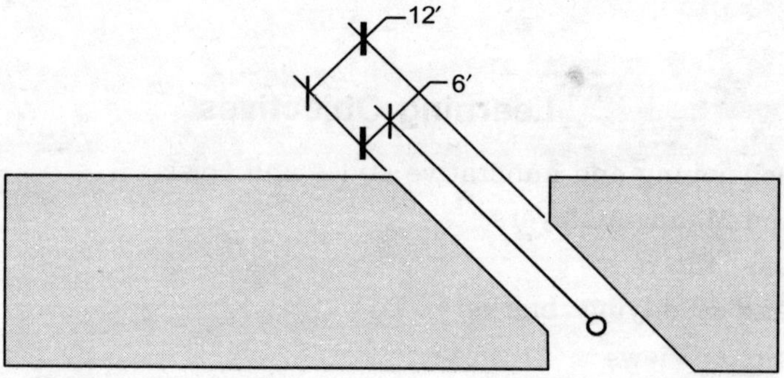

Note: You can also create linear dimensions with extension lines that are not perpendicular to their dimension lines. These are called oblique dimensions and they are most commonly used with isometric drafting. In this case, the dimension line angles will be oriented at 30 and 60 degrees depending on the current isoplane.

Radial Dimensions

A radial dimension measures the radius or diameter of arcs and circles with an optional centerline or center mark. Several options are displayed in the illustration.

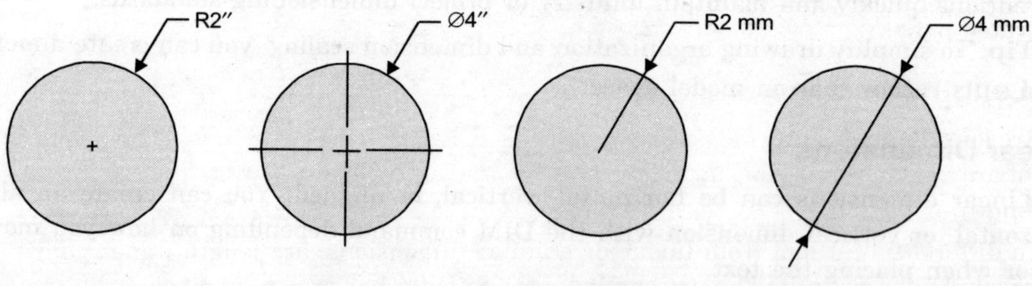

Note: When part of the dimension is located within the dimensioned arc or circle, the non-associative centerline or center mark is automatically suppressed.

Angular Dimensions

Angular dimensions measure the angle between two selected geometric objects or three points. From left to right, the example shows angular dimensions created using a vertex and two points, an arc, and two lines.

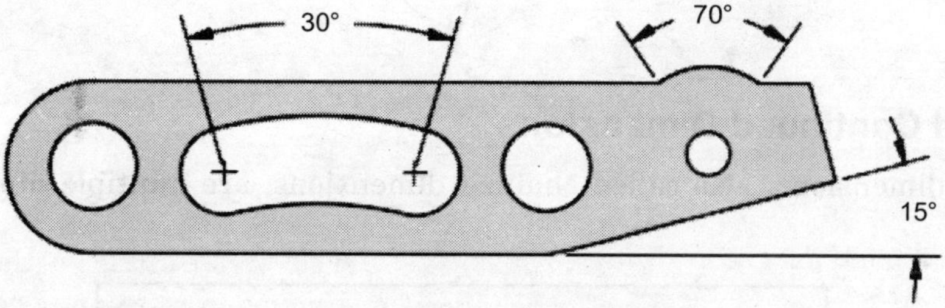

Ordinate Dimensions

Ordinate dimensions measure the perpendicular distances from an origin point called the *datum*, such as a hole in a part. These dimensions prevent escalating errors by maintaining accurate offsets of the features from the datum.

Important: The datum is established by the current location of the UCS origin.

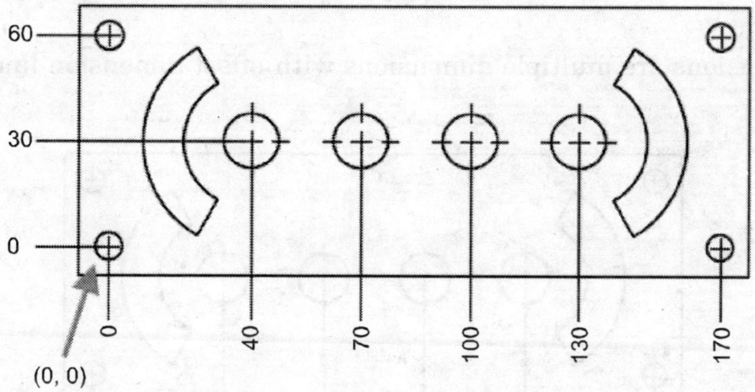

In this example, the datum (0, 0) is indicated as the hole in the lower-left corner of the illustrated panel.

Arc Length Dimensions

Arc length dimensions measure the distance along an arc or polyline arc segment. Typical uses of arc length dimensions include measuring the travel distance around a cam or indicating the length of a cable.

To differentiate them from linear or angular dimensions, arc length dimensions display an arc symbol by default. The arc symbol, also called a hat or cap, is displayed either above the dimension text or preceding the dimension text.

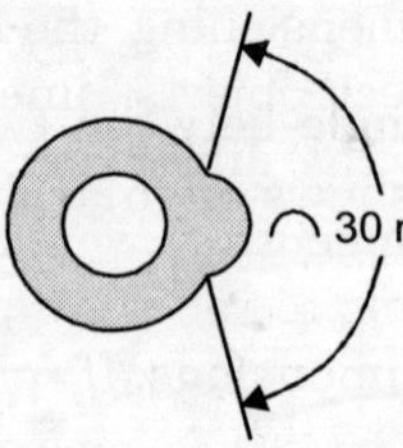

Baseline and Continued Dimensions

Continued dimensions, also called chained dimensions, are multiple dimensions placed end-to-end.

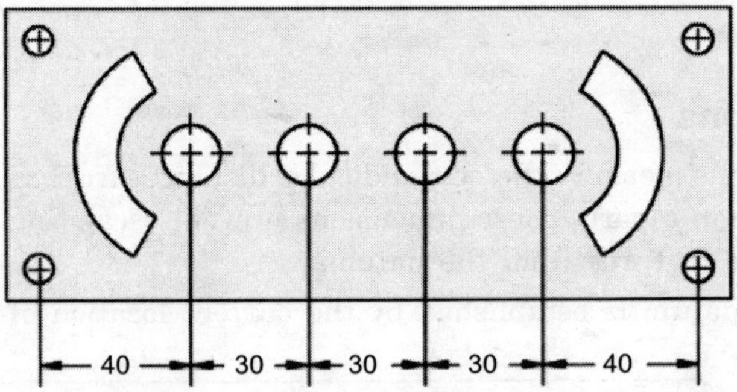

Baseline dimensions are multiple dimensions with offset dimension lines measured from the same location.

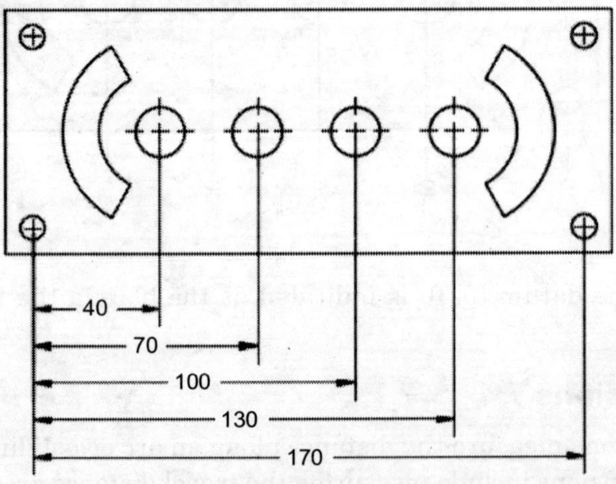

Creates multiple types of dimensions within a single command session.

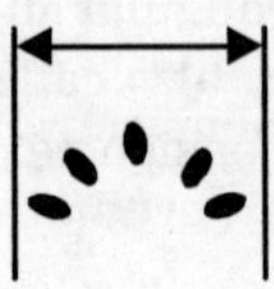

When you hover over an object for dimensioning, the DIM command automatically previews a suitable dimension type to use. Select objects, lines, or points to dimension and click anywhere in the drawing area to draw the dimension.

The supported dimension types range from vertical, horizontal, aligned, and rotated linear dimensions, to angular dimensions, to radius, diameter, jogged radius, and arc length dimensions, to baseline and continued dimensions. If required, you can change the dimension type using command line options.

The following prompts are displayed.

Select objects: Automatically selects an applicable dimension type for the objects you select and displays the prompts corresponding to that dimension type.

Selected object type	Action
Arc	Defaults the dimension type to radius dimensions.
Circle	Defaults the dimension type to radius dimensions.
Line	Defaults the dimension type to linear dimensions.
Dimension	Displays options to modify the selected dimension.
Ellipse	Defaults to options set for selecting a line.

First extension line origin: Creates a linear dimension when you select two points.

Angular: Creates an angular dimension showing the angle between three points by or the angle between two lines (same as the DIMANGULAR command).

- **Vertex:** Specifies the point to use as the vertex of the angle.
- **Specify first side of angle:** Specifies one of the lines that form the angle.
- **Specify second side of angle:** Specifies the other line that forms the angle.
- **Angular dimension location:** Specifies where to place the dimension arc line. Depending on the position of the dimension, the quadrant for the dimension changes.
- **Mtext:** Displays the Text Editor contextual tab, which you can use to edit the dimension text.
- **Text:** Customizes the dimension text at the Command prompt. The generated dimension is displayed within angle brackets.
- **Text angle:** Specifies the angle of the dimension text.
- **Undo:** Returns to the previous prompt.
- **Undo:** Returns to the previous prompt.

Baseline: Creates a linear, angular, or ordinate dimension from the first extension line of the previous or selected dimension (same as the DIMBASELINE command).

Note: By default, the last created dimension is used as the base dimension.

- **First extension line origin:** Specifies the first extension line of the base dimension as the extension line origin for the baseline dimension.
- **Second extension line origin:** Specifies the next edge or angle to dimension.
- **Feature Location:** Uses the endpoint of the base dimension (ordinate dimension) as the endpoint for the baseline dimension.
- **Select:** Prompts you select a linear, ordinate, or angular dimension to use as the base for the baseline dimension.

- **Offset:** Specifies the offset distance from which the baseline dimensions are created.
- **Undo:** Undoes the last baseline dimension created.

Continue: Creates a linear, angular, or ordinate dimension from the second extension line of a selected dimension (same as the DIMCONTINUE command).

- First extension line origin: Specifies the first extension line of the base dimension as the extension line origin for the continued dimension.
- Second extension line origin: Specifies the next edge or angle to dimension.
- Feature location: Uses the endpoint of the base dimension (ordinate dimension) as the endpoint for the continued dimension.
- Select: Prompts you select a linear, ordinate, or angular dimension to use as the base for the continued dimension.
- Undo: Undoes the last baseline dimension created.

Ordinate: Creates an ordinate dimension (same as DIMORDINATE command).

- **Feature location:** Prompts for a point on a feature such as an endpoint, intersection, or center of an object.
- **Leader endpoint:** Uses the difference between the feature location and the leader endpoint to determine whether it is an Xor a Y ordinate dimension. If the difference in the Y ordinate is greater, the dimension measures the X ordinate. Otherwise, it measures the Y ordinate.
- **Xdatum:** Measures the X ordinate and determines the orientation of the leader line and dimension text.
- **Ydatum:** Measures the Y ordinate and determines the orientation of the leader line and dimension text.
- **Mtext:** Displays the Text Editor contextual tab, which you can use to edit the dimension text.
- **Text:** Customizes the dimension text at the Command prompt. The generated dimension is displayed within angle brackets.
- **Angle:** Specifies the angle of the dimension text.
- **Undo:** Returns to the previous prompt.
- **Undo:** Returns to the previous prompt.

Align: Aligns multiple parallel, concentric, or same datum dimensions to a selected base dimension.

Base dimension: Specifies a dimension to use as basis for the dimensions alignment.

- **Dimensions to align:** Selects the dimensions to align to the selected base dimension.
- **Distribute:** Specifies the method on how to distribute a group of selected isolated linear or ordinate dimensions.
- **Equal:** Equally distributes all selected dimensions. This method requires a minimum of three dimension lines.
- **Offset:** Distributes all selected dimensions at a specified offset distance.

Layer: Assigns new dimensions to the specified layer, overriding the current layer. Enter Use Current or " · " to use the current layer. (DIMLAYER system variable)

Undo: Reverses the last dimension operation.

The following options are displayed when you place a dimension in such a way that it overlaps an existing dimension.

Move away: Arranges the existing dimension and the newly inserted dimension into a baseline dimension type.

Break up: Splits up the existing dimension into two dimensions, and arranges those dimensions into a continued dimension type.

Replace: Deletes the existing dimension and replaces it with the one you insert.

None: Dimension Style manager

Creates new styles, sets the current style, modifies styles, sets overrides on the current style, and compares styles.

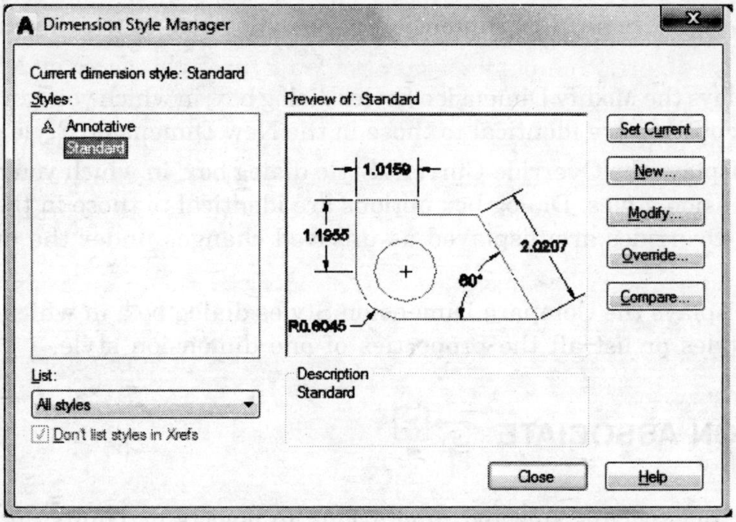

List of Options

The following options are displayed.

Current Dimension Style: Displays the name of the dimension style that is current. The default dimension style is STANDARD. The current style is applied to dimensions you create.

Styles: Lists dimension styles in the drawing. The current style is highlighted. Right-click in the list to display a shortcut menu with options to set the current style, rename styles, and delete styles. You cannot delete a style that is current or in use in the current drawing.

The  icon before the style name indicates that the style is annotative.

Unless you select Don't List Styles in Xrefs, dimension styles are displayed in externally referenced drawings using the syntax for externally referenced named objects. Although you cannot change, rename, or make current externally referenced dimension styles, you can create new styles based on them.

The item selected in List controls the dimension styles displayed.

List: Controls the display of styles in the Styles list. Select All Styles if you want to see all dimension styles in a drawing. Select Styles in Use if you want to see only the dimension styles currently used by dimensions in the drawing.

Don't List Styles in Xrefs: When selected, suppresses display of dimension styles in externally referenced drawings in the Styles list.

Preview: Shows a graphic representation of the style selected in the Styles list.

Description: Describes the style selected in the Styles list relative to the current style. If the description is longer than the space provided, you can click in the pane and use arrow keys to scroll down.

Set Current: Sets the style selected under Styles to current. The current style is applied to dimensions you create.

New: Displays the Create New Dimension Style dialog box, in which you can define a new dimension style.

Modify: Displays the Modify Dimension Styles dialog box, in which you can modify dimension styles. Dialog box options are identical to those in the New Dimension Style dialog box.

Override: Displays the Override Current Style dialog box, in which you can set temporary overrides to dimension styles. Dialog box options are identical to those in the New Dimension Style dialog box. Overrides are displayed as unsaved changes under the dimension style in the Styles list.

Compare: Displays the Compare Dimension Styles dialog box, in which you can compare two dimension styles or list all the properties of one dimension style.

10.2 DIMENSION ASSOCIATE 

Associates or reassociates selected dimensions to objects or points on objects.

Each selected dimension is highlighted in turn, and prompts for association points appropriate for the selected dimension are displayed.

A marker is displayed for each association point prompt.

- If the definition point of the current dimension is not associated to a geometric object, the marker appears as an X.
- if the definition point is associated, the marker appears as an X inside a box.

Press Esc to end the command without losing the changes that were already specified. Use UNDO to restore the previous state of the changed dimensions.

Note: The marker disappears if you pan or zoom with a wheel mouse.

The following prompts are displayed.

Select Objects: Specifies one or more non-associative dimension or leader objects to reassociate them manually to objects or points on objects.

Disassociated: Specifies all non-associative dimension or leader objects for manual reassociation.

10.3 LAYER

Create, rename, and remove layers. Set the current layer. Change layer properties such as color, linetype, lineweight, and transparency.

Use the Layer Properties Manager to work with layers. Click Home tab ➤ Layers panel ➤ Layer Properties.

Create a Layer

1. In the Layer Properties Manager, click New Layer.

   A layer name is added to the layer list.

2. Enter a new layer name by typing over the highlighted layer name.
   - Layer names can be up to 255 characters long (double-byte or alphanumeric), and include letters, numbers, spaces, and several special characters.
   - Layer names cannot include the following characters: < > / \ “: ; ? * | =’
3. For complex drawings with many layers, enter descriptive text in the Description column.
4. Specify the settings and default properties of the new layer by clicking in each column.

Rename a Layer

1. In the Layer Properties Manager, click to select a layer.
2. Click the layer name or press F2.
3. Enter a **new** name.

Remove a Layer

1. In the Layer Properties Manager, click to select a layer.
2. Click Delete Layer.

The following layers cannot be deleted:

- Layers 0 and Defpoints
- Layers containing objects, including objects in block definitions
- The current layer
- Layers used in an external reference

Note: To remove all unused layers, use PURGE.

Set the Current Layer

1. In the Layer Properties Manager, click to select a layer.
2. Click Set Current.

Change the Properties Assigned to Layers

1. If you want to change multiple layers, use one of the following methods in the Layer Properties Manager:
   - Press and hold Ctrl, and choose several layer names.
   - Press and hold Shift, and choose the first and last layers in a range.
   - Right-click, and click Show Filters in Layer List. Choose a layer filter from the list of layers.
2. Click the current setting in the column that you want to change.
   The dialog box for that property displays.
3. Choose the setting that you want to use.

When changing layer properties:

- If the linetype you want is not displayed, click Load and use one of the following methods:
  - ♦ In the Load or Reload Linetypes dialog box, choose the linetypes to load.
  - ♦ In the Load or Reload Linetypes dialog box, click File to open an additional linetype definition (LIN) file. Choose the linetypes to load and click OK.
- Lineweights are not displayed automatically. If you want to display or hide lineweights, click Show/Hide Lineweight on the status bar.

  If no change is visible, it's probably due to a combination of the thickness of the line compared to the display resolution of your monitor.
- Transparency is not displayed automatically. If you want to display or hide the transparency of objects, click Show/Hide Transparency on the status bar.

Layer Properties Manager: You can add, delete, and rename layers, change their properties, set property overrides in layout viewports, and add layer descriptions.

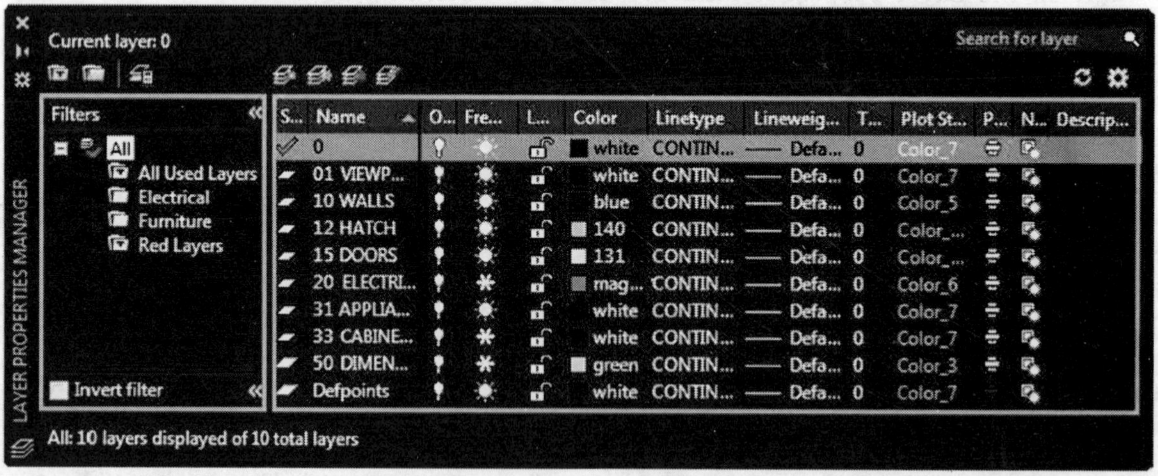

Filter List Filter List

Hover over the image to find clickable areas which jump to the Help section for those controls.

Layers and Layer Properties

New Layer 

Creates a layer with a default name that you can immediately change. The new layer inherits the properties of the currently selected layer in the layer list.

New Layer Frozen VP In All Viewports 

Creates a layer and freezes it in all existing layout viewports. This button is accessible from the Model tab or layout tabs.

Delete Layer 

Deletes selected layers. The following layers cannot be deleted:

- Layers 0 and Defpoints
- Layers containing objects, including objects in block definitions
- The current layer
- Layers used in an external reference
- Layers in a partially opened drawing (Not applicable to AutoCAD LT.)

Note: Be careful about deleting layers if you are working on a drawing in a shared project or one based on a set of layering standards.

Set Current 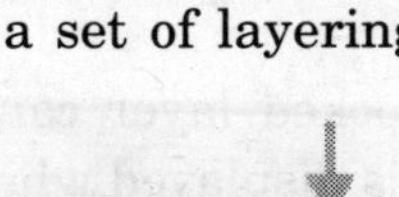

Sets the selected layer as the current layer. New objects are automatically created on the current layer. (CLAYER system variable)

Layer List: Modify the layer properties using the layer list. Click the current setting to change the layer property for the selected layer or group of layers.

Note: The layer list can be filtered and sorted to make it easier to find and select the layers that you want to change.

Sort: Click a column label to sort by that column.

Column Order: Drag a column to a new location in the list to change the column order.

Status

- {blank} Layer filter

- This layer is the current layer
- This layer contains objects
- This layer does not contain any objects

Note: To improve performance, all layers are indicated as containing objects, , by default. You can turn on this feature in layer settings.

- This layer contains objects, and a property override is turned on in a layout viewport
- This layer does not contain any objects, and a property override is turned on in a layout viewport

Name: Displays the name of the layer or filter. Press F2 to enter a new name.

On

Turns the selected layers on and off. When a layer is on, it is visible and available for plotting. When a layer is off, it is invisible and not plotted, even if the setting in the Plot column is turned on.

Freeze

Freezes the selected layers. You can freeze layers to improve performance and reduce regeneration time in complex drawings. Objects on frozen layers are not displayed, plotted, or regenerated.

In drawings that support 3D modeling, frozen layers are not rendered. (Not applicable to AutoCAD LT.)

Tip: Freeze the layers that you want to remain invisible for long periods. If you plan to switch visibility settings frequently, use the On/Off setting to avoid regenerating the drawing.

Lock

Locks and unlocks the selected layers. Objects on a locked layer cannot be modified. Objects on locked layers appear faded and a small lock icon is displayed when you hover over the object.

Note: Set the fade level for locked layers to see which objects are on locked layers.

Color ■□

Displays the Select Color dialog box, where you can specify a color for the selected layers.

Linetype: Displays the Select Linetype dialog box, where you can specify a linetype for the selected layers.

Lineweight: Displays the Lineweight dialog box, where you can specify a lineweight for the selected layers.

Transparency: Displays the Transparency dialog box, where you can specify a transparency for the selected layers. Valid values are from 0 to 90. The higher the value, the more transparent the objects appear.

Plot Style: Displays the Select Plot Style dialog box where you can change the plot style for the selected layers. For color-dependent plot styles (the PSTYLEPOLICY system variable is set to 1), you cannot change the plot style associated with a layer.

Plot

Controls whether the selected layers are plotted. If you turn off plotting for a layer, the objects on that layer are still displayed. Layers that are off or frozen are not plotted, regardless of the Plot setting.

New VP Freeze

Freezes selected layers in new layout viewports. For example, freezing the DIMENSIONS layer in all new viewports restricts the display of dimensions in any newly created layout viewports but does not affect the DIMENSIONS layer in existing viewports. If you later create a viewport that requires dimensions, you can override the default setting by changing the current viewport setting.

VP Freeze (available only from a layout tab)

Freezes selected layers only in the current layout viewport. If a layer is already frozen or turned off in the drawing, you can't thaw the layer in the current layout viewport.

VP Color (available only from a layout tab) ■□

Sets an override for the color associated with the selected layer for the current layout viewport.

VP Linetype (available only from a layout tab): Sets an override for the linetype associated with the selected layer for the current layout viewport.

VP Lineweight (available only from a layout tab): Sets an override for the lineweight associated with the selected layer for the current layout viewport.

VP Transparency (available only from a layout tab): Sets an override for transparency associated with the selected layer for the current layout viewport.

VP Plot Style (available only from a layout tab): Sets an override for the plot style associated with the selected layer for the current layout viewport. Override settings are not visible in the viewport or plotted when the visual style in the drawing is set to Conceptual or Realistic. For color-dependent plot styles (the PSTYLEPOLICY system variable is set to 1), you cannot set a plot style override.

Description:(Optional) Describes the layer or the layer filter.

Manage the Layer List

Search for Layer

Filters the layer list by name as you enter characters in the box.

The following wildcards are available:

Character	Definition
# (pound)	Matches any numeric digit
@ (at)	Matches any alphabetic character
. (period)	Matches any non-alphanumeric character
* (asterisk)	Matches any string and can be used anywhere in the search string
? (question mark)	Matches any single character; for example, ?BC matches ABC, 3BC, and so on
~ (tilde)	Matches anything but the pattern; for example, ~*AB* matches all strings that don't contain AB
[ ]	Matches any one of the characters enclosed; for example, [AB]C matches AC and BC
[~]	Matches any character not enclosed; for example, [~AB]C matches XC but not AC
[-]	Specifies a range for a single character; for example, [A-G]C matches AC, BC to GC, but not HC
` (reverse quote)	Reads the next character literally; for example, `~AB matches ~AB

New Property Filter

Displays the Layer Filter Properties dialog box, where you can create a layer filter. Layer filters limit the layers listed in the Layer Properties Manager to those with specified settings and properties. For example, you can limit the list of layers to only those that are turned on and thawed.

New Group Filter

Creates a layer filter that includes only the layers that you drag to the filter.

Invert Filter: Displays all layers that do not meet the criteria in the selected layer filter.

Filters List: Displays the list of layer filters in the drawing. Click » « to expand or collapse the filter list. When the Filters list is collapsed, use the Layer Filter button (▾) at the bottom-left corner of the Layer Properties Manager to display a list of filters.

There are six predefined filters:

- **All:** Lists all the layers in the current drawing.
- **All non-Xref Layers:** Lists all the layers that are not referenced from an xref drawing.
- **All Used Layers:** Lists all the layers that contain objects.
- **Xref:** Lists all the layers being referenced from an xref drawing.
- **Viewport Overrides:** Lists all the layers containing property overrides in the current layout viewport.
- **Unreconciled New Layers:** Lists all new unreconciled layers that were added since the drawing was last opened, saved, reloaded, or plotted. When the layer notification feature is turned on, a new layer is considered unreconciled until you accept the layer as reconciled.

Miscellaneous Tools

Layer States Manager

Displays the Layer States Manager, where you can save, restore, and manage sets of layer settings that are called layer states.

Refresh

Refreshes the order of the layer list and the layer status information.

Settings

Displays the Layer Settings dialog box, where you can set various display options.

Column Label Shortcut Menu

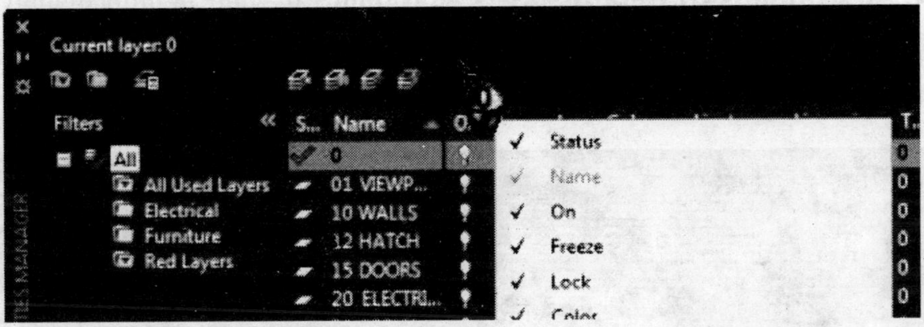

Column Names: Lists all columns by name. A check mark indicates that the column is included in the display. Click a column name to display or hide the column. VP Freeze, VP Color, VP Linetype, VP Lineweight, and VP Plot Style are available only when a layout viewport is active.

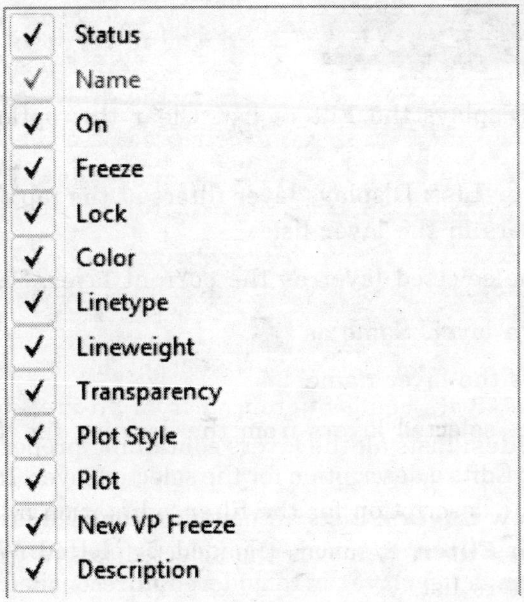

Customize: Displays the Customize Layer Columns dialog box, where you can specify which columns are hidden or displayed, or change the column order.

Maximize All Columns: Changes the width of all columns to fit the column headers and the data content.

Maximize Column: Changes the width of the selected column to fit the column header and the data content of that column.

Optimize all columns: Changes the width of all columns to fit the content in each column.

Optimize column: Changes the width of the selected column to fit the content in that column.

Freeze column (or Unfreeze column): Freeze keeps the column, and any columns to the left, visible while scrolling. Unfreeze so that all columns scroll.

Restore All Columns to Defaults: Sets all columns to their default display and width settings.

Layer List Shortcut Menu

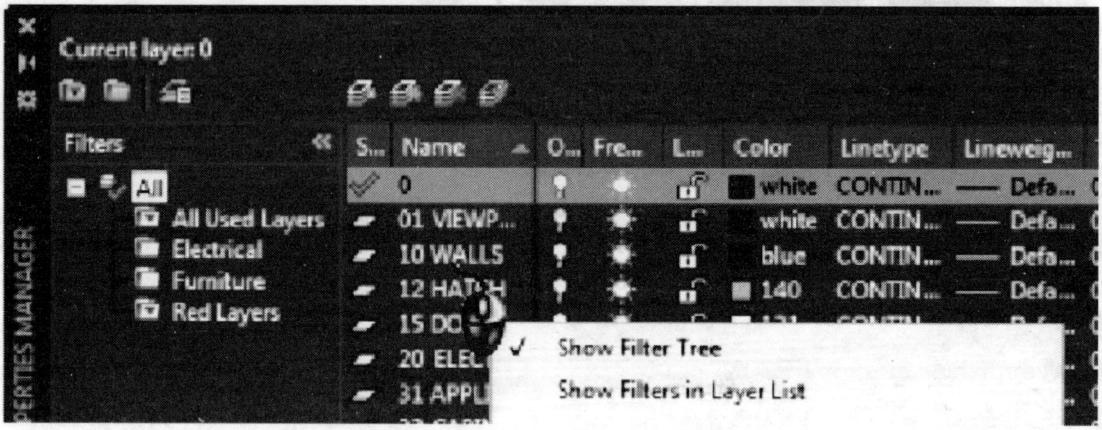

Show Filter Tree: Displays the Filters list. Clear this option to hide the list. Same as » «.

Show Filters in Layer List: Displays layer filters at the top of the layer list. Clear this option to display only layers in the layer list.

Set Current: Sets the selected layer as the current layer. Same as .

New Layer: Creates a layer. Same as .

Rename Layer: Edits the layer name.

Delete Layer: Deletes selected layers from the drawing file. Same as .

Change Description: Edits a description for the selected layer. If layer filters are displayed in the layer list, editing the description for the filter, edits it on all layers in the filter.

Remove from Group Filter: Removes the selected layers from the layer group filter that is selected in the Filters list.

Reconcile Layer: Removes new layers from the Unreconciled New Layers filter. This option is available only when unreconciled layers are selected. When the layer notification feature is turned on, a new layer is considered unreconciled until you accept the layer as reconciled.

Remove Viewport Overrides For: This option is only available in layout viewports. Removes a single override property or all property overrides on selected layers (or all layers) for the current viewport or all viewports. Different options display in a flyout menu depending on where the cursor is located when the shortcut menu is opened. To remove a single property override, right-click the property override.

New Layer Frozen in All Viewports: Creates a layer and freezes it in all existing layout viewports and new viewports. Same as .

VP Freeze Layer: Freezes the selected layer in all new and existing layout viewports.

VP Thaw Layer in All Viewports: Thaws the selected layer in all new and existing layout viewports.

Isolate Selected Layers: Turns off all layers except the selected layers.

Merge Selected Layer(s) To: Combines the selected layers into a specified layer. Objects on the selected layers are moved to the new layer and inherit the properties for that layer.

Select All: Selects all the layers displayed in the layer list.

Clear All: Deselects all layers in the layer list.

Select All But Current: Selects all the layers displayed in the layer list except the current layer.

Invert Selection: Selects everything displayed in the layer list except the items that are currently selected.

Invert Layer Filter: Displays all layers that do not meet the criteria in the active layer property filter.

Layer Filters: Displays a list of layer filters, including default layer filters. Click a filter to apply it to the layer list.

Save Layer States: Saves the current layer settings as a layer state.

Restore Layer State: Displays the Layer States Manager, where you can select a layer state to restore. This operation restores only those settings that were specified in the Layer States Manager when the layer state was saved.

Filter List Shortcut Menu

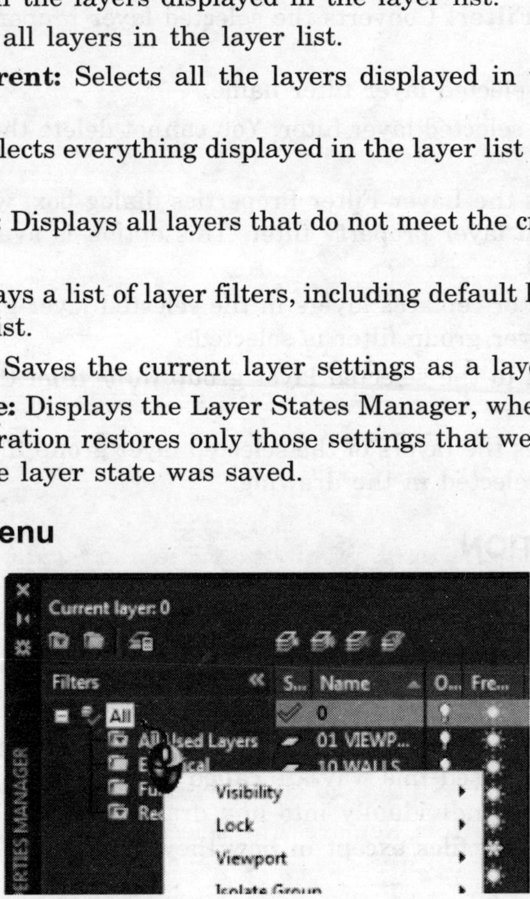

Visibility: Changes the visibility of the layers in the selected filters.

Lock: Controls whether layers are locked or unlocked in the selected filters.

Viewport: In the current layout viewport, controls the VP Freeze setting of the layers in the selected filters.

Isolate Group: Freezes all layers that are not included in the selected filters.

- **All Viewports.**
  - ♦ In all layout viewports, sets VP Freeze for all layers that are not included in the selected filters.
  - ♦ In model space, freezes all layers that are not in the selected filter, except for the current layer.
- **Active Viewport Only.**
  - ♦ In the current layout viewport, sets VP Freeze for all layers that are not included in the selected filters.
  - ♦ In model space, turns off all layers that are not included the selected filters, except for the current layer.

New Properties Filter: Displays the Layer Filter Properties dialog box. Same as .

New Group Filter: Creates a layer group filter. Same as .

Convert to Group Filter: Converts the selected layer property filter to a layer group filter.

Rename: Edits the selected layer filter name.

Delete: Removes the selected layer filter. You cannot delete the All, All Used Layers, or Xref layer filters.

Properties: Displays the Layer Filter Properties dialog box, where you can modify the definitions of the selected layer property filter. This option is available only when a layer property filter is selected.

Select Layers: Adds or replaces layers in the selected layer group filter. This option is available only when a layer group filter is selected.

- **Add:** Adds layers to the selected layer group filter from the objects selected in the drawing.
- **Replace:** Replaces the layers of the selected layer group filter with the layers of the objects that are selected in the drawing.

10.4 BLOCK DEFINITION

Create Block Libraries

A block library is a collection of block definitions stored in a single drawing file. You can use block libraries supplied by Autodesk or other vendors or create your own.

You can organize a set of related block definitions by creating the blocks in the same drawing file. Drawing files used this way are called block, or symbol, libraries. These block definitions can be inserted individually into any drawing. Block library drawings are not different from other drawing files except in how they are used.

When you define a block in the block library drawing (using BLOCK), you can include a short description of the block.

You can also creating a visual index of each block definition by inserting it in the library drawing and including text that provides the block name, the date of creation, the date of the last modification, and any special instructions or conventions.

Use DesignCenter to view and copy block definitions individually from block library drawings (or from any existing drawing) to your current drawing. DesignCenter does not overwrite an existing block definition in a drawing with one that comes from another drawing.

Use Tool Palettes to Organize Blocks

Once you've added a block tool to a tool palette, you can insert the block reference by dragging it from the tool palette to the drawing or by clicking and placing it in the drawing.

Create a Block Within a Drawing

A block is essentially a block definition that includes the block name, the block geometry, the location of the base point to be used for aligning the block when you insert it, and any associated attribute data. You can define a block from drawing geometry either in the Block Definition dialog box or by using the Block Editor. Once the block definition is created, you can reference it in the same or a different drawing.

The block definition in the following illustration comprises a name, four lines, and a base point at the intersection of the two diagonal lines. In this (and in the second example), when the block is saved, the geometry is removed from the drawing area.

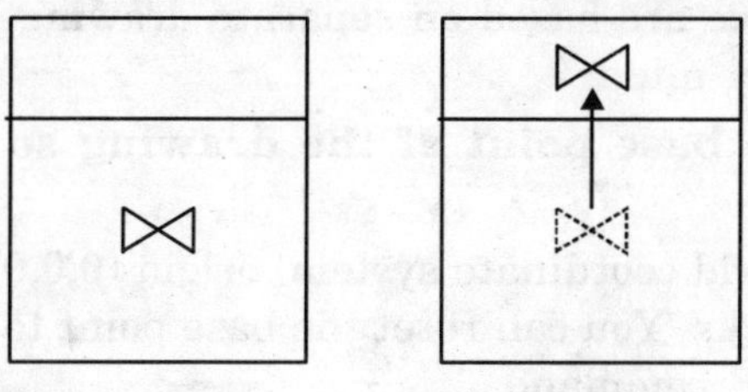

Objects in drawing area — Block definition created from selected objects

The following illustration shows a typical sequence for creating a block definition within a drawing.

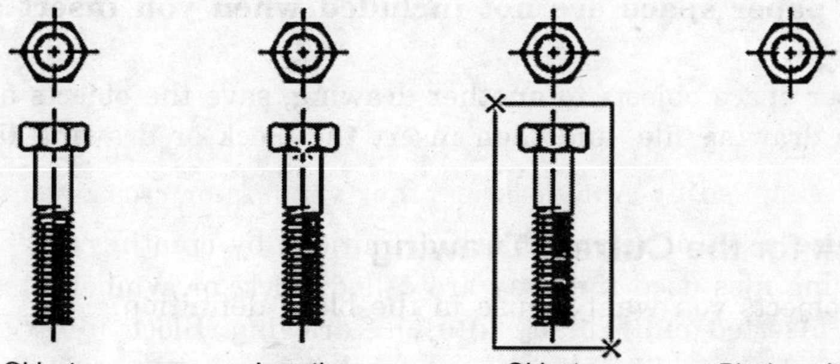

Objects created — Insertion point selected — Objects selected — Blocks written to block table

Create a Drawing File for Use as a Block

You can create drawing files for the purpose of inserting them into other drawings as blocks. Individual drawing files are easy to create and manage as the source of block definitions. Collections of symbols can be stored as individual drawing files and grouped in folders.

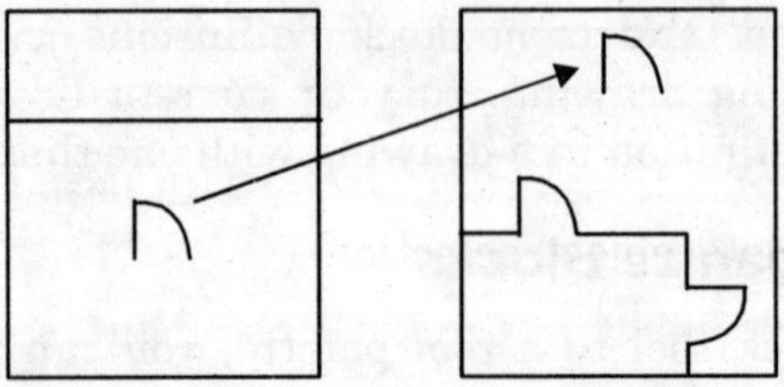

Methods for creating drawing files for use as blocks include

- Create and save a complete drawing file.
- Create and save only selected objects to a new drawing using EXPORT or WBLOCK (Write Block).

With either method, you create an ordinary drawing file that can be inserted as a block into any other drawing file. Use WBLOCK when you need to save several versions of a symbol as separate drawing files, or when you want to create a drawing file without leaving the current drawing.

If the blocks you reference are based on separate drawings (rather than a named set of objects), keep these points in mind:

- **You can change the base point of the drawing so that it works better as a block.**

By default, the WCS (world coordinate system) origin (0,0,0) is used as the base point for drawing files inserted as blocks. You can reset the base point to specify a different base point for insertion using the BASE command.

- **If you expect the original drawing to change, consider attaching the drawing as an xref instead of inserting it as a block.**

  If the original drawing is updated after you insert it as a block, the changes are not reflected in the destination drawing.

- **Objects in paper space are not included when you insert a drawing as a block.**

To transfer paper space objects to another drawing, save the objects as a block or save them in a separate drawing file, and then insert the block or drawing file into the other drawing.

To Define a Block for the Current Drawing

1. Create the objects you want to use in the block definition.
2. Click Insert tab Block Definition panel Create Block. 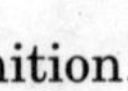

3. In the Block Definition dialog box, enter a block name in the Name box.
4. Under Objects, select Convert to Block.

   Note: If Delete is selected, the original objects are erased from the drawing when the block is created. If necessary, you can use OOPS to restore them.
5. Under Base Point and Objects, make sure Specify On-screen is not selected.
6. Select Objects.
7. Use your pointing device to select the objects to be included in the block definition. Press Enter to complete object selection.
8. In the Block Definition dialog box under Base Point, specify the block insertion point using one of these methods.
   - Click Pick Point to specify a point using the pointing device.
   - Enter the X,Y,Z coordinate values of the point.
9. In the Description box, enter a description for the block definition. This description is displayed in DesignCenter™ (ADCENTER).
10. Click OK.

    The block is defined in the current drawing and can be inserted at any time.

10.5 FILTER (OBJECT SELECTION METHOD)

Selects, edits, and names filters for object selection.

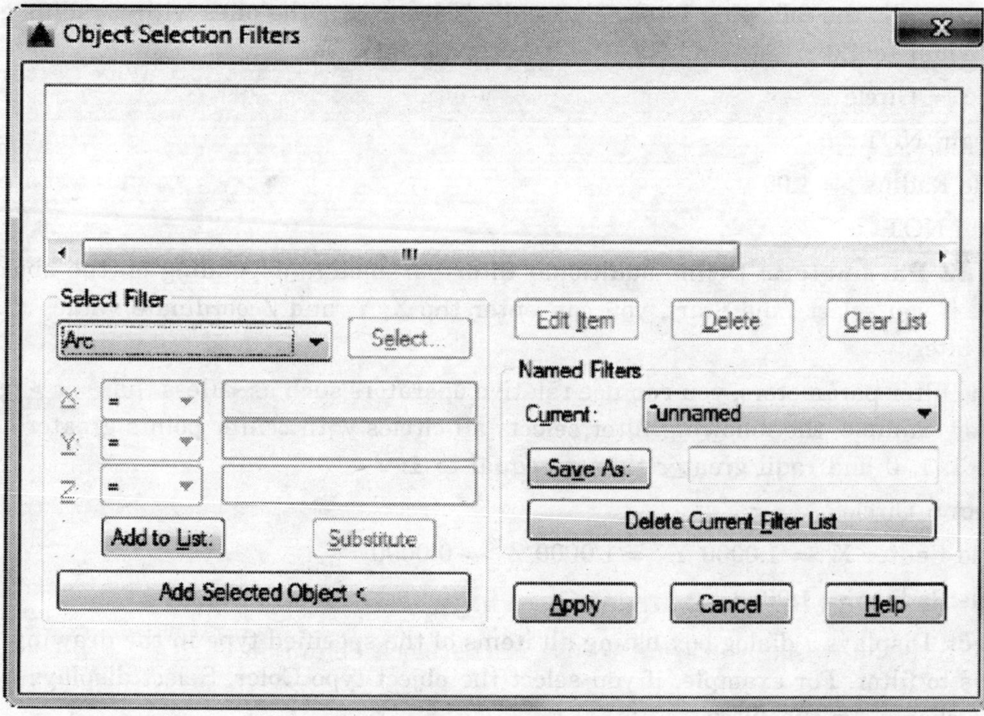

List of Options

The following options are displayed.

Filter Property List: Displays a list of the filter properties that compose the current filter. The current filter is the filter that you select in Current in the Named Filters area.

Select Filter: Adds filter properties to the current filter.

Object Types and Logical Operators: Lists object types that you can filter and logical operators (AND, OR, XOR, and NOT) for grouping the filter expressions.

If you use logical operators, make sure that you pair and balance them correctly in the filter list. The number of operands you can enclose depends on the operation.

Logical operators		
Starting operator	**Encloses**	**Ending operator**
Begin AND	One or more operands	End AND
Begin OR	One or more operands	End OR
Begin XOR	Two operands	End XOR
Begin NOT	One operand	End NOT

For example, the following filter selects all circles except the ones with a radius greater than or equal to 1.0:

Object = Circle

**Begin NOT

Circle Radius >= 1.00

**End NOT

X, Y, Z Parameters: Define additional filter parameters depending on the object. For example, if you select Line Start, you can enter the X, Y, and Zcoordinate values that you want to filter.

In the filter parameters, you can use relative operators such as < (less than) or > (greater than). For example, the following filter selects all circles with center points greater than or equal to 1, 1, 0 and radii greater than or equal to 1:

Object = Circle

Circle Center X >= 1.0000 Y >= 1.0000 Z >= 0.0000

Circle Radius >= 1.0000

Select: Displays a dialog box listing all items of the specified type in the drawing. Select the items to filter. For example, if you select the object type Color, Select displays a list of colors to choose for the filter.

Add to List: Adds the current Select Filter property to the filter list. Filter properties that you add to the unnamed filter remain available during the current work session unless you manually delete them.

Substitute: Replaces the filter property selected in the filter property list with the one displayed in Select Filter.

Add Selected Object: Adds one selected object in the drawing to the filter list.

Edit Item: Moves the selected filter property into the Select Filter area for editing. The edited filter replaces the selected filter property.

Delete: Deletes a selected filter property from the current filter.

Clear List: Deletes all the listed properties from the current filter.

Named Filters: Displays, saves, and deletes filters.

Current: Displays saved filters. Select a filter list to make it current. The named filter and its list of properties are loaded from the default file, *filter.nfl*.

Save As: Saves a filter and its list of properties. The filter is saved in the *filter.nfl* file. Names can contain up to 18 characters.

Delete Current Filter List: Deletes a filter and all its properties from the default filter file.

Apply: Exits the dialog box and displays the Select Objects prompt, where you create a selection set. The current filter is used on the objects you select.

10.6 ORTHOGRAPHIC PROJECTION VIEWS

Create orthogonal, and isometric projected views from an existing model documentation drawing view.

1. Click Layout tab ➤ Create View panel ➤ Projected View.
2. Click the drawing view you want to use as the parent view. A preview of a projected view appears at the cursor.
3. Move the preview to the desired location and click to place the view.

   Note: As you move the preview the orientation of the projected view changes to reflect its relationship to the parent view.
4. Repeat step 3 until all the required projected view are created.
5. Press ENTER.

To specify the projection angle for drawing view: You can set up the defaults for model documentation drawing views to use either the first angle or third angle projection.

1. Click Layout tab ➤ Styles and Standards panel ➤ Dialog box launcher.
2. In the Drafting Standards dialog box, in the Projection type section, click the desired projection angle.

   Note: The settings you specify have no effect on any existing drawing views. They apply to the next drawing view you create.

10.7 SECTION VIEW

When you create a section view, it uses the settings of the current section view style. You can apply a different section view style to it after the section view is created.

1. In the drawing area, select the section line corresponding to the section view to which you want to apply the style.
2. Click View tab ➤ Palettes panel ➤ Properties.
3. In the Properties palette, under Annotation, in the Style list, select the section view style to apply.

The section view style is applied to the section line and the corresponding section view.

Section View Style Manager

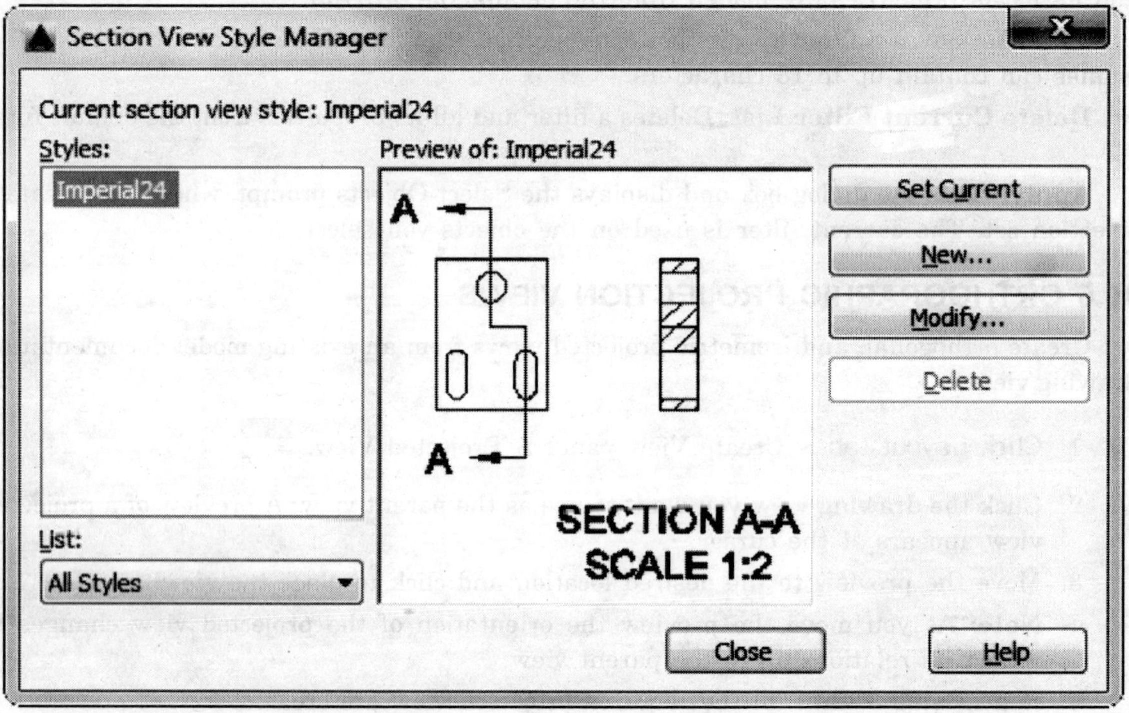

Section view styles control the appearance of model documentation section views. The style specify the formatting of the section line, view label and hatching.

List of Options: The following options are displayed.

Current section view style: Displays the name of the section view style that is applied to section views you create.

Styles: Displays a list of section view styles available in the current drawing file. The current style is highlighted.

List: Filters the contents of the Styles list. Click All Styles to display all section view styles available in the drawing file. Click Styles in use to display only the section view styles that are referenced by section views in the current drawing.

Preview of: Displays a sample image depicting the effects of the style that is selected in the Styles list.

Set Current: Sets the section view style selected in the Styles list as the current style. All new section views are created using this section view style.

New: Displays the Create New Section View Style dialog box, in which you can define new section view styles.

Modify: Displays the Modify Section View Style dialog box, in which you can modify section view styles.

Note: You cannot rename the section view styles Imperial24 and Metric50.

Delete: Deletes the section view style selected in the Styles list. This button is not available if the style selected in the Styles list is currently in use by a section view.

EXERCISE

1. By selecting which one of the following, the objects on the layer are not visible and are not considered when the drawing is regenerated?

   (a) When you off the layer (b) When you froze the layer

   (c) When you lock the layer (d) None

2. If a layer is locked:

   (a) Details can be added to a locked layer

   (b) Details can be erased from a locked layer

   (c) Details cannot be added or erased from a locked layer

   (d) Details can be added to a locked layer but they disappear when the drawing file is saved.

3. When a layer is turned off:

   (a) File space is saved when saving the file

   (b) It makes no real difference. Details can still be added to the layer

   (c) Details cannot be erased from the layer

   (d) Details on the layer cannot be seen.

4. When dimensioning from the command line what do the following abbreviations mean?

   (a) hor? horizontal (b) ver? vertical

   (c) int? intersect (d) per? perpendicular

5. An aligned dimension is one which:
   (a) Is aligned to the previous dimension;
   (b) Is in line with the object being dimensioned
   (c) Can only be used when placing a dimension with a radius
   (d) Is lined up with the edges of the drawing area of AutoCAD.
6. Which of the following is correct?
   (a) When a block has been constructed and saved the data is saved as file with a name in a named directory
   (b) When a block has been constructed and saved, the file data is saved as part of the file in which the block was constructed
   (c) When a block has been constructed and saved, the file data is saved as a separate file from the drawing in which the block was constructed
   (d) When a block has been constructed and saved, the file data is saved as new file.
7. Which of the following is correct?
   (a) When a wblock has been constructed and saved the data is saved as file with a name in a named directory
   (b) When a wblock has been constructed and saved, the file data is saved as part of the file in which the block was constructed
   (c) When a wblock has been constructed and saved, the file data is saved as a separate file from the drawing in which the block was constructed
   (d) When a wblock has been constructed and saved, the file data is saved as new file.
8. In first angle orthographic projection an end view is:
   (a) Always placed to the right of the front view
   (b) Viewed from the one side and placed on the other side of the front view
   (c) Always placed to the left of the front view
   (d) Placed either side of the front view irrespective of the viewing position
9. In third angle orthographic projection a plan is:
   (a) Placed either above or below the front view depending upon whether the viewing is from above or below
   (b) Always placed above the front view
   (c) Always placed below the front view
   (d) Placed either above or below no matter which viewing direction is chosen.
10. When hatching a sectional view in an engineering drawing
    (a) Every detail cut by a section plane must be hatched
    (b) It is up to the operator constructing the drawing which parts are hatched
    (c) There is no need to hatch any part of the drawing
    (d) A bolt within a sectional view is not hatched

11. By Using Different Layers draw the following

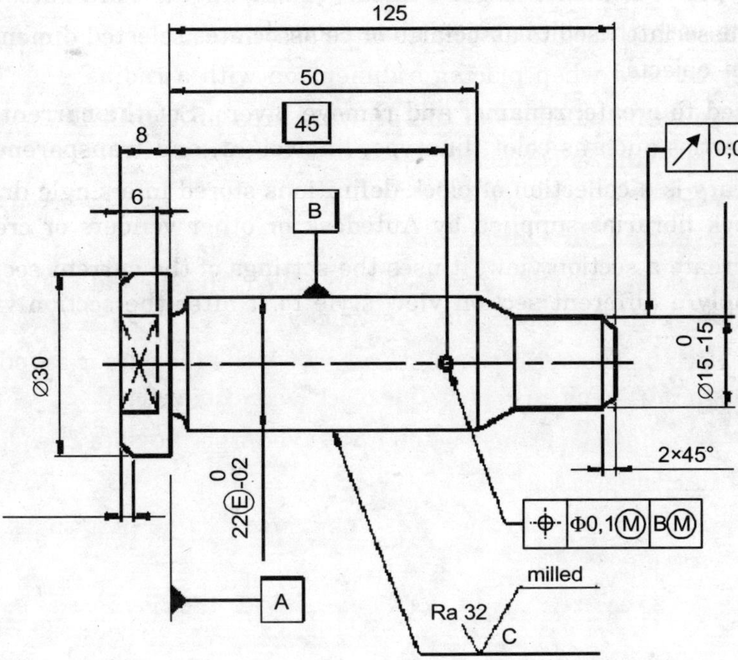

12. Draw Hatching in the following pattern.

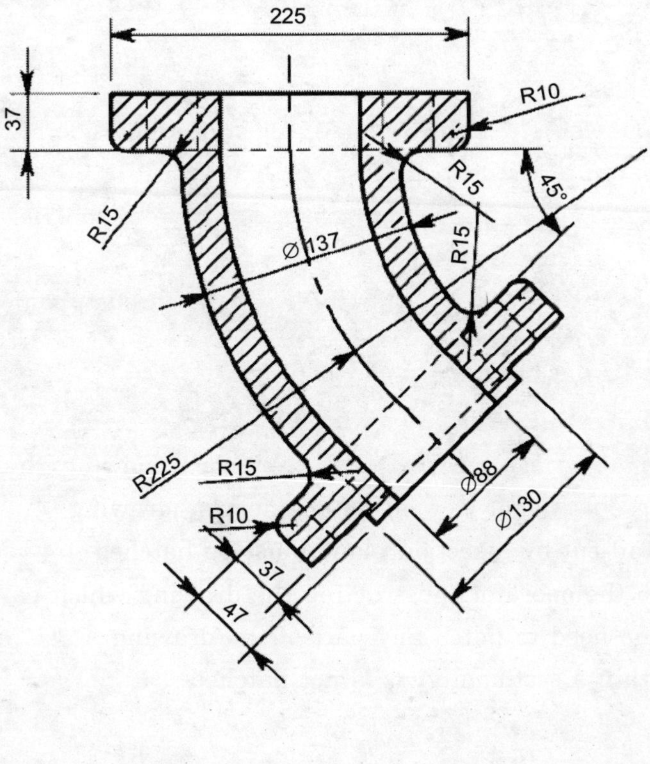

SUMMARY

- The basic types of dimensioning are linear, radial, angular, ordinate, and arc length.
- Dimension associate used to associates or re associates selected dimensions to objects or points on objects.
- Layer is used to create, rename, and remove layers. Set the current layer. Change layer properties such as color, linetype, lineweight, and transparency.
- A block library is a collection of block definitions stored in a single drawing file. You can use block libraries supplied by Autodesk or other vendors or create your own.
- When you create a section view, it uses the settings of the current section view style. You can apply a different section view style to it after the section view is created.

CHAPTER

11

SOLID MODELLING

Learning Objectives

- Working with simple and composite solids
- Using 3D viewing techniques
- Creating complex solids and surfaces
- Modifying objects in 3D space
- Editing solids
- Converting 3D objects

11.1 GEOMETRIC MODELING

Generally speaking, a model is nothing but the copy of a real or imagined object, mapping thereof using limited information. A computerized geometric model maps up the shape and dimensions of an object.

From the topological point of view, geometric modelling systems can be classified into two basic groups:

- Manifold modelling systems: they include modelling systems suitable for modelling forms that can be mapped into a manifold of 2D points.
- Objects of a non-manifold topology are not realistic in general; they cannot be mapped into a manifold of 2D points. This usually arises from the fact that a model includes basic units of different dimensions (1D, 2D or 3D) or the latter are interconnected within a model.

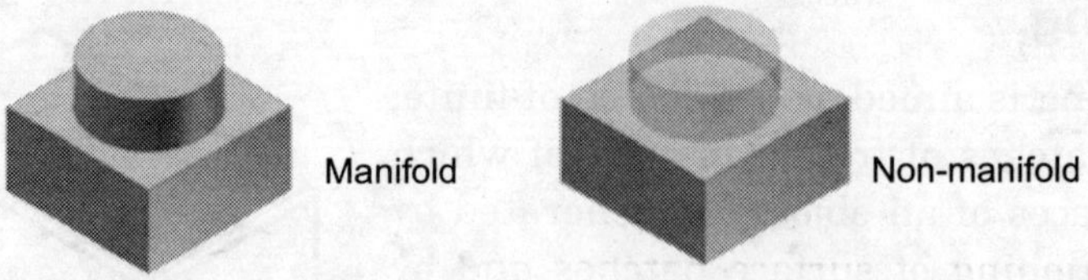

Manifold modelling systems Manifold modelling systems can be broken down into two further groups based on the completeness of information on features of shape:

- Modelling systems of other than full value include:
  - (i) Wireframe modelling
  - (ii) Surface modeling
- Modelling systems of full value include:
  - (i) Mantle modelling
  - (ii) Solid modeling

Wireframe modelling A wireframe model depicts the edges delimiting the surfaces of the object modelled. These edges can consist of lines, arcs, and curves. Disadvantages of the modelling method: all edges are shown on the image displayed; visibility cannot be depicted; volume and mass characteristics cannot be specified; data provision is lengthy and difficult; □ not suitable for designing shapes and specifying more complex forms.

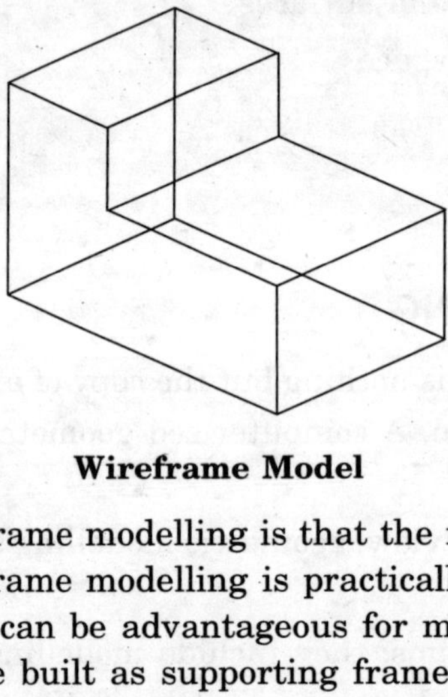

Wireframe Model

A basic shortcoming of wireframe modelling is that the model displayed does not clearly show the object modelled. Wireframe modelling is practically out of use today. However, in many cases, a wireframe image can be advantageous for model design in mantle and solid modelling; wireframe models are built as supporting frames for surface modelling.

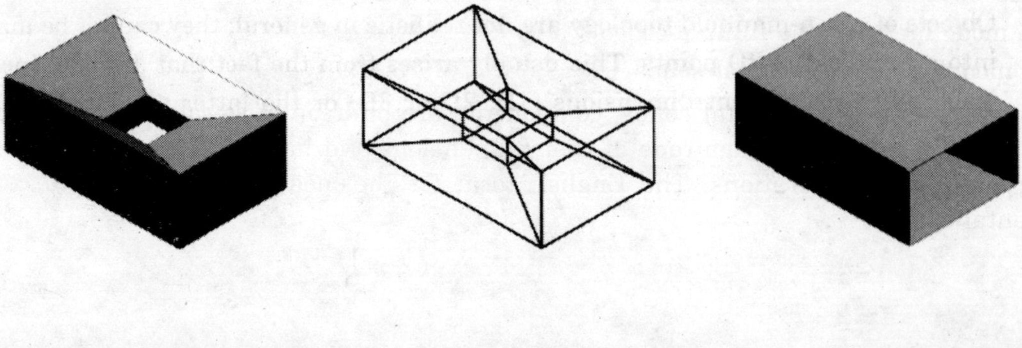

Surface Modelling

Surface modelling is aimed at the design of finite, non-open surface patches of free forms, out of which the delimiting surfaces of an object are generated by the geometric positioning of surface patches and by the stipulation of various continuity restrictions. This modelling method does not manage topological information. The non-contacting surfaces on the surface model shown in the figure below are intended to illustrate that surfaces are interconnected only at "sight" level. Characteristics of surface modelling: a surface model is suitable for hide and shade displays; not suitable for calculating volume or mass characteristics; not suitable for producing numerical models for engineering calculations.

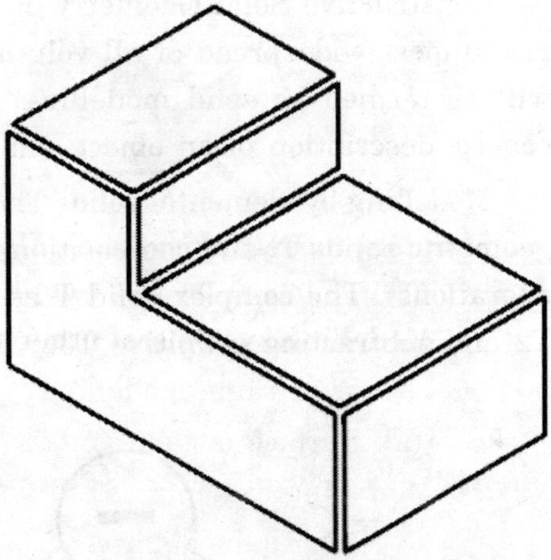

Surface Model

Mantle modelling describes the finite and closed cover of an object (the mantle) by a polyeder approach or realistic geometry. Mantle modelling makes methodological use of the basic assumption that each physical object has an unambiguously determinable delimiting surface. Geometrically, this delimiting surface is the mantle, which is a continuous closing set of surface patches. In addition to other information, this modelling method also provides a comprehensive topological characterization of the model. Characteristics of surface modelling: A mantle model is suitable for hide and shade displays; It is suitable for calculating volume or mass characteristics; It is suitable for completing production technology designs.

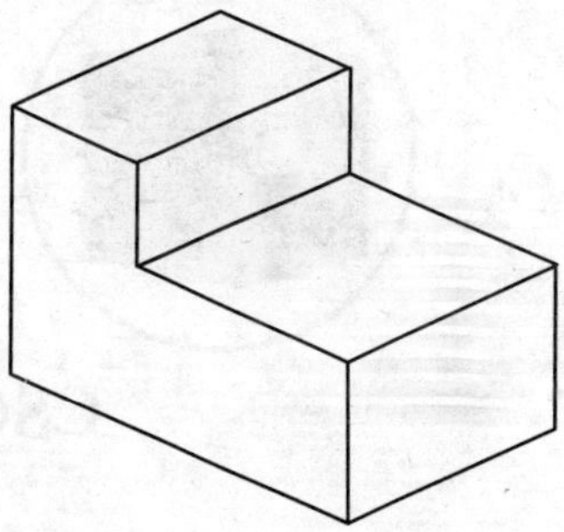

Mantle Model

Solid modelling: Solid or volume modelling describes objects as finite, closed, regular sets of points. A solid model provides a complete, characteristic and concise description of the object. The data structure includes descriptions of the basic units building up the solid and their relationships as well. solid modelling is much more simple than wireframe, surface, or mantle modelling. Many varieties of solid-oriented modelling systems have been developed: volume breakdown methods: prism breakdown modelling; half space modelling; volume fillup methods: modelling by elemental cells; modelling by elemental solids.

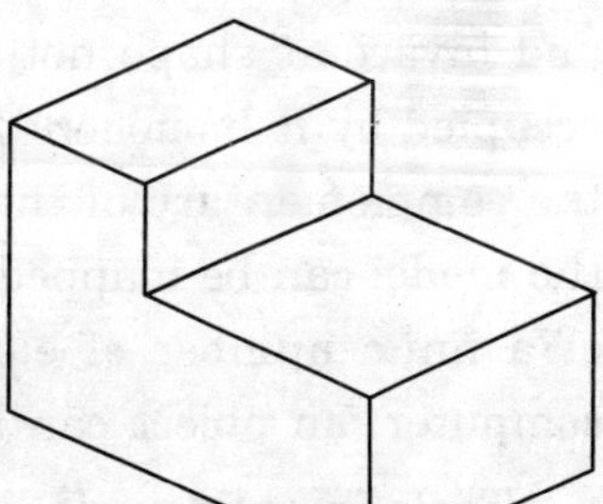

Solid model

In modelling by elemental solids, components are built up of so-called solid primitives, falling into the order of magnitude of their dimensions and having a specific geometry, by using composition operations. The English term for the modelling procedure to combine elemental solids is:

Constructive Solid Geometry or CSG modelling for short. Modelling by elemental solids is the most widespread of all volumetric modelling methods. Later on, this modelling form will be termed as solid modelling. A solid model provides a complete, characteristic and concise description of an object, enabling integrated and automated design.

Modelling by elemental solids The two basic toolset groups of solid modelling are elemental geometric solids Ti and composition operations (is a summary indication of composition (set) operations). The complex solid T as in the figure is generated by adding primitives T1 and T2 and subtracting primitive T3.

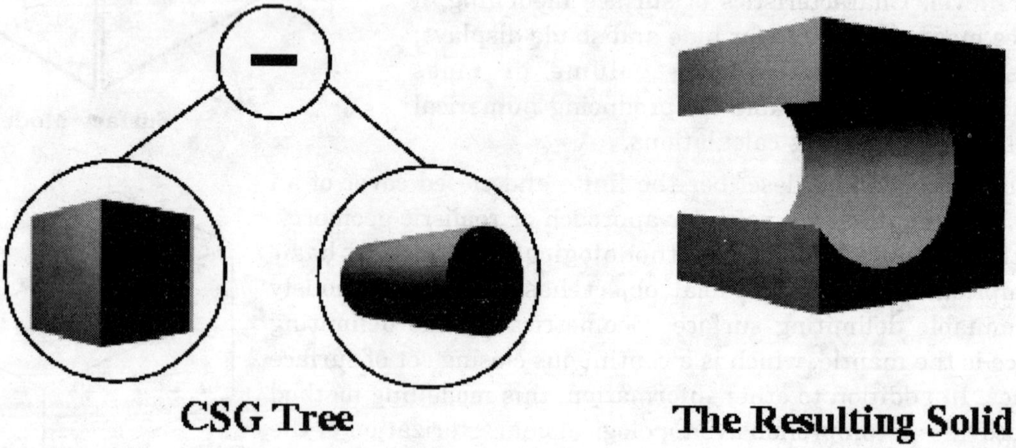

CSG

Assumptions in solid modelling an object is a rigid solid, meaning that it has a concrete and invariant shape not affected by spatial location or position; □ an object fills the space occupied by it homogeneously, meaning that the inside of the object is always connected to the complementary of the model through the cover; object extension is finite, meaning that the model can be mapped for computer display; an object can be generated as a composition of a finite number of elemental solids, meaning that an object model can be stored in a computer; an object can be modelled as a closed set in terms of rigid solid motion.

Tools for solid modelling The toolset for solid modelling includes the following: □ creation of solid primitives; □ composition operations; □ manipulation of solid primitives and solids; □ illustration. Creation of solid primitives Elemental solids, also named as solid primitives can be pre-defined or created by the user. Pre-defined solid primitives include: □ rectangle □ wedge; □ cylinder; □ cone; □ torus; □ sphere.

11.2 TOOLS FOR SOLID MODELLING

Tools for Solid Modelling

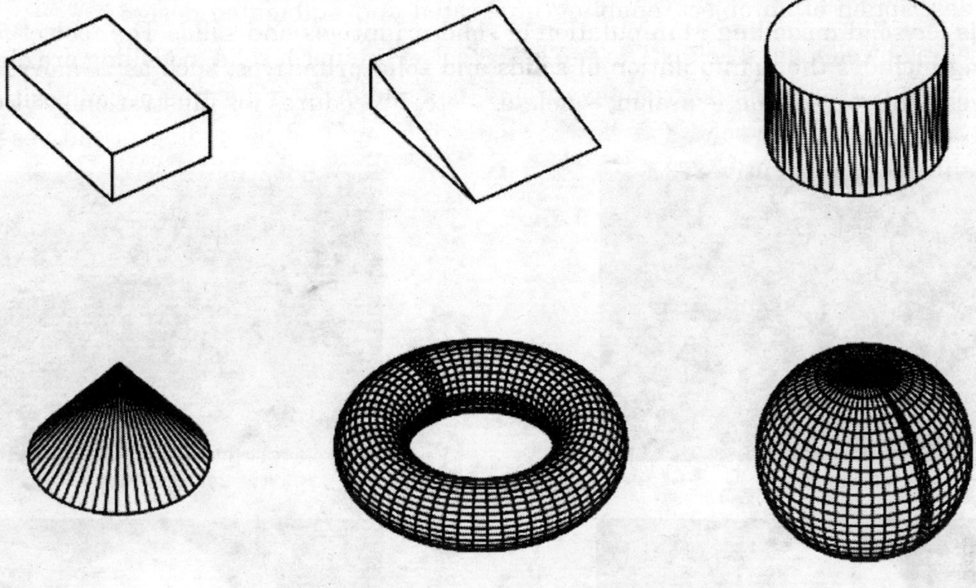

Pre-defined solid primitives

Some softwares may include additional solid primitives; e.g. cone, convex arc, concave arc, etc.

Tools for solid modelling A common feature of user-created solid primitives is that they can be created by moving surfaces.

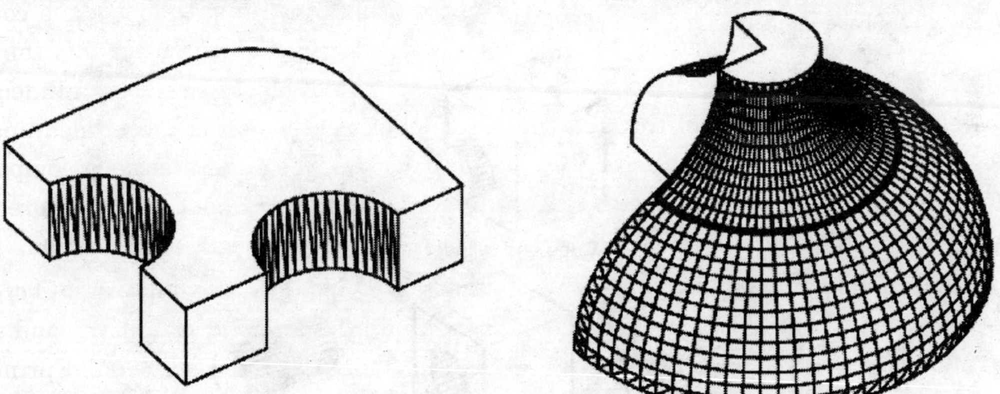

User-created solid primitives

User-created solid primitives: ~ extrude; ~ revolve; ~ (sweep); ~ (loft). User-created solid primitives.

Tools for solid modelling: Composition operations of two operands include union (□), combining the sets of points of two discrete solids; difference (\), producing the difference of two sets of points; and intersection (□), defining the common set of points to be found in both solids. Composition operations. Cube and cylinder Union Difference Intersection

Tools for solid modelling Manipulation of solid primitives and solids The toolset for solid modelling includes the manipulation of solids and solid primitives, such as: ~ move; ~ copy; ~ revolve; ~ mirror; ~ scale; ~ assign; ~ delete; ~ etc. Procedures for illustration ~ wireframe; ~ hide; ~ shade.

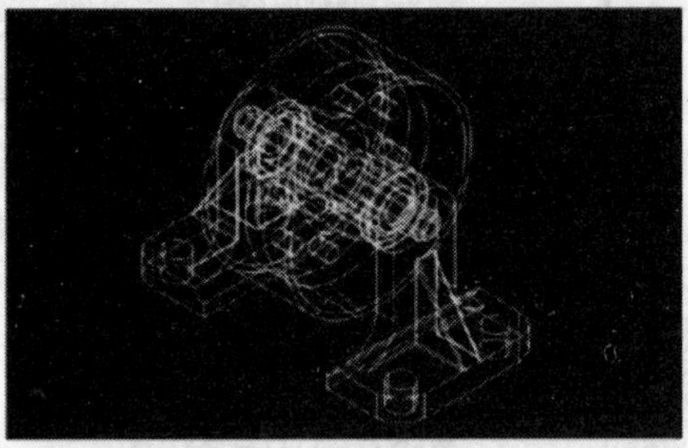

Wireframe

Hide

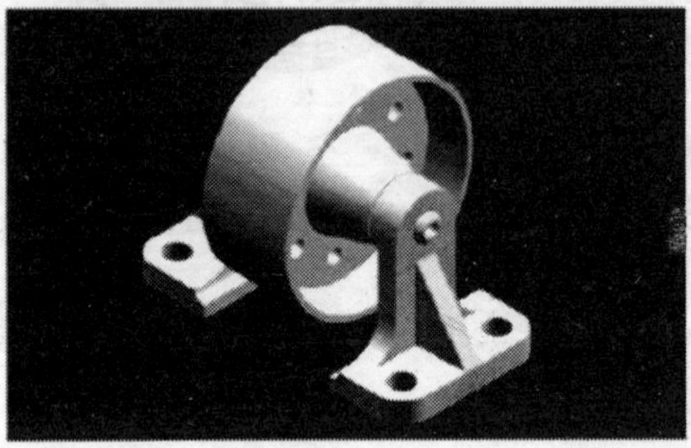

Shade

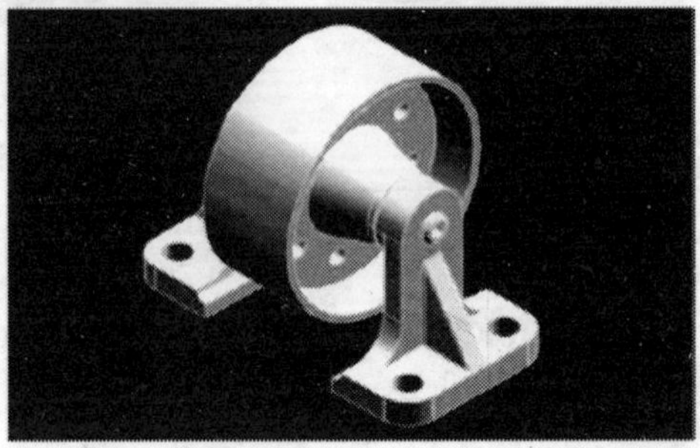

Shade + hide

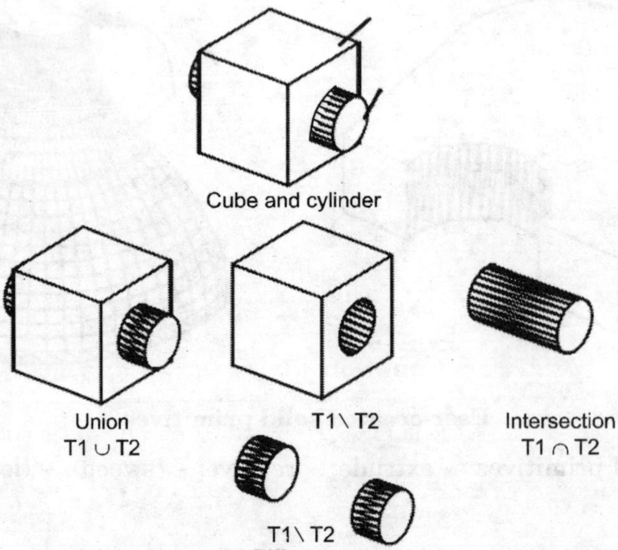

11.3 CREATING PARTS, ASSEMBLY (SOLIDWORKS)

A Design Automation Software Package Used to Produce

- Parts
- Assemblies
- Drawings

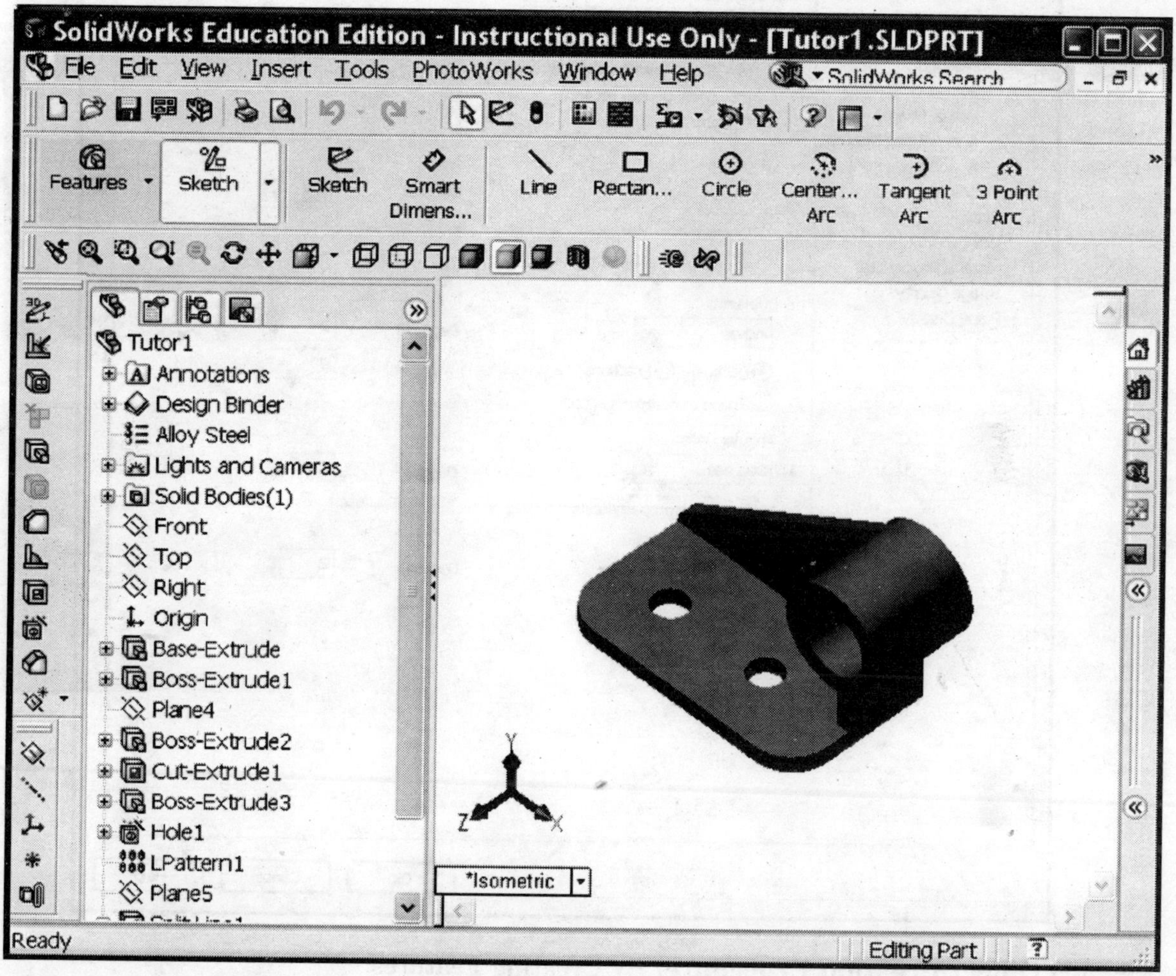

SolidWorks Model Types

Type	Function	Data File
Part	3-D Object	*.SLDPRT
Assembly	Many Parts	*.SLDASM
Drawing	Multi-views	*.SLDDRW

Part Modeling

(1) Setting Up Unit: Menu: Tools->Options

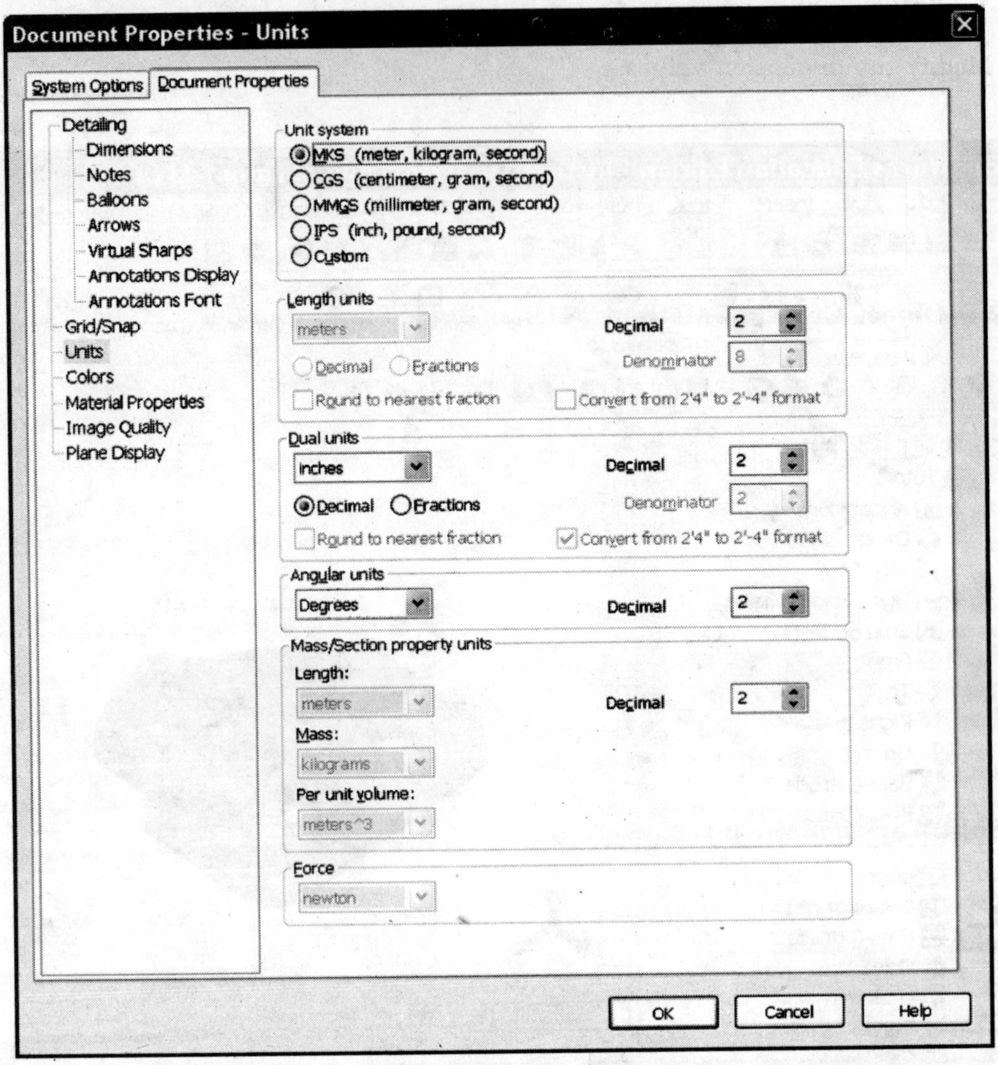

3-D Object Creation Procedure: By Creating Features

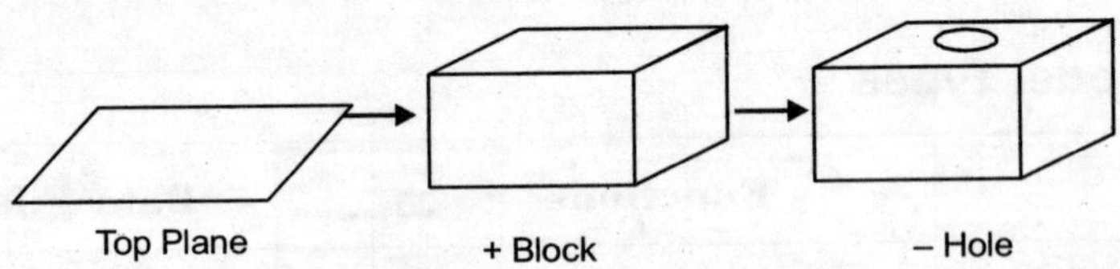

Each Feature

- 2-D Sketching
- 3-D Formation

2-D Sketching: Parametric Modeling

(a) Procedure

- Sketch the geometry
- Dimension the geometry

Modify the dimension values

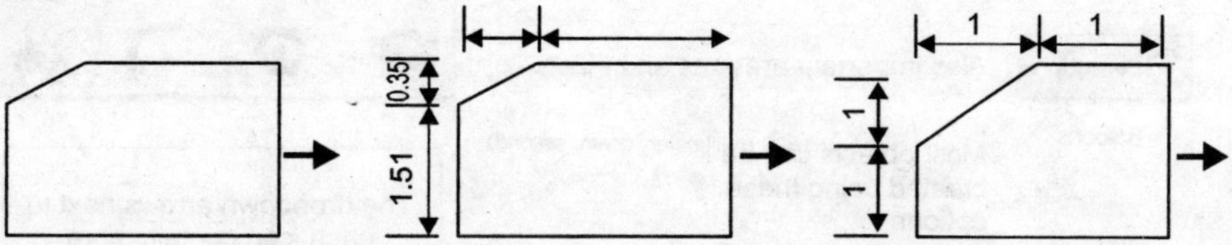

2-D Object Creation Methods: Menu: Tools->Sketch Entities

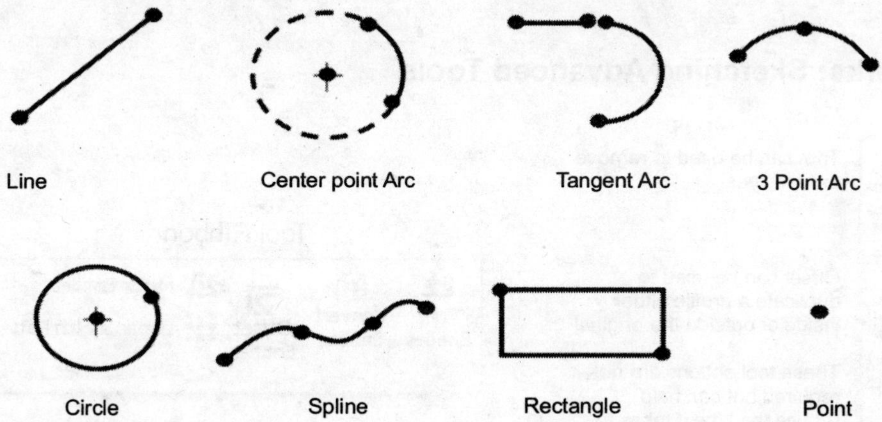

SolidWorks: Main Interface

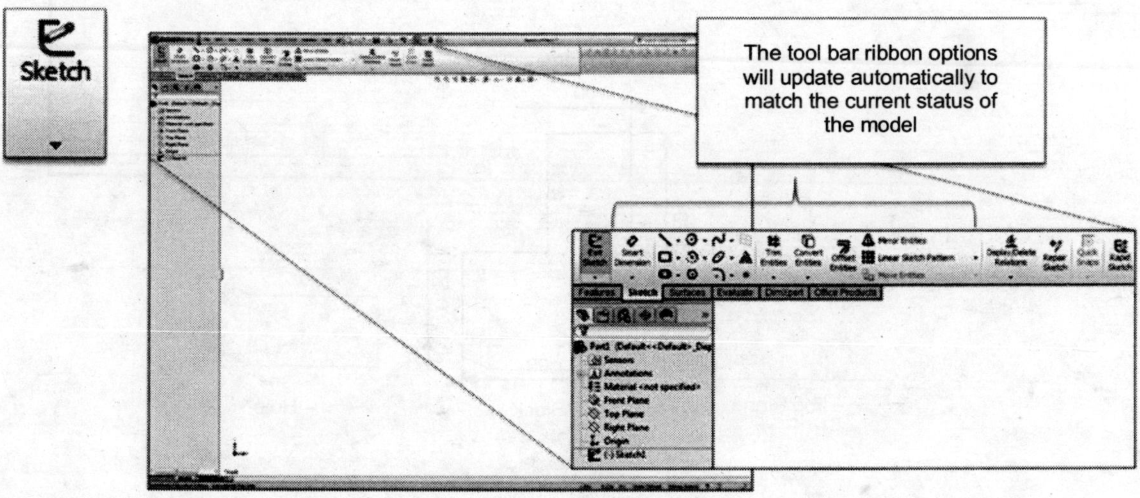

SolidWorks: Sketching Basic Tools

Tool Ribbon

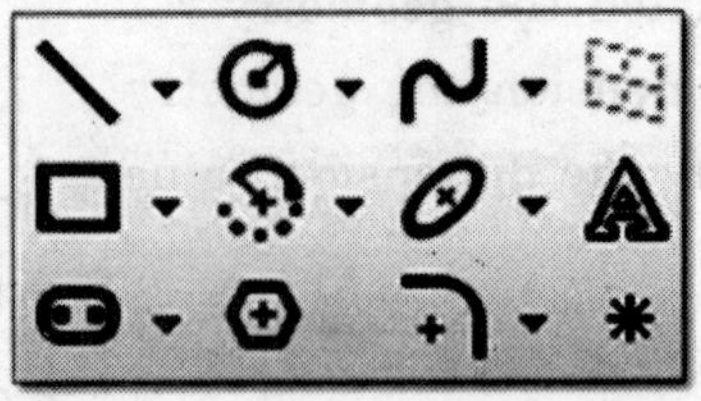

Lines and circles are the most important tools

Also important are Arcs and Fillets

Most objects can be created using these options

The dropdown arrows next to each icon contain more options for each tools type

SolidWorks: Sketching Advanced Tools

Trim can be used to remove excess lines

Tool Ribbon

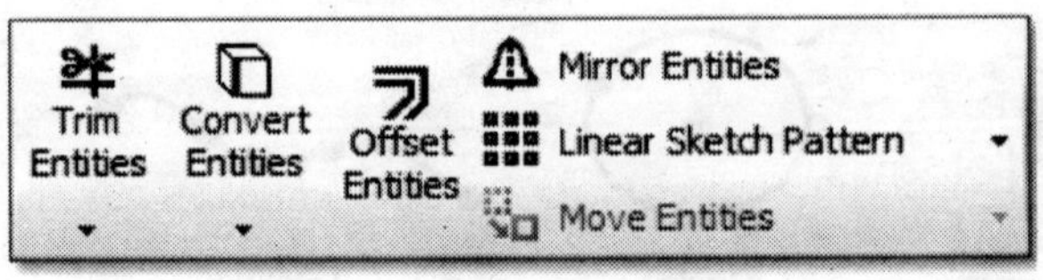

Offset can be used to duplicate a profile either inside or outside the original

These tool options are not required but can help reduce the time it takes to create a sketch

Part

Top Plane

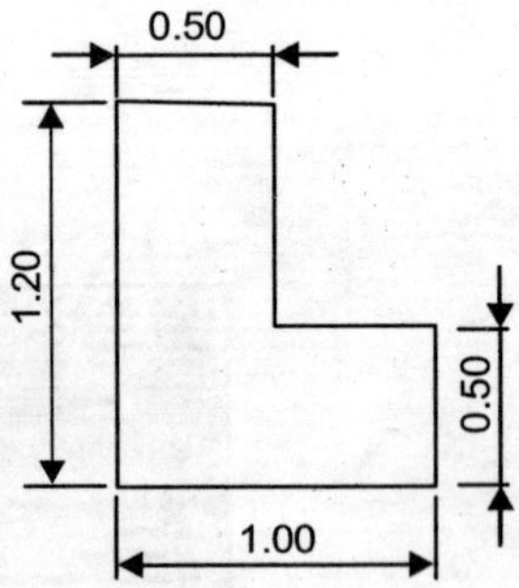

Part is created by sketch.

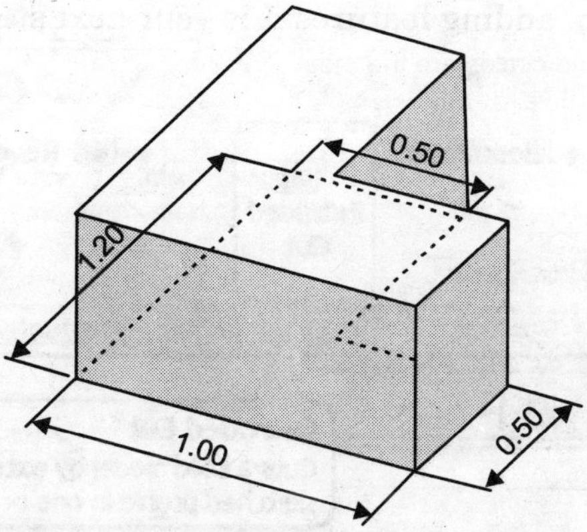

Sketch is the base to define your part, form and features.

Before you start creating sketches you must select plane or face where the sketch will be place on.

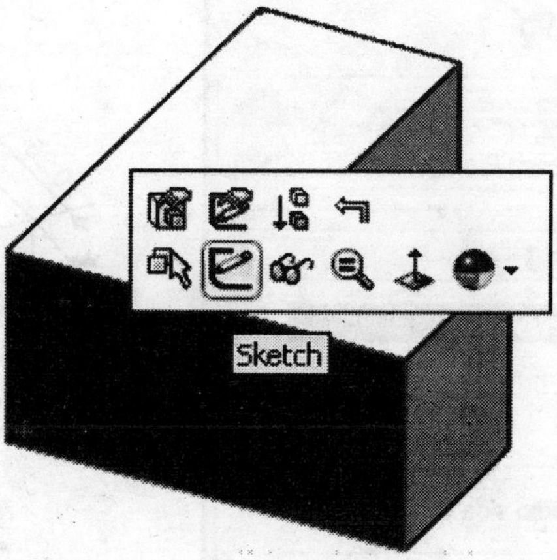

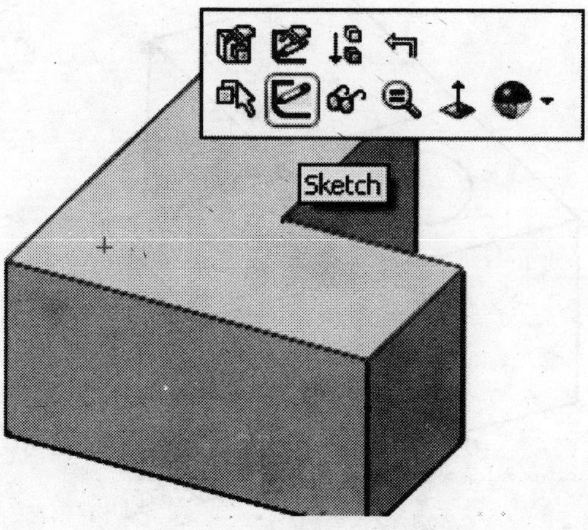

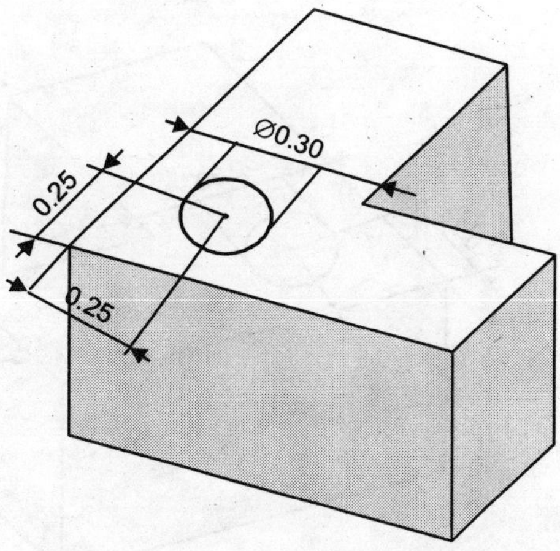

After select plane or face the sketch will be, sketch on it!

When you done with sketch, adding features it is your next step. Select Feature>Extruded Cut

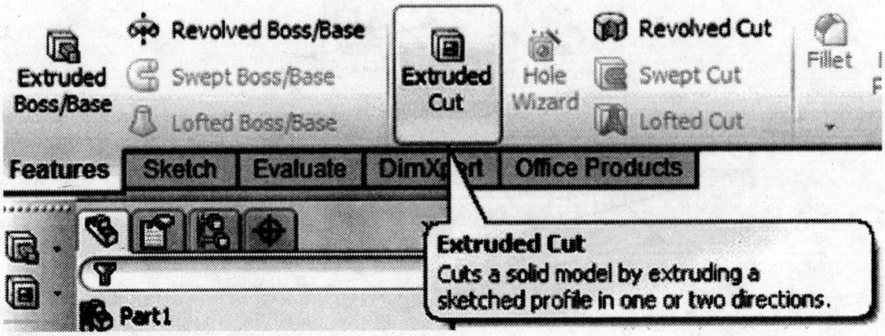

Select Through All and OK.

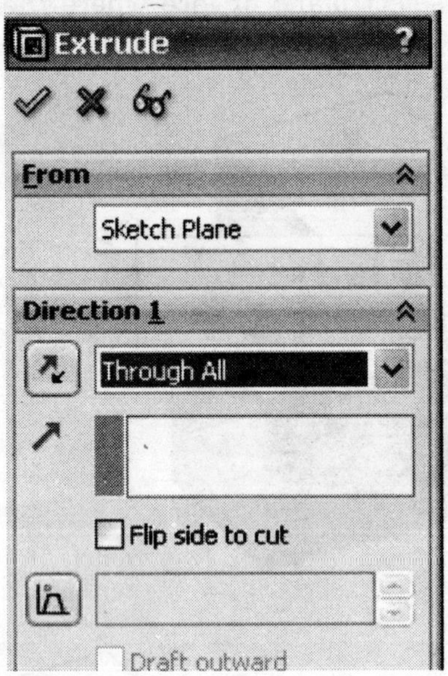

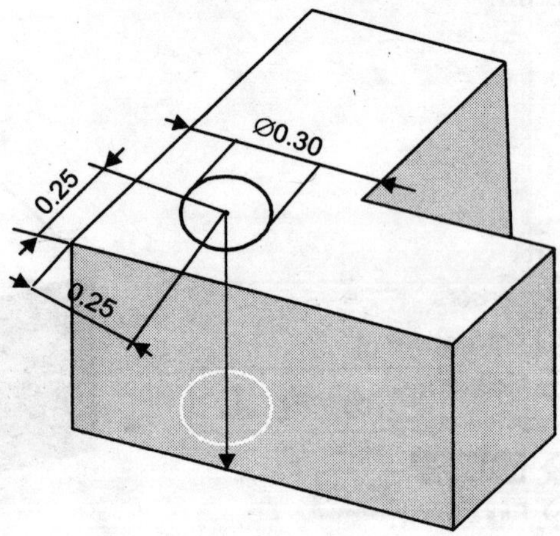

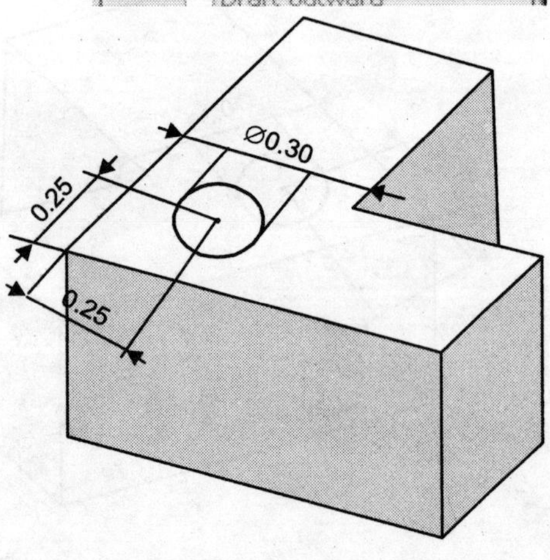

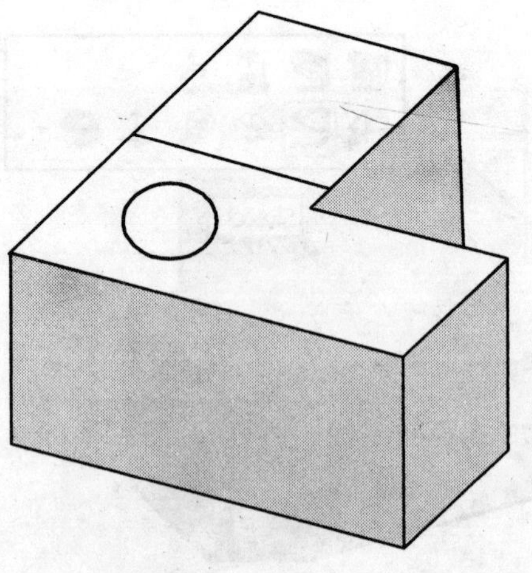

Assembly

Assembly is how all parts works together in assembly, checking for clashes and it functionality. First all parts inserted in assembly by Insert Component tool.

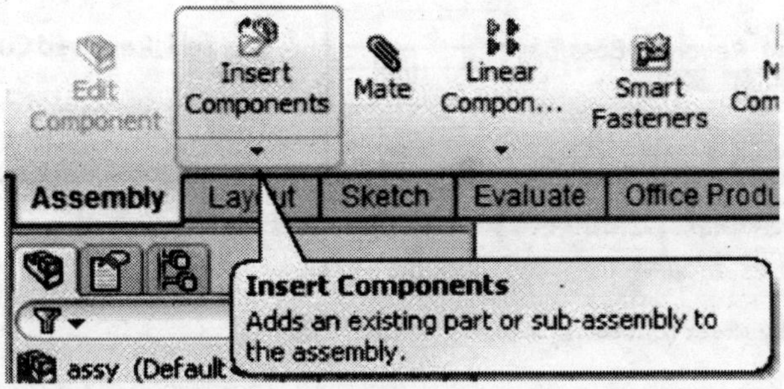

When all parts inserted into workspace, Mate is command to define how parts mate with each other.

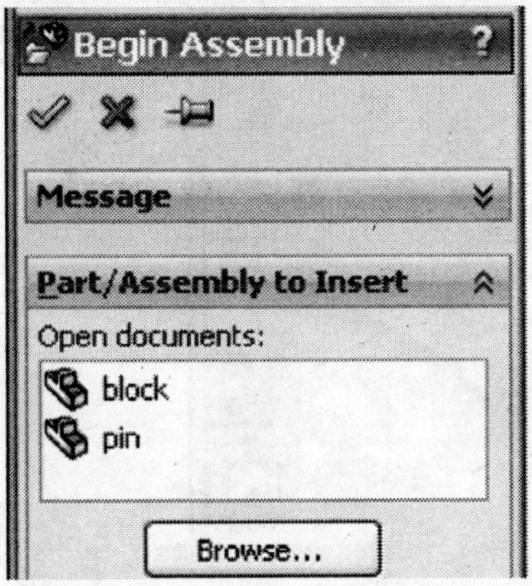

Let's mate this block and pin together, click Mate and select pin face and hole face, OK.

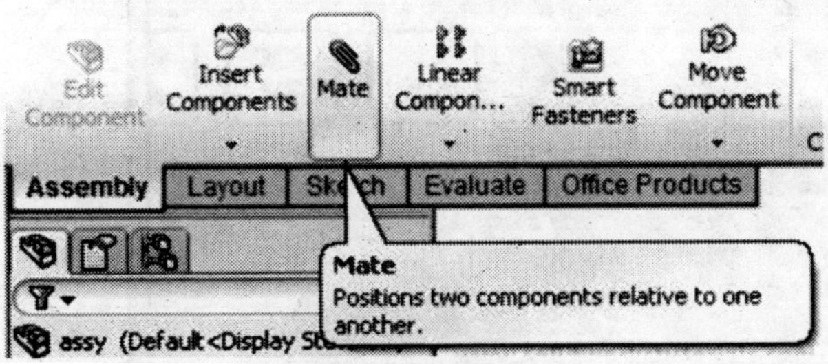

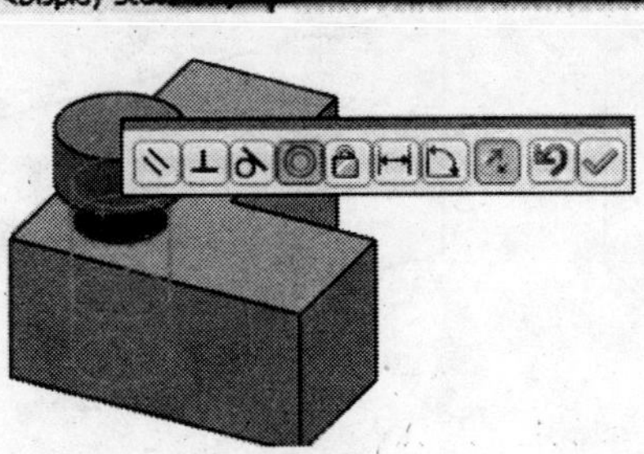

Drawing: Drawing is use for detailing part by adding dimension to it. To create a drawing first you need to select drawing block.

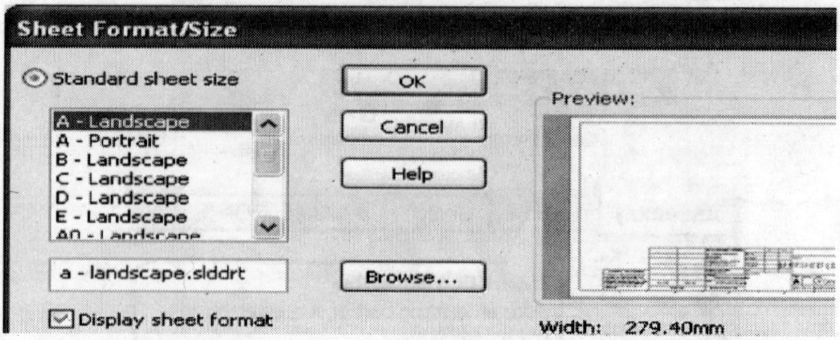

When block inserted, select click view palette to add drawing view.

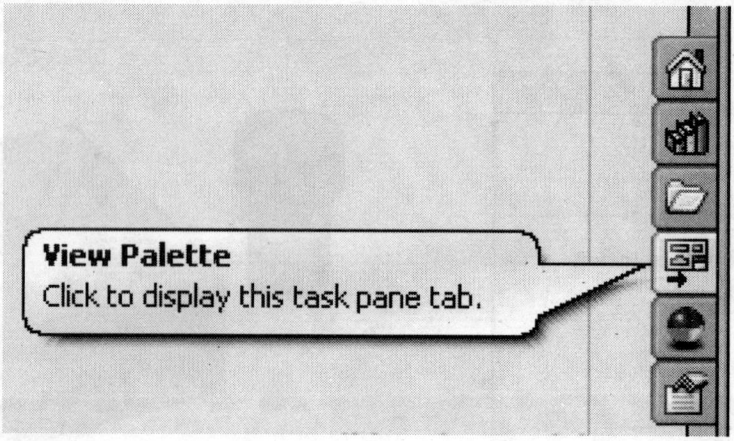

Choose the part you wish to make drawing.

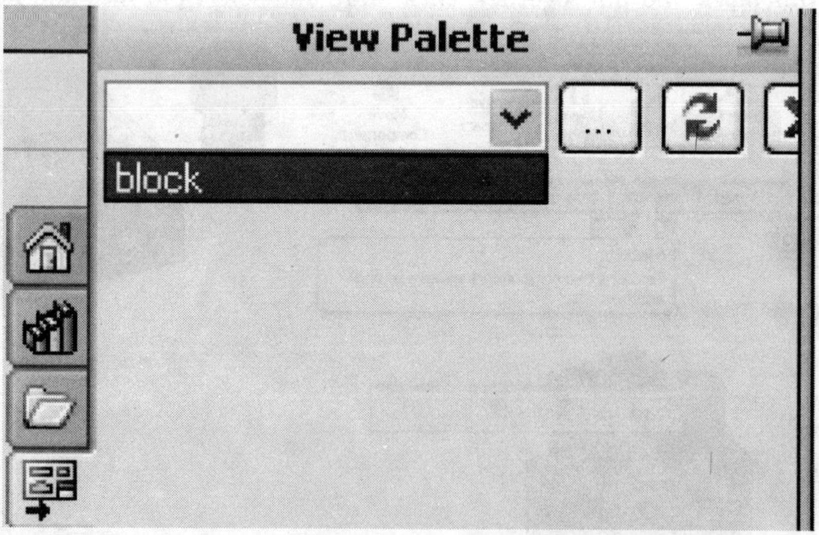

Now just drag and drop the part view on drawing block and add dimensions.

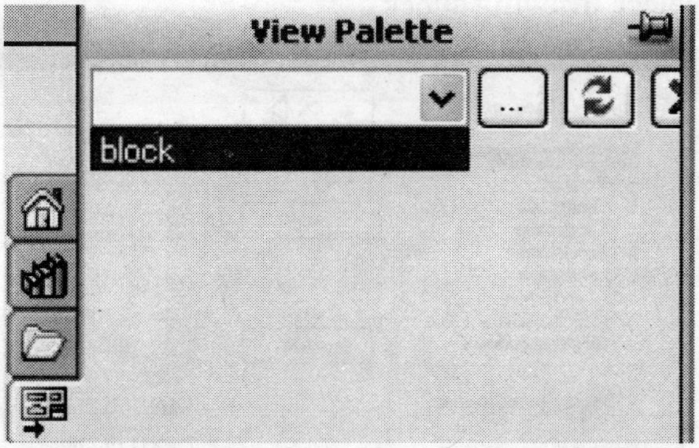

11.4 A PARAMETRIC AND NON-PARAMETRIC MODELING

A parametric model contains information like dimensions, constraints, and relationships between various entities like edges, sketches and features. You can easily make changes to the design, and it updates and responds to those changes and A non parametric model does not contain such relationships. It is essentially a "dumb model" which often happens when a CAD model is imported from another program.

Dumb models can be modified, but they do not have the additional constraints and relationships to allow the update to affect other design elements.

11.5 DRAWINGS USING SOLIDWORKS

Fortunately, Solidworks makes it very easy for us to create drawings from a part or assembly file. In fact, if built properly, Solidworks will also dimension the entire part and assembly for us...something that saves a lot of time! Once we open or finish the part, we choose "make drawing from part" from the file menu at the top which will automatically create a drawing file from our part:

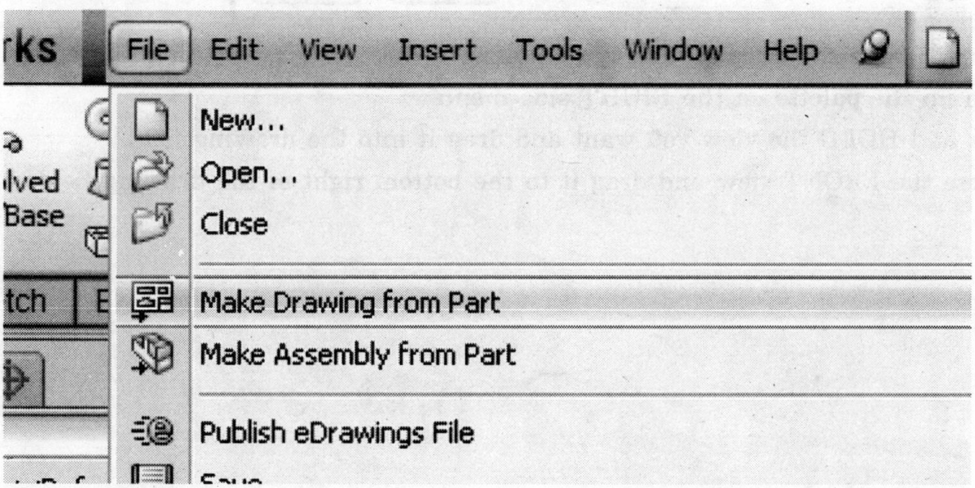

At first you see several different formats that are set up for you. Choose A-Landscape.

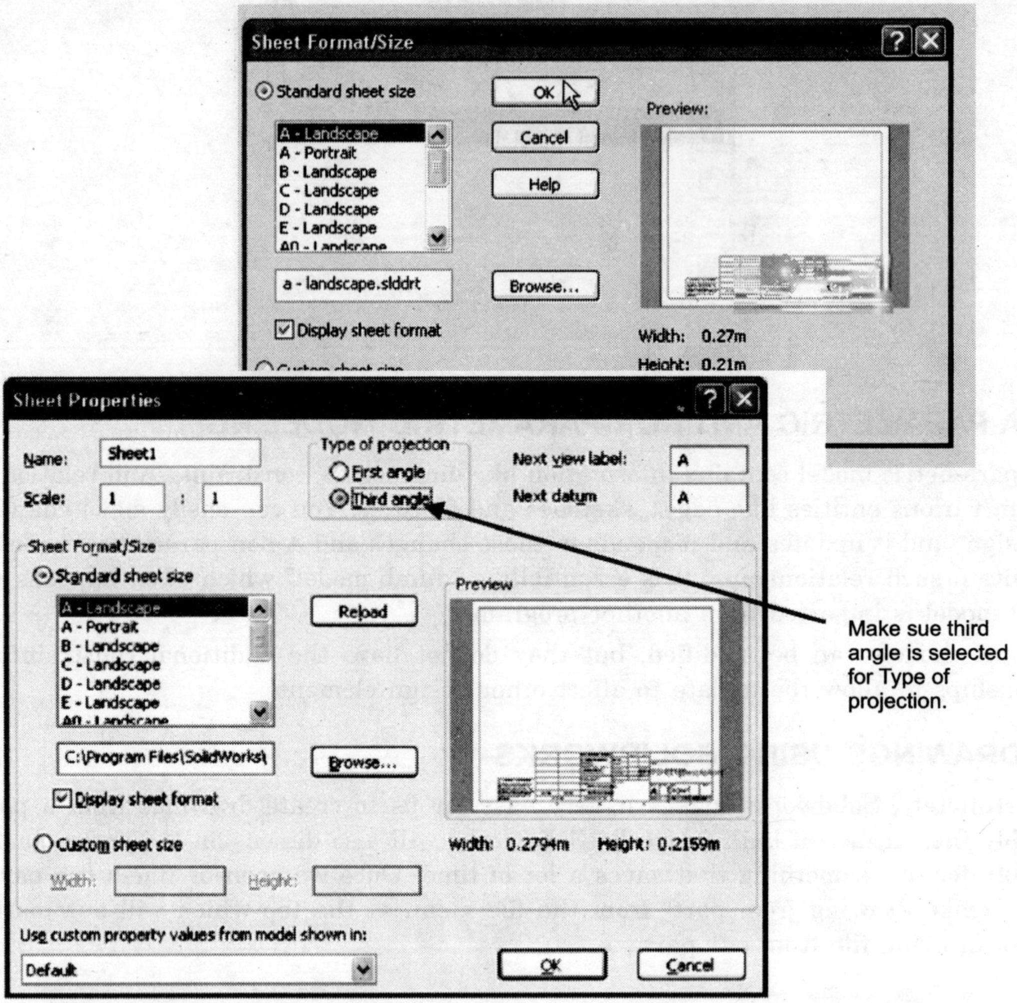

This will open the Sheet Properties window:

Open up the palette on the RIGHT side menu

Click and HOLD the view you want and drag it into the drawing field.

Choose the FRONT view and drag it to the bottom right of the drawing field.

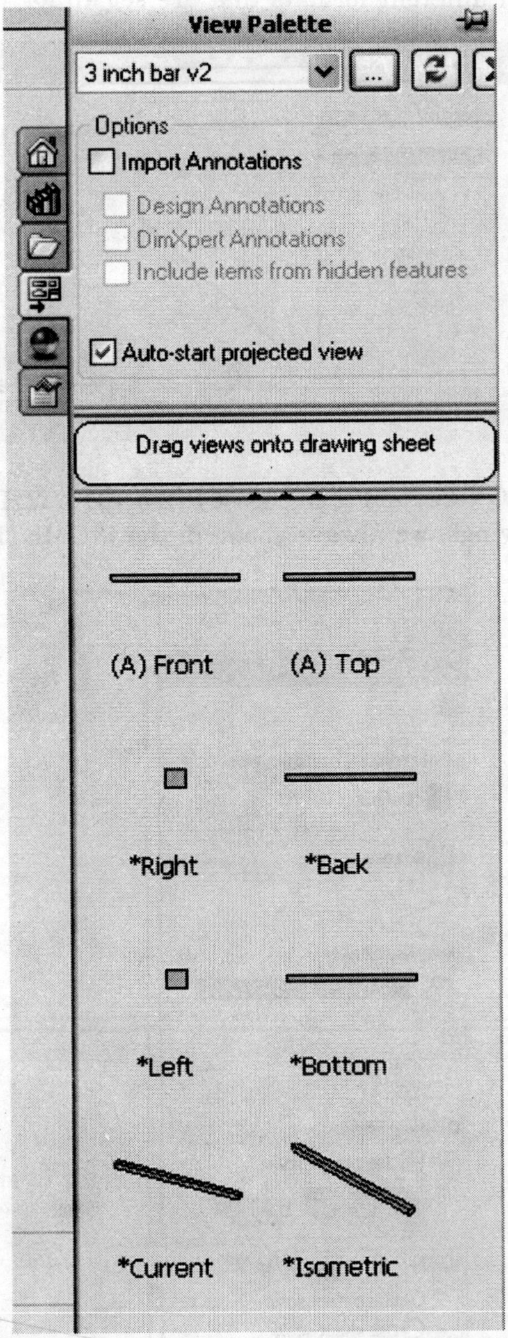

As soon as the front view is in position move the mouse UPWARD. You will see another design; this is actually the TOP view. Click to place the top view then return to the front view and move the mouse sideways. Repeat, then go diagonal.

Click the green check when finished.

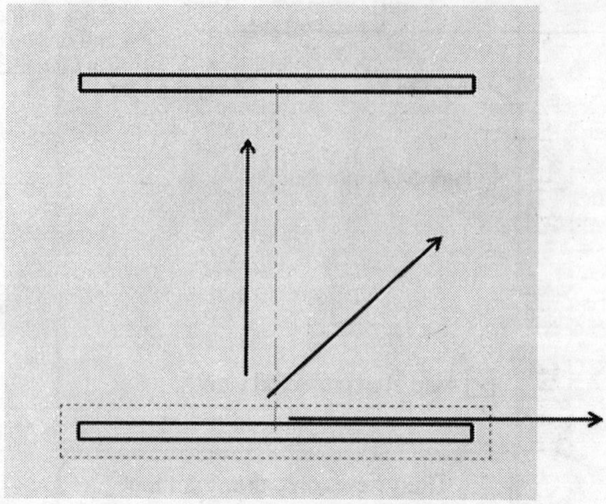

Click on the ISOMETRIC view, and then choose SHADED WITH EDGES under DISPLAY STYLE on the left. In drawings, we always shade in the ISOMETRIC view.

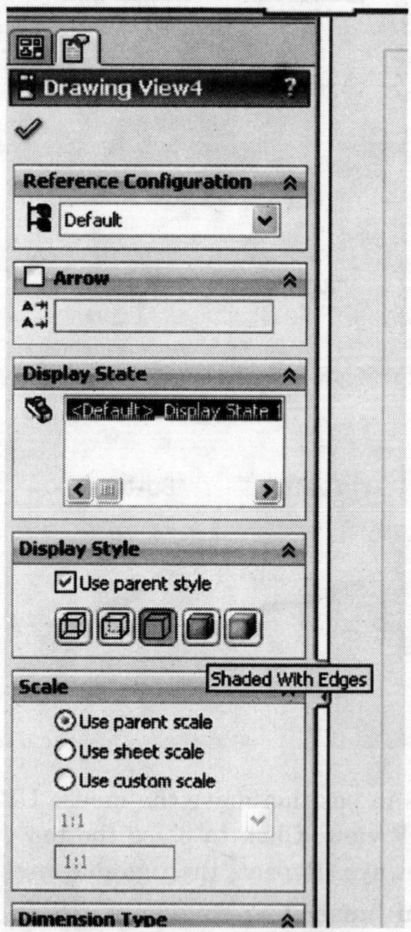

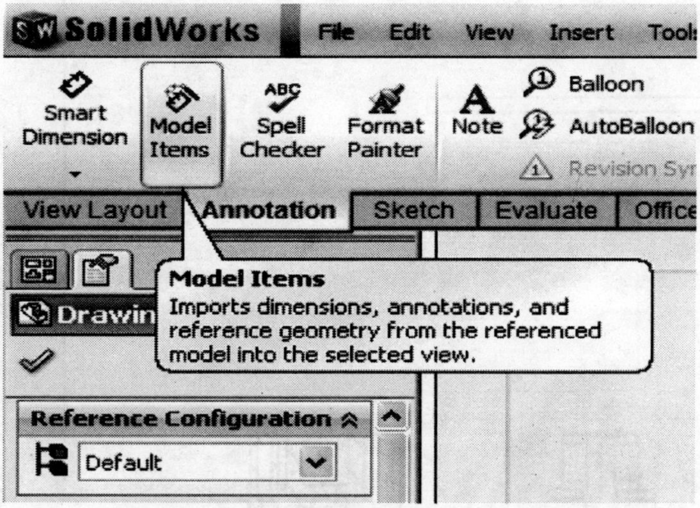

Click on the FRONT VIEW, then within the ANNOTATIONS tab at the top click on MODEL ITEMS.

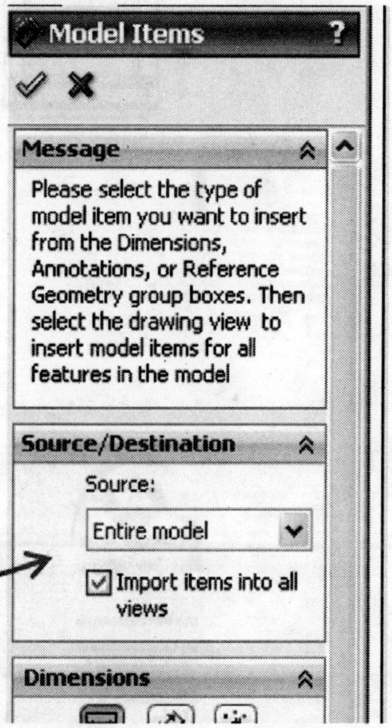

Under source/destination choose "entire model" from the drop down menu. Then check the box, import items into all views. Dimensions will then be added to the drawing. If you don't see dimensions on each view, try adding them separately.

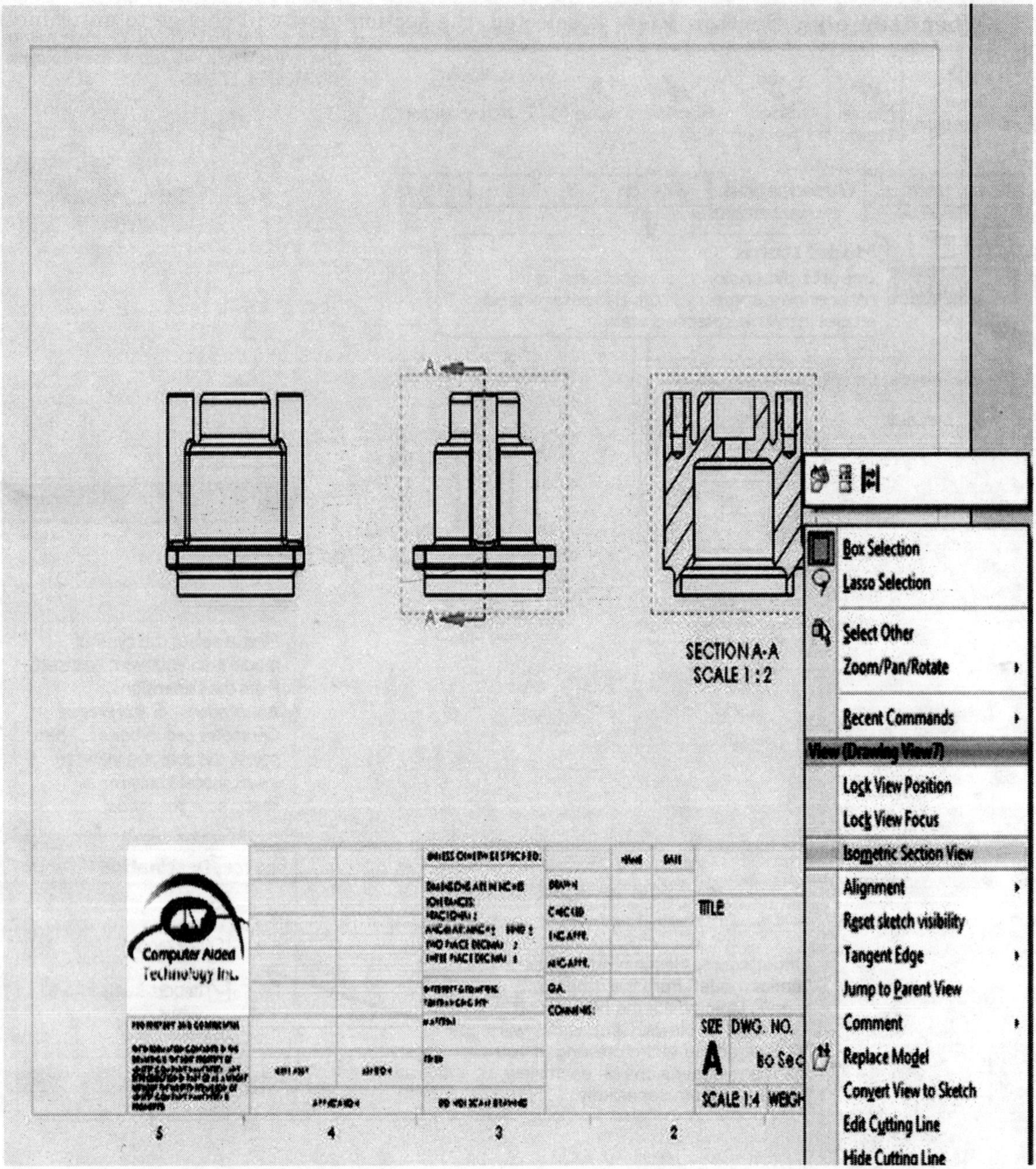
A
A
SECTION A-A
SCALE 1 : 2
Box Selection
Lasso Selection
Select Other
Zoom/Pan/Rotate
Recent Commands
View (Drawing View7)
Lock View Position
Lock View Focus
Isometric Section View
Alignment
Reset sketch visibility
Tangent Edge
Jump to Parent View
Comment
Replace Model
Convert View to Sketch
Edit Cutting Line
Hide Cutting Line
Computer Aided
Technology Inc.
TITLE
SIZE DWG. NO.
A
SCALE 1:4
5
4
3
2

When **Isometric Section View** is selected, the section view will change to an isometric view.

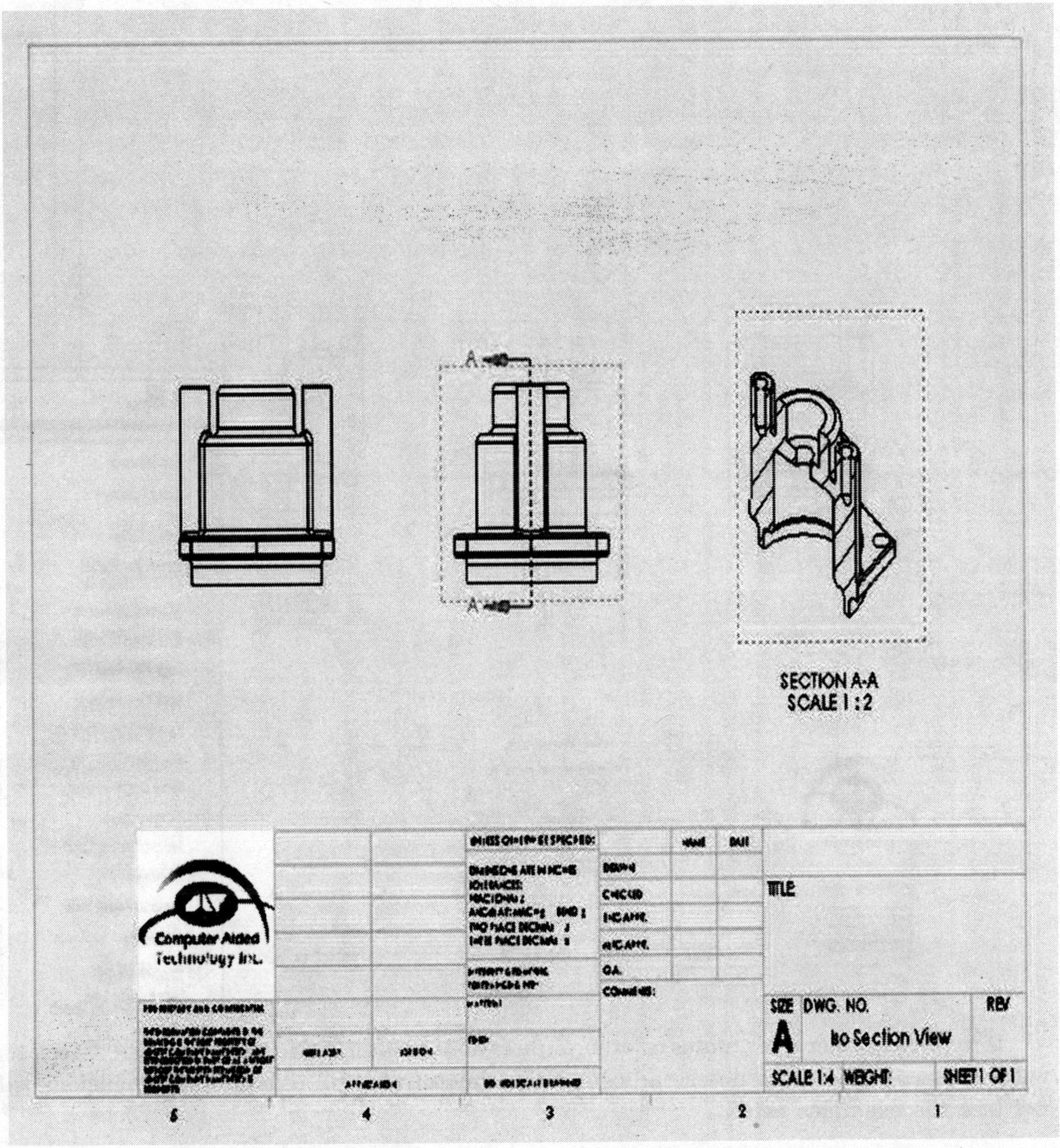

To change it back to its original state, right mouse click and select **Remove Isometric View**.

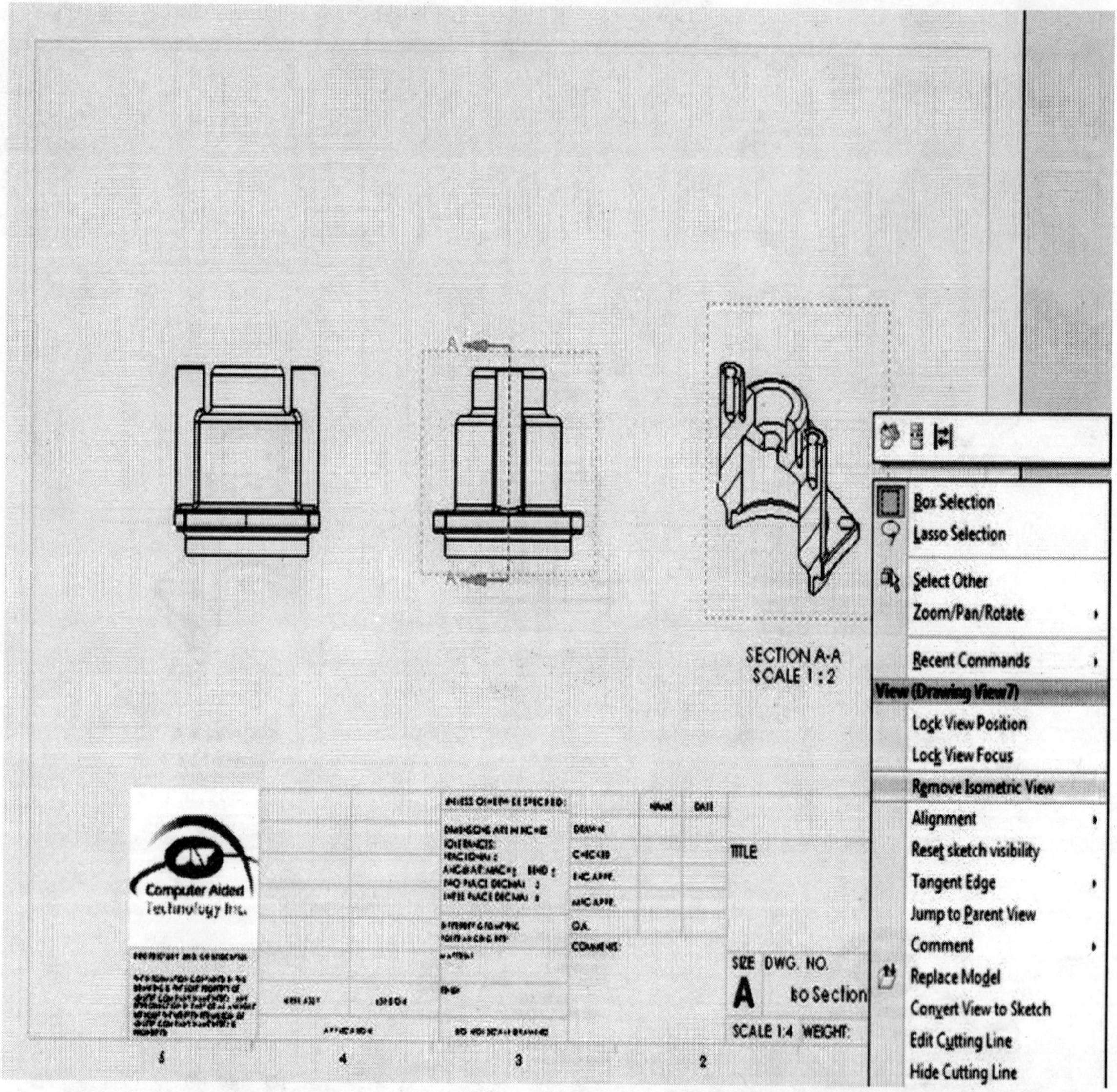

If there are dimensions, notes or other annotations (centerline in this example) associated with the section view, the notes and centerline annotations will remain but dimensions will not (see the examples below.)

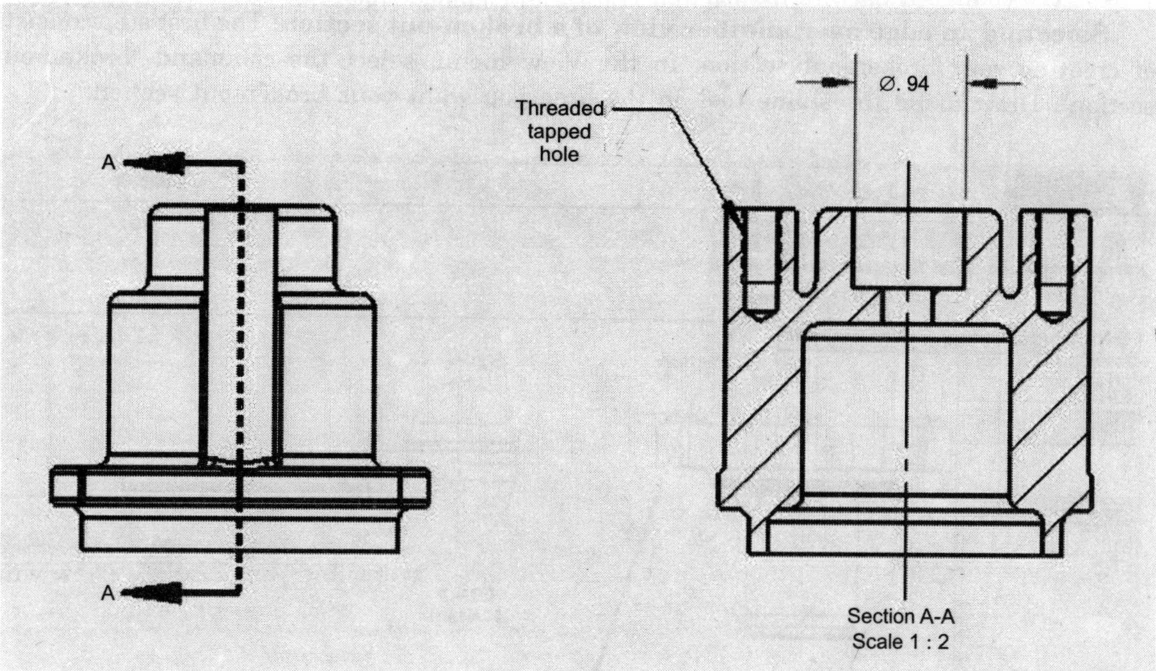

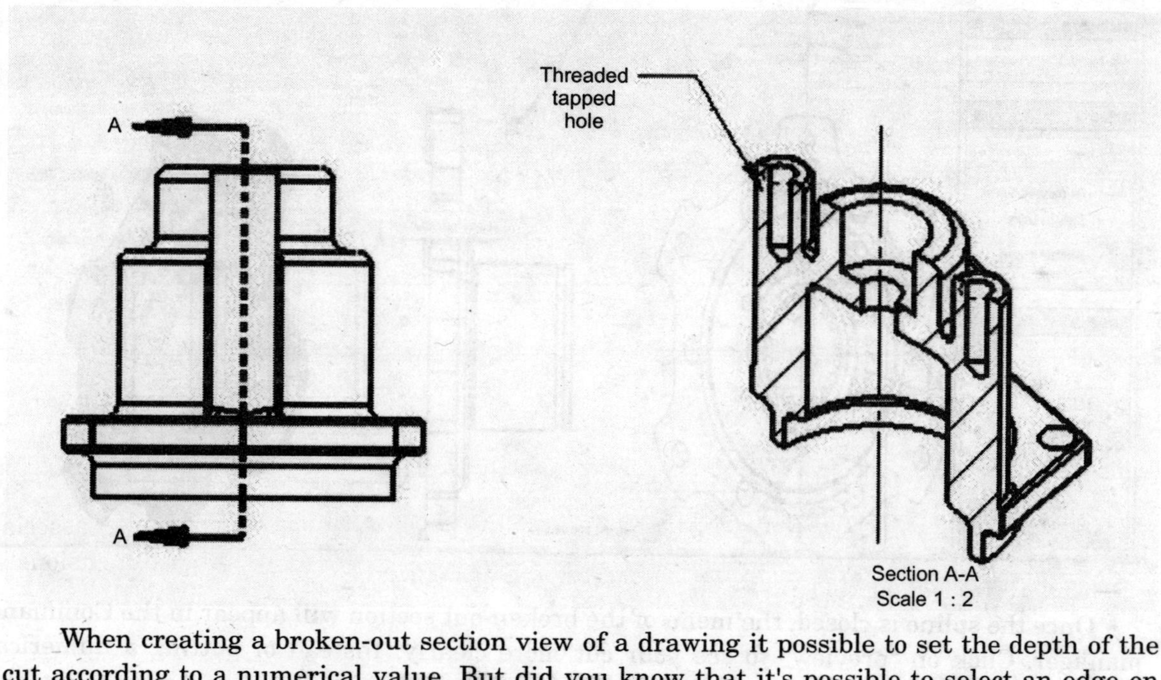

When creating a broken-out section view of a drawing it possible to set the depth of the cut according to a numerical value. But did you know that it's possible to select an edge on another view to define the depth? Or that you can put your 3D drawing in temporary 3D view mode and select the depth on this one?

Selecting an edge over another view of a broken-out section: The first step consists of creating your broken-out section. In the View menu, select the command "broken-out section". Draw using the spline tool on the area you want your broken-out section.

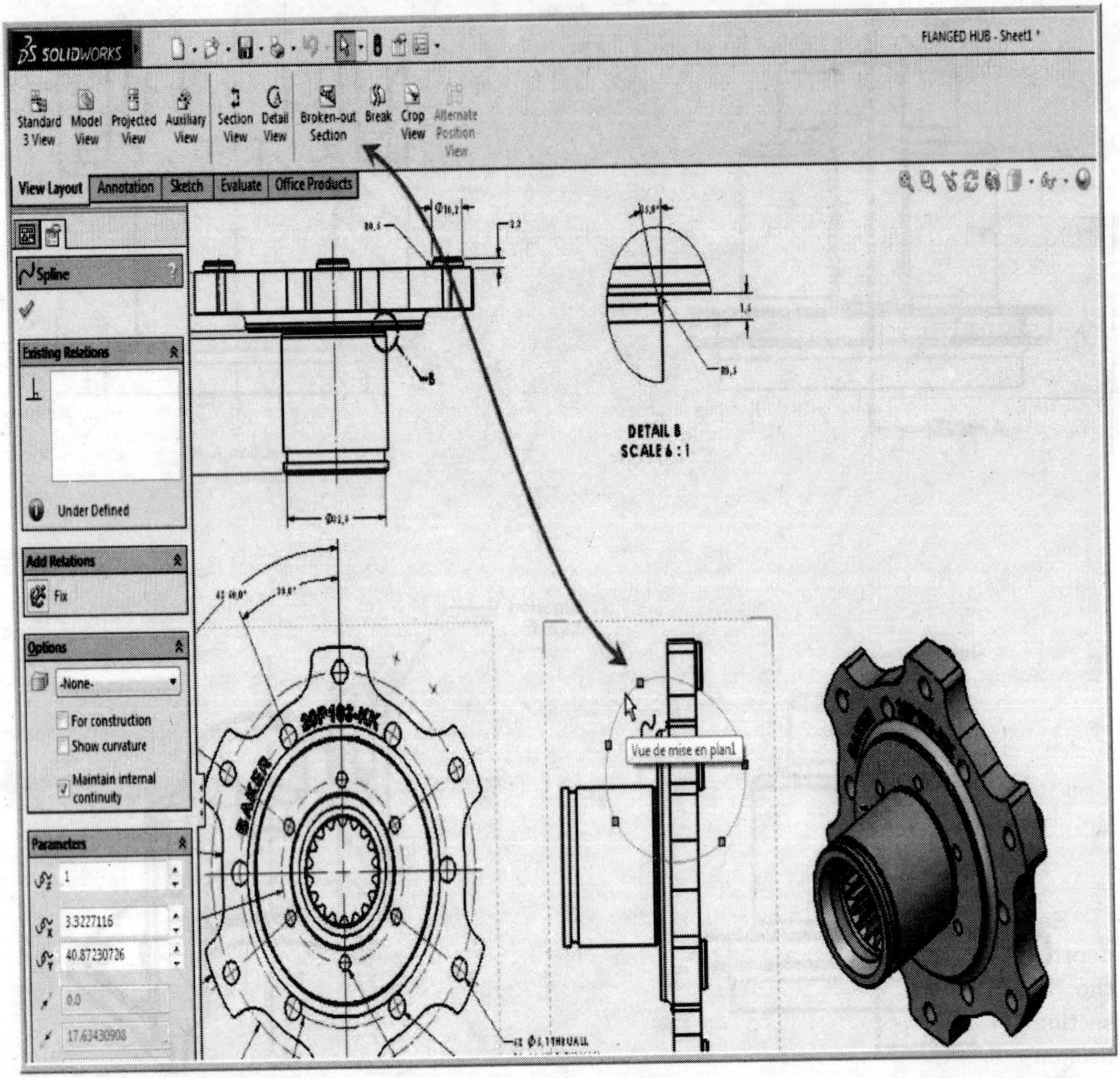

Once the spline is closed, the menu of the broken-out section will appear in the Command manager. Click on "preview" to see your cut more clearly. Instead of putting a numerical value, activate the "selection" box. Note that in the other view the line becomes visible. Simply select a straight edge or circular edge on the other view. The depth of the cut will adjust according to your selection.

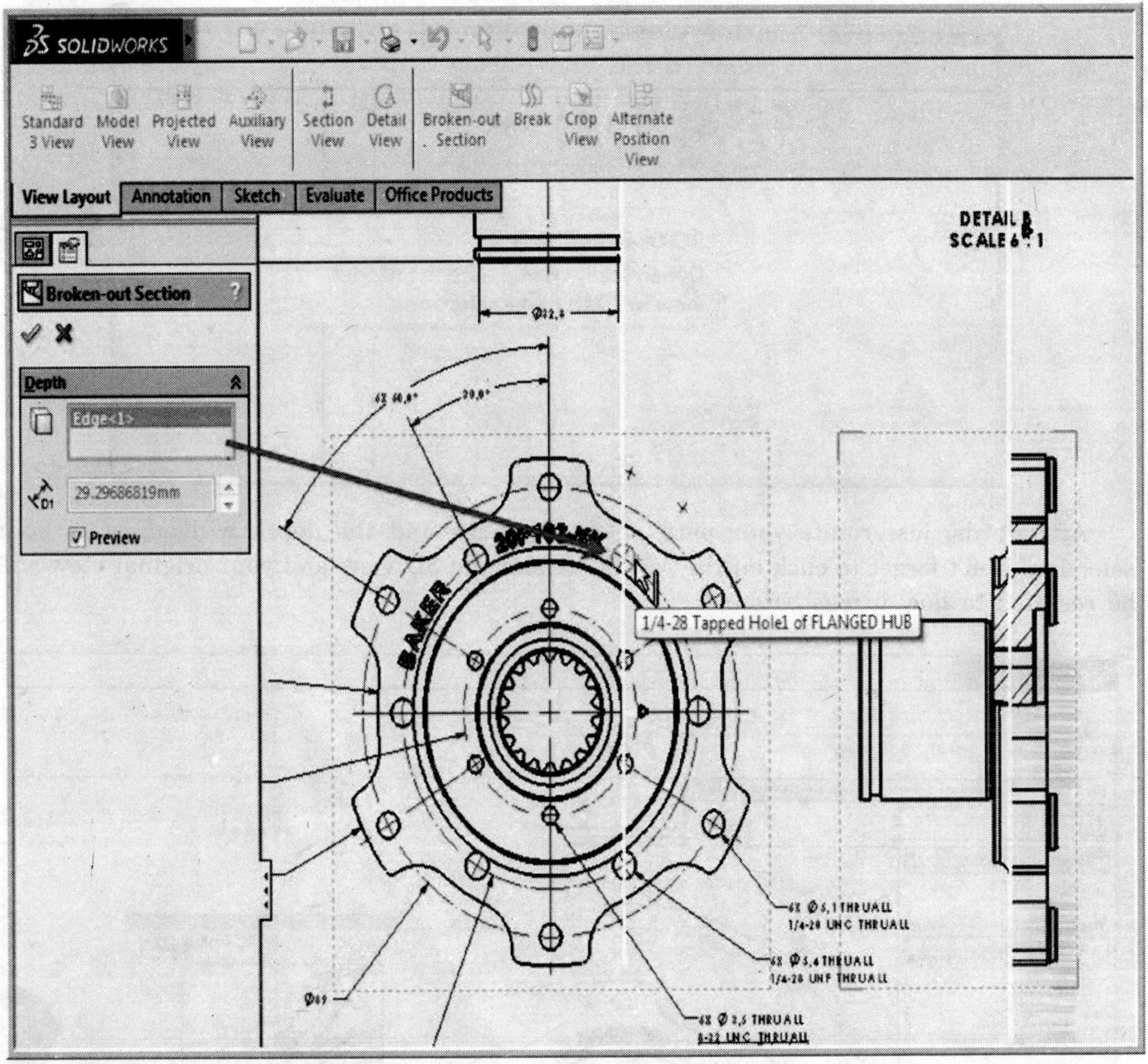

Selecting an edge on the same view as the broken-out section: How to select the depth of the straight section when there is no adequate projected view? The answer is to use the "3D Drawing View" feature and select the edge in the same view where the straight section has been produced. To do this, you must perform the exact previous step to your broken-out section with the spline tool. Then, click on the selection menu, and instead of selecting an edge, click on the "3D Drawing View."

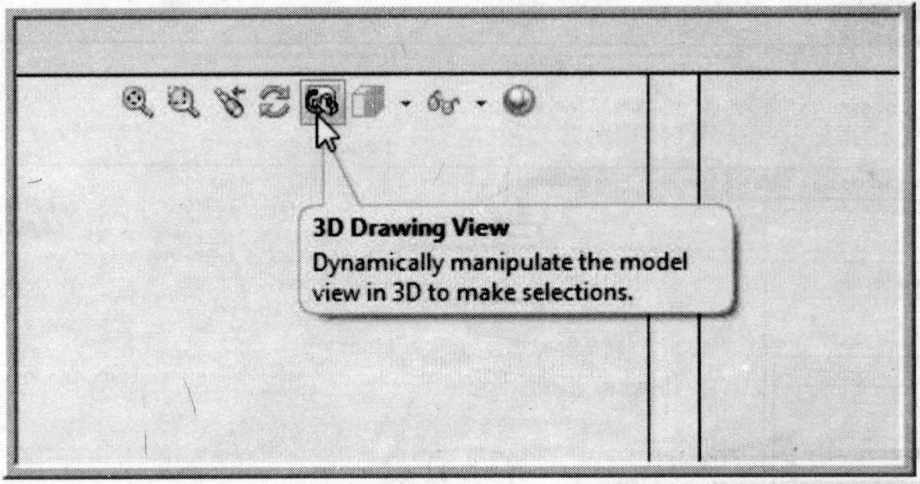

Afterwards, just rotate your part; select the edge and the depth will adjust to your selection. Don't forget to click on the red "X" to exit the 3D view and your original view will be restored to how it was before.

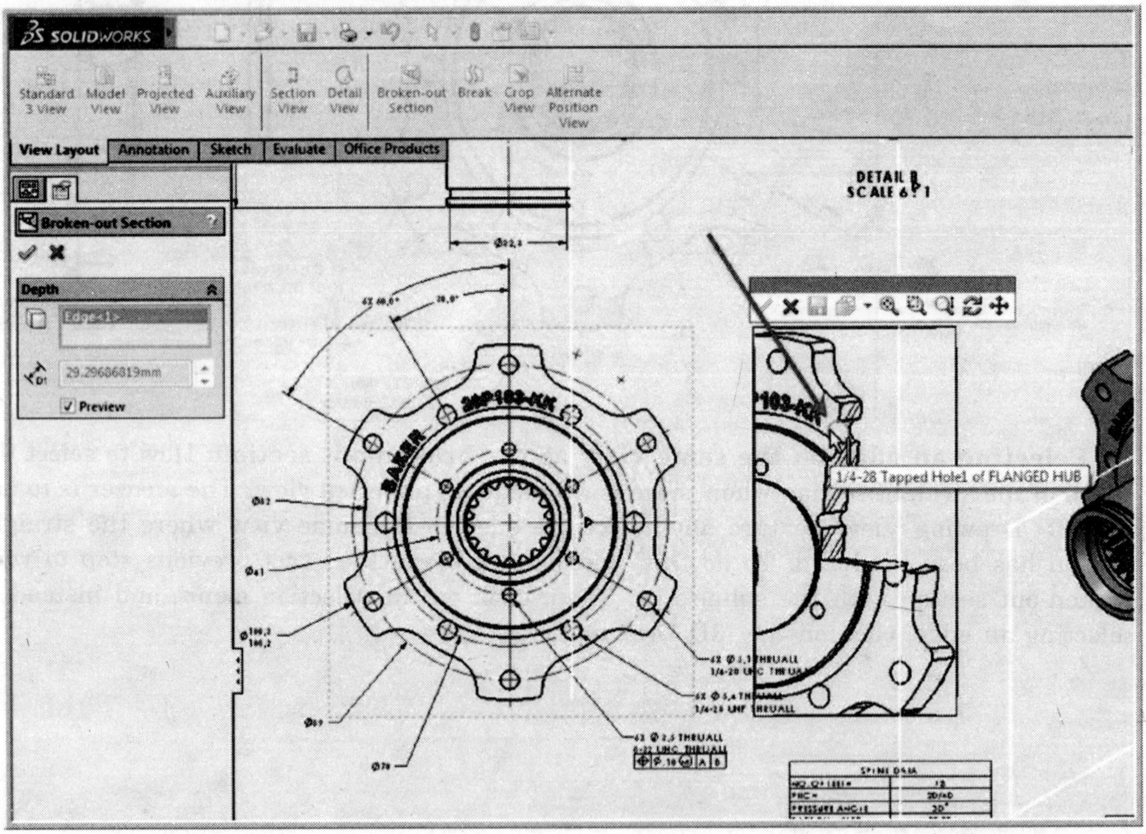

Now we have all the necessary views and we can add some dimensions. Click on the annotation tab.click on the smart dimension.

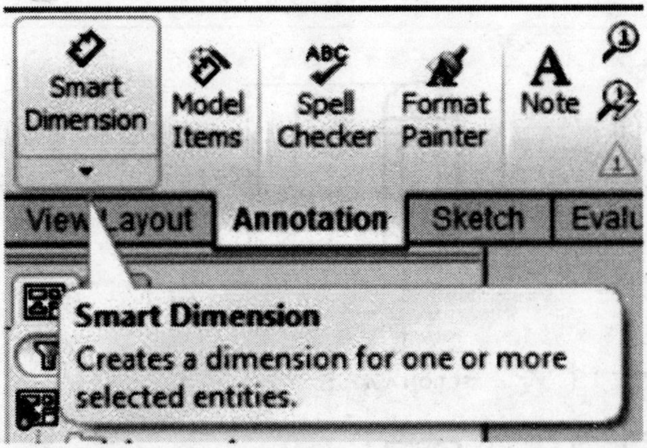

For creating dimensions you need to choose the start and end points:

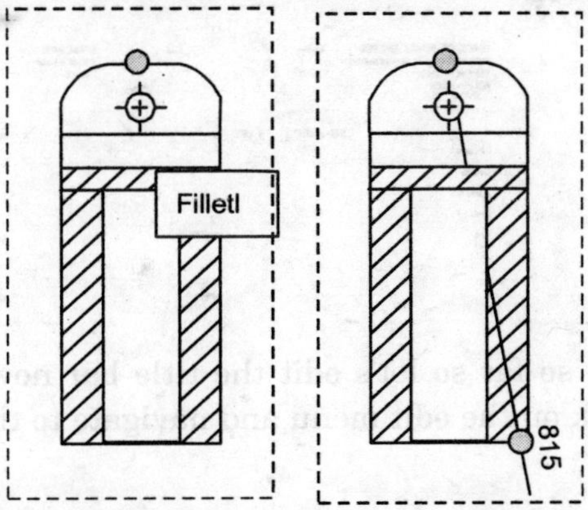

Then drag it to a side:

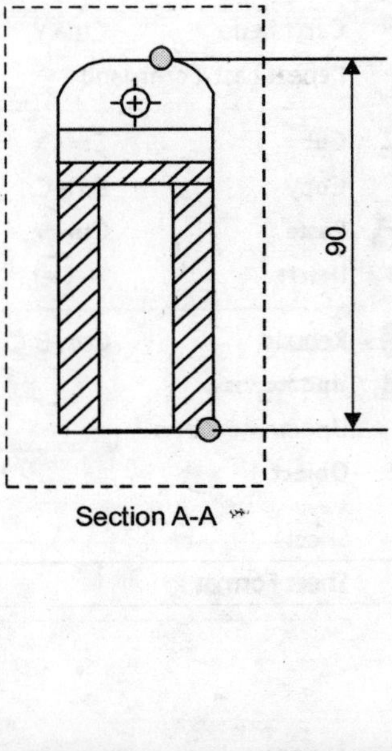

Add some necessary dimensions like below:

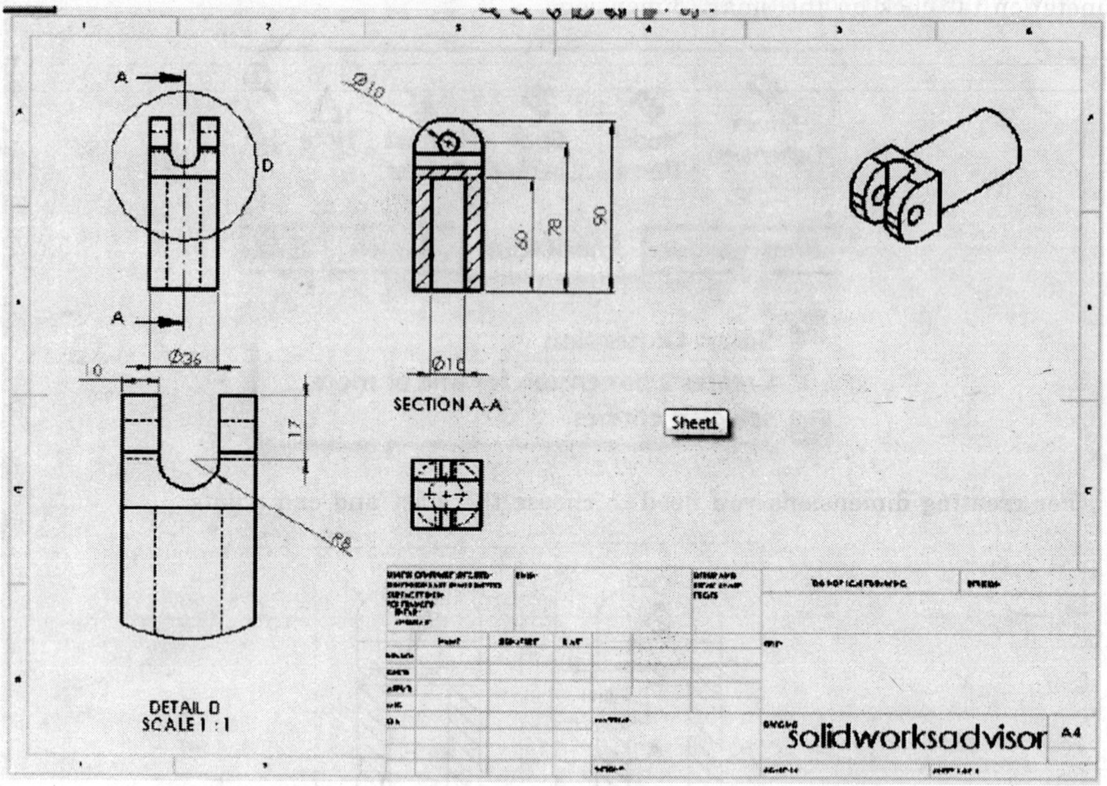

We were creating views so far so let's edit the title bar now. We should edit the sheet format for this purpose. Click on the edit menu and navigate to the "sheet format" option and select it.

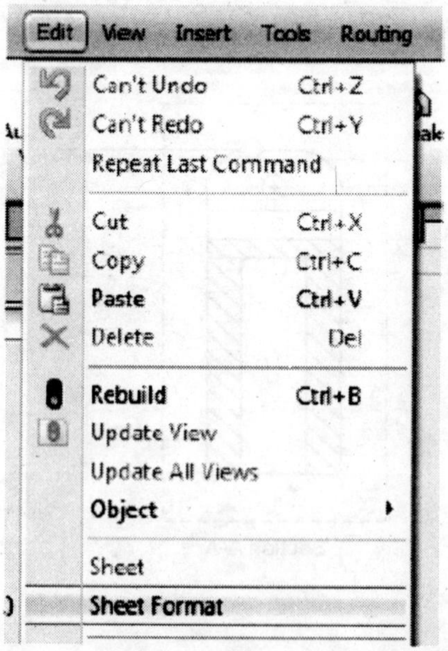

Now we can edit the title block:

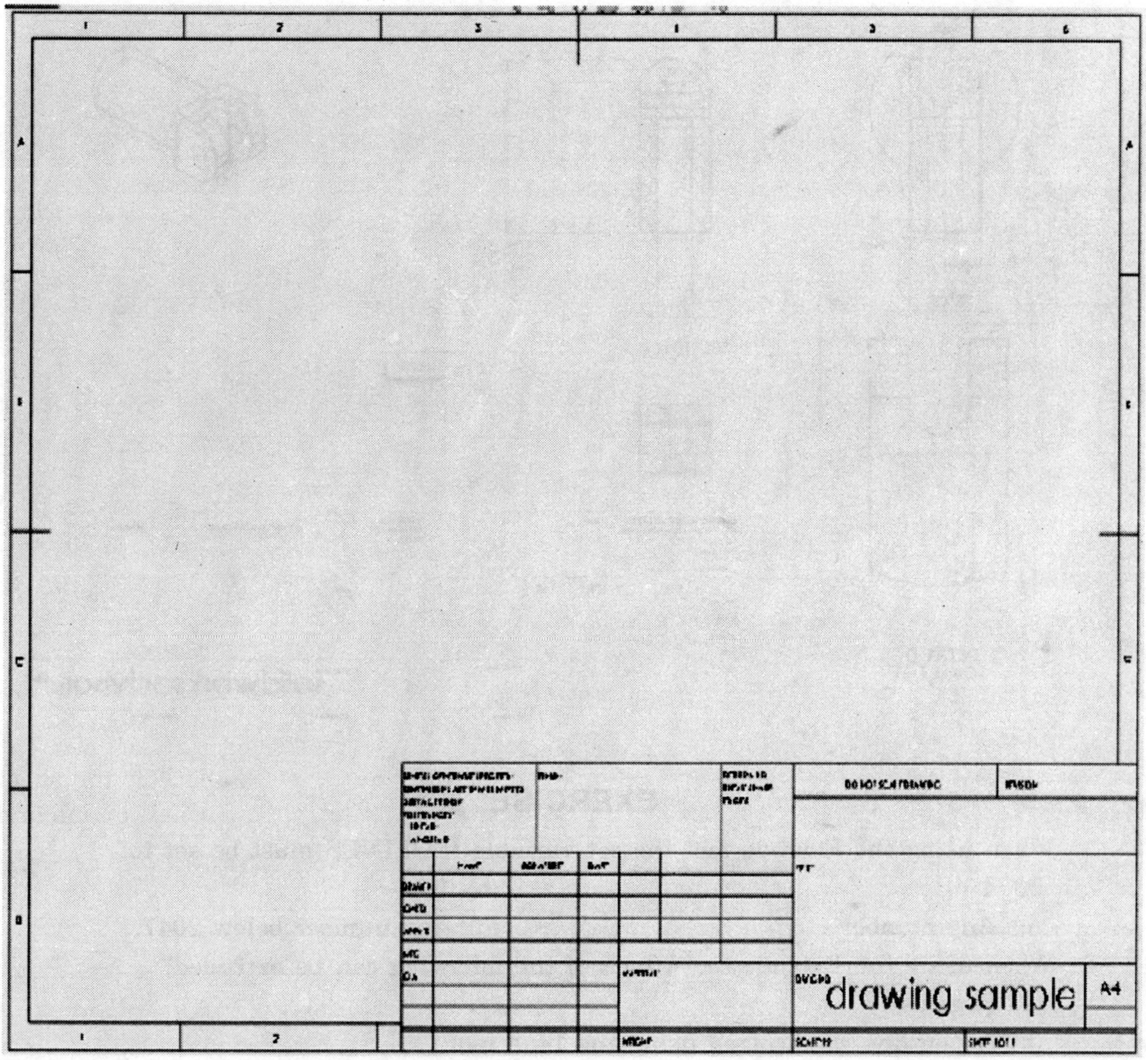

Double click any part of the table for editing for example double click on the title ((drawing sample)) and type your desired title.

Now you know the basics go on and train what you just learnt. If you need to edit the views again just select "sheet" from edit menu.

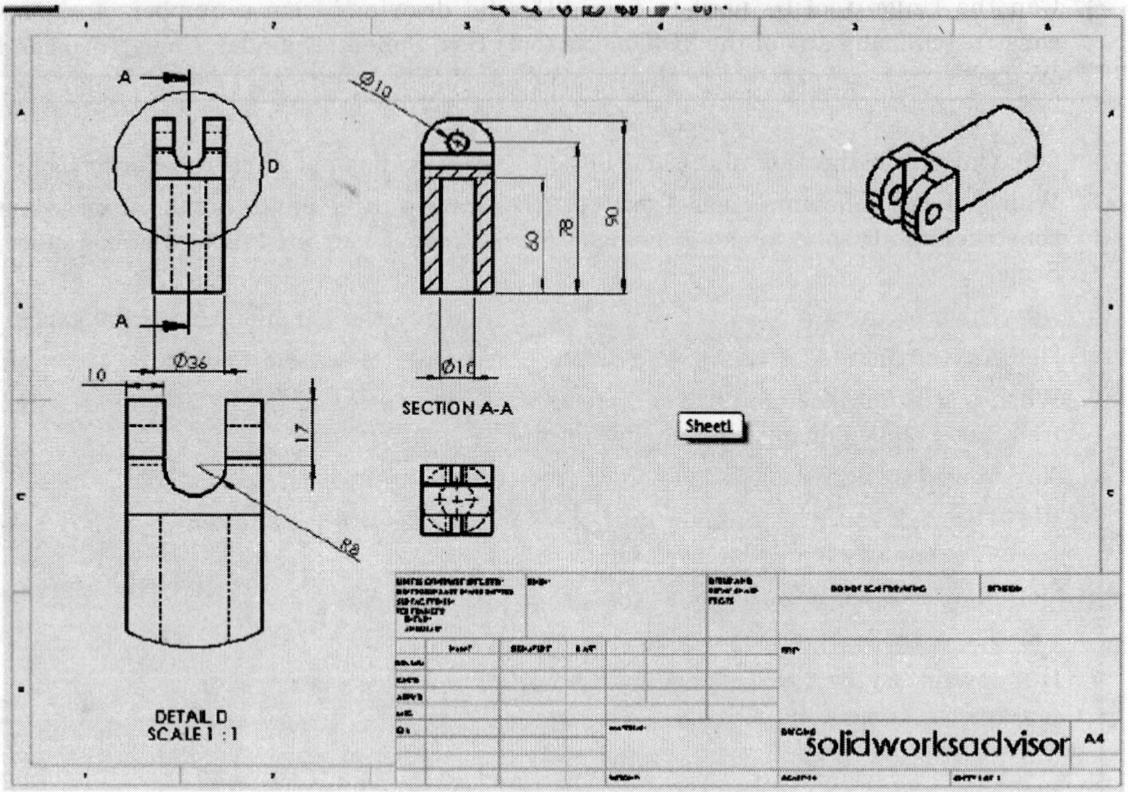

EXERCISE

1. When using the Extrude tool the set variable ISOLINES must be set to:
   (a) 4 (b) 16
   (c) Any number (d) Any number below 2047.
2. When using the Extrude tool which of the following can be extruded?
   (a) Regions
   (b) An outline constructed using the Line tool
   (c) Circles
   (d) Closed plines
   (e) Ellipses
   (f) Any outline
   (g) Another extrusion.
3. When using the Revolve tool which of the following can be extruded?
   (a) Regions
   (b) An outline constructed using the Line tool
   (c) Circles
   (d) Closed plines
   (e) Ellipses

4. Can the Union tool be used to form 3D solid drawings from a number of objects constructed using any of the 3D Objects tools Box, Sphere, Cylinder, Cone, Torus and Wedge?
   - (a) Yes
   - (b) No
   - (c) Only from the Box and Cone tools
   - (d) Yes, but not from the Torus tool.
5. Which of the following tools from the 2D Draw control panel can be used when constructing 3D solid model drawings - Move, Copy, Chamfer, Fillet, Mirror, Rotate, Scale?
   - (a) None of them
   - (b) Only the Chamfer and Fillet tools
   - (c) Any of them
   - (d) Only Move and Copy.
6. When a selected face of a 3D solid model drawing is selected by the Move Faces tool from the Solids Editing toolbar, the face is:
   - (a) Moved away from the 3D solid as a separate entity
   - (b) Moved in such a manner that the body of the 3D solid to which the face is attached is extruded
   - (c) Moved to be placed on another 3D model
   - (d) The whole of the 3D model is moved.
7. Is it possible to use the Extrude faces tool from the Solids Editing toolbar on a region constructed from a closed polyline?
   - (a) Yes, any region can have its face extruded using this tool
   - (b) Yes, if the region has not been obtained by using any of the Boolean operators
   - (c) No - a region of any type cannot be acted upon by this tool
   - (d) Yes, providing the line of extrusion is a straight line or polyline
8. Draw using 3D Software :

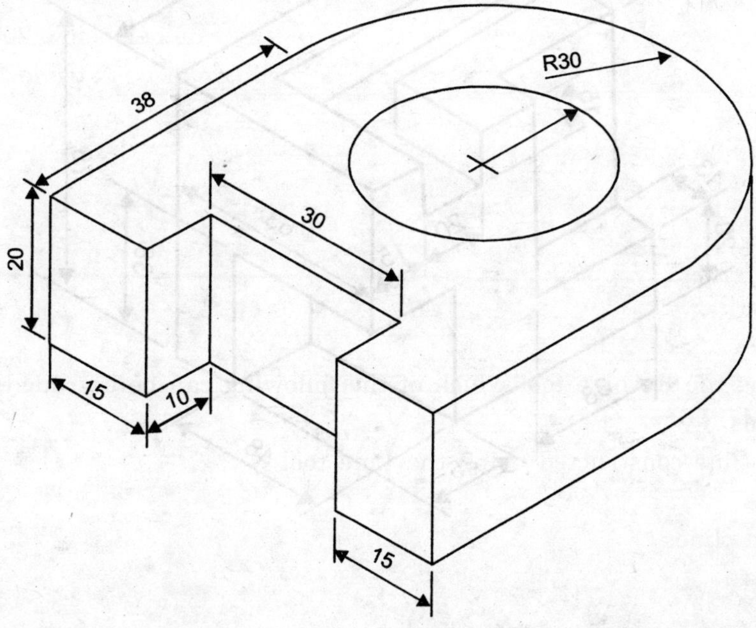

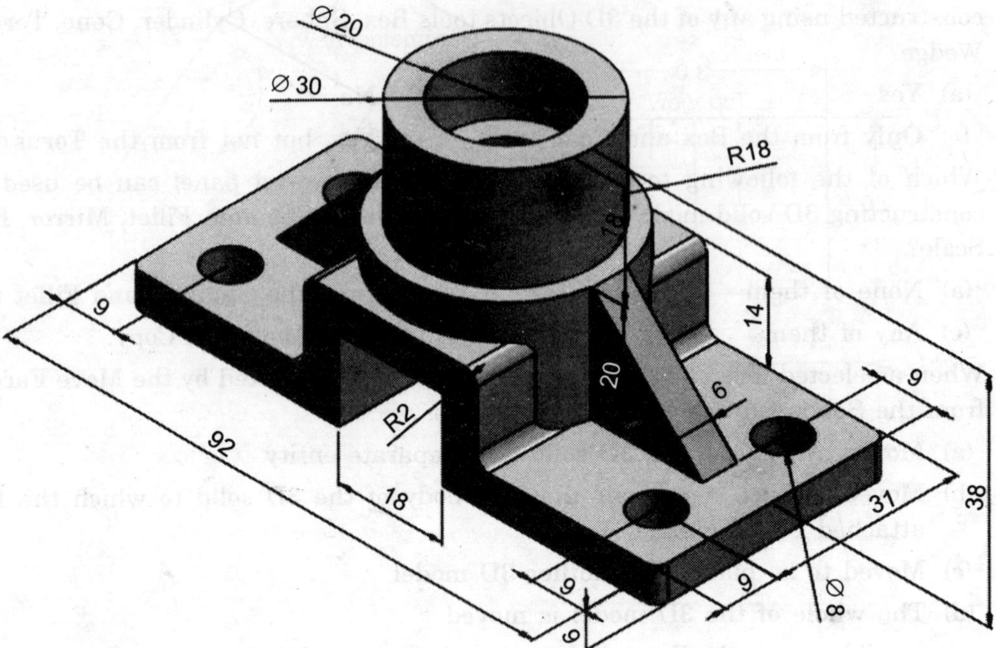
Ø 20
Ø 30
R18
18
9
14
20
6
9
R2
92
18
31
38
9
9
8 Ø
9

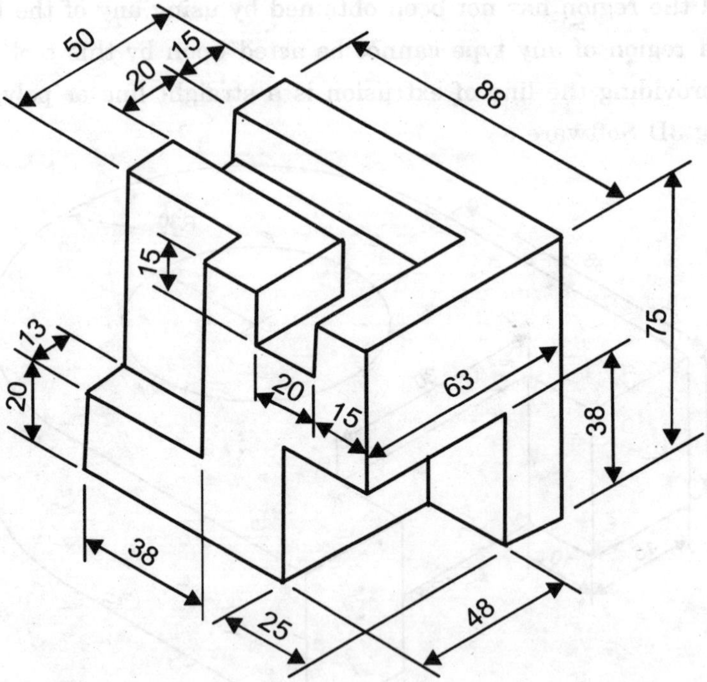
50
15
20
88
15
75
73
20
20
15
63
38
38
25
48

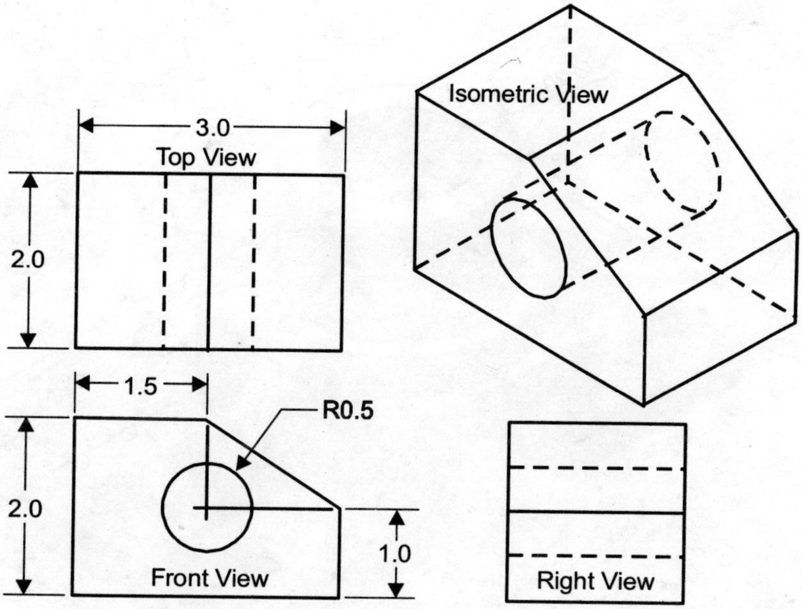

SUMMARY

- Solid modelling Solid or volume modelling describes objects as finite, closed, regular sets of points. A solid model provides a complete, characteristic and concise description of the object.
- In modelling by elemental solids, components are built up of so-called solid primitives, falling into the order of magnitude of their dimensions and having a specific geometry, by using composition operations.
- Modelling systems of other than full value include: (i) Wireframe modelling (ii) Surface modeling
- Modelling systems of full value include: (i) Mantle modelling (ii)Solid modeling
- A parametric model contains information like dimensions, constraints, and relationships between various entities like edges, sketches and features. and A Non parametric model does not contain such relationships. .
- SOLIDWORKS offers an option that allows the user to show an existing section view as an isometric section view. A right-mouse click on an existing section view will display the **Isometric Section View** option.